Introduction to
Bed, Bank and Shore Protection

Gerrit J. Schiereck

Spon Press
Taylor & Francis Group

LONDON AND NEW YORK

First published by Delft University Press
on behalf of Vereniging voor Studie- en Studentenbelangen te Delft
Leeghwaterstraat 42, 2628 CA Delft, The Netherlands
tel. +31 15 27 82124, telefax +31 15 27 87585, e-mail: hlf@vssd.nl
internet: http://www.vssd.nl/hlf

Published 2004 by Spon Press
11 New Fetter Lane, London EC4P 4EE
Simultaneously published in the USA and Canada
by Spon Press
29 West 35th Street, New York, NY 10001

Spon Press is an imprint of the Taylor & Francis Group

Printed and bound in Great Britain by TJ International Ltd, Padstow, Cornwall

Every effort has been made to ensure that the advice and information in this book is
true and accurate at the time of going to press. However, neither the publisher nor the
author can accept any legal responsibility or liability for any errors or omissions that
may be made. In the case of drug administration, any medical procedure or the use of
technical equipment mentioned within this book, you are strongly advised to consult
the manufacturer's guidelines.

British Library Cataloguing in Publication Data
A catalogue record for this book is available from the British Library
Library of Congress Cataloging in Publication Data
Schiereck, Gerrit J. (Gerrit Jan)
    Introduction to bed, bank and shore protection/ Gerrit J. Schiereck.
        p. cm.
    Includes bibliographical references and index.
    ISBN 0-415-33177-5 (alk. paper)
    1. Shore protection. 2. Riparian areas. 3. River channels. 4 River engineering. 5.
Hydrodynamics. 6. Hydraulic structures. I. Title.

TC330.S35 2003

ISBN 0-415-33177-3

Introduction to
Bed, Bank and Shore Protection

Contents

A little learning is a dangerous thing;
Drink deep or taste not the Pierian spring.
Alexander Pope (1688-1744)

Preface

Every book is unique. This one is because of a combination of two things:
*   the coverage of subjects from hydraulic, river and coastal engineering, normally treated in separate books
*   the link between theoretical fluid mechanics and practical hydraulic engineering.

On the one side, many fine textbooks on fluid motion, wave hydrodynamics etc. are available, while on the other side one can find lots of manuals on hydraulic engineering topics. The link between theory and practice is seldom covered, making the use of manuals without understanding the backgrounds a "dangerous thing". Using a cookbook without having learned to cook is no guarantee for a tasty meal and distilling whisky without a thorough training is plainly dangerous. Manuals are often based on experience, either in coastal or river engineering, or they focus on hydraulic structures, like weirs and sluices. This way, the overlap and analogy between the various subjects is lacking, which is a pity, especially in non-standard cases where insight into the processes is a must.

This book tries to bridge the gap between theoretical hydrodynamics and the design of protections. Understanding of what happens at an interface between soil and water is one of the key notions. However, this can only partly be derived from a textbook. Using one's eyes every time one is on a river bank, a bridge or a beach is also part of this process. In the same sense, a computer program never can replace experimental research completely and every student who wants to become a hydraulic engineer should spend time doing experiments whenever there is a possibility to do so.

Gerrit J. Schiereck
Dordrecht, June 2003

Acknowledgement
It is impossible to compile a book like this without the help of many people. In alfabetical order, I want to thank for their major or minor, but always important,

contributions: Kees d'Angremond, Alice Beurze, Rob Booij, Henri Fontijn, Pieter van Gelder, Mark Lindo, Jelle Olthof, Jacques Oostveen, Kristian Pilarczyk, Hermine Schiereck, Jacques Schievink, Wijnand Tutuarima, Wim Venis, Henk Jan Verhagen, Arnold Verruijt, Mark Voorendt, Dick de Wilde.

Reminder I

Things you should remember before studying this book

1. $H = z + \dfrac{p}{\rho g} + \dfrac{u^2}{2g} = h + \dfrac{u^2}{2g}$ energy head = location + pressure head + velocity

    head = piezometric head + velocity head (Bernoulli)

2. $p = \rho g h$ hydrostatic pressure

3. $\Delta h = \dfrac{u^2}{2g} \rightarrow u = \sqrt{2g\Delta h}$ (Torricelli)

4. $\dfrac{p}{\rho g} = \dfrac{u^2}{2g} \rightarrow p = \dfrac{\rho u^2}{2}$ transfer of velocity into pressure

5. $\bar{u} = C\sqrt{RI} \left(C = 18\log\dfrac{12R}{k_r} \right)$ (Chezy)

6. $u_f = kI$ (Darcy)

7. $\sigma' = \sigma - p$ (effective stress = total stress – water pressure)

8. $\left.\begin{array}{l} L = cT \\ c_0 = \dfrac{gT}{2\pi} \end{array}\right\} \rightarrow L_0 = c_0T = \dfrac{gT^2}{2\pi} \approx 1.56T^2$ (deep water wavelength)

9. $c_s = \sqrt{gh} \rightarrow L_s = \sqrt{gh}\,T$ (shallow water wavelength)

10. $\text{Re} = \dfrac{uL}{\nu}$ (Reynolds number, inertia versus viscosity) for open

    channel flow, with $L = h,$ flow is turbulent for Re > 2000-3000

11. $Fr = \dfrac{u}{\sqrt{gh}}$ (Froude number, kinetic versus potential energy)

    Fr < 1: sub-critical flow
    Fr = 1: critical flow (u = celerity shallow water wave)
    Fr > 1 super-critical flow

1 INTRODUCTION

Eroded coast (North Holland), courtesy Rijkswaterstaat

1.1 How to look at protections

1.1.1 *Why and when*

The interface of land and water has always played an important role in human activities; settlements are often located at coasts, river banks or deltas. When the interface consists of rock, erosion is usually negligible, but finer material can make protection necessary. In a natural situation, the interface moves freely with erosion and sedimentation. Nothing is actually wrong with erosion, unless certain interests are threatened. Erosion is somewhat like weed: as long as it does not harm any crop or other vegetation, no action is needed or even wanted. There should always be a balance between the effort to protect against erosion and the damage that would occur otherwise.

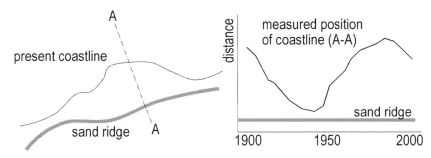

Figure 1-1 To protect or not to protect, that's the question

Figure 1-1 shows cyclic sedimentation and erosion of silt (with a period of many decades) seaward of a natural sand ridge. In a period of accretion people have started to use the new land for agricultural purposes. When erosion starts again, the question is whether the land should be protected and at what cost. Sea-defences are usually very costly and if the economic activities are only marginal, it can be wise to abandon the new land and consider the sand ridge as the basic coastline. If a complete city has emerged in the meantime, the decision will probably be otherwise. With an ever increasing population, the pressure on areas like these also increases. Still, it is good practice along a natural coast or bank to build only behind some set-back line. This set-back line should be related to the coastal or fluvial processes and the expected lifetime of the buildings. For example, a hotel has a lifetime of, say 50 years. It should then be built at a location where erosion will not threaten the building within 50 years, see Figure 1-2. So, in fact the unit for a set-back line is not meters but years! These matters are Coastal Zone Management issues and are beyond the scope of this book.

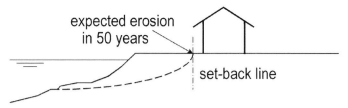

Figure 1-2 Building code in eroding area

Besides erosion as a natural phenomenon, nature can also offer protection. Coral reefs are excellent wave reductors. Vegetation often serves as protection: reed along river banks and mangrove trees along coasts and deltas reduce current velocities and waves and keep the sediment in place. Removal of these natural protections usually mark the beginning of a lot of erosion trouble and should therefore be avoided if possible. So, a first measure to fight erosion, should be the conservation of vegetation at the interface. Moreover, vegetation plays an important role in the ecosystems of banks. Chapter 12 deals with these aspects and with the possibilities of nature-friendly protections.

Finally, it should be kept in mind that, once a location is protected along a coast or riverbank that has eroded on a large scale, the protected part can induce extra erosion and in the end the whole coast or bank will have to be protected. So, look before you leap, should be the motto.

A lot of cases remain where protection is useful. Figure 1-3 gives some examples of bed, bank and shore protections. Along canals, rivers and estuaries, bank protection is often needed to withstand the loads caused by flow, waves or ships. Shore protection structures include seawalls, revetments, dikes and groynes. Bed protection is necessary where bottom erosion could endanger structures, like bridge piers, abutments, in- or outlet sluices or any other structures that let water pass through.

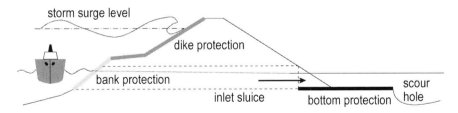

Figure 1-3 Examples of protection

1.1.2 Design

Protections of the interface of land or soil and water are mostly part of a larger project: e.g. a navigation channel, a sea defence system, an artificial island or a bridge. Therefore, the design of a protection should be tuned to the project as a whole, as part of an integrated design process, see De Ridder,1999. In general it can be said that the resulting design should be *effective* and *efficient*. Effective means that the structure should be functional both for the user and the environment. This implies that the structure does what it is expected to do and is no threat for its environment. Efficient means that the costs of the (effective) structure should be as low as possible and that the construction period should not be longer than necessary.

A design that combines effectiveness and efficiency can be said to be "*value for money*". The intended value becomes manifest in the *terms of reference* (ToR) which contains the demands for a structure. This ToR has to be translated into concepts (possible solutions). Demands and concepts do not match one to one and a fit between the two is to be reached with trial and error. Promising concepts are engineered and compared. One comparison factor, of course, is costs. The designer's task to get value for money can be accomplished by compromising between four elements, see Figure 1-4.

Figure 1-4 Value for money

The design process is of a cyclic nature because it is impossible to go directly from left to right in Figure 1-4. In the first phase, the designer works with a very general notion of the ToR and with some concepts in mind, based on his own or others' experiences. An integrated design process starts with a rough approach to all four elements in Figure 1-4, refining them in subsequent design phases. Effectivity can be evaluated in terms of functionality, environment and technology, while efficiency is expressed in terms of costs and construction although, of course, there are several overlaps and links between these aspects. They all play a role in each of the design phases, but the focus gradually shifts as indicated in Figure 1-5.

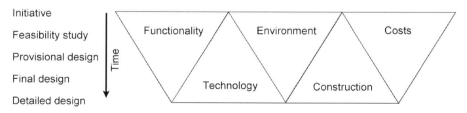

Figure 1-5 Focus during design process

Level of detail

In any project it is possible to discern various levels of detail. It is good to be aware of the level of detail one is working on and to keep an eye on the adjacent levels. An example of these levels (other divisions are, of course, possible):

1. System   (Macro level)
2. Components   (Meso level)
3. Parts   (Mini level)
4. Elements   (Micro level)

Examples of the macro level are e.g. a coastal zone, a watersystem (river, lake etc.) a harbour or a polder. On the meso level, one can think of components like a sea defence (dike, sea wall etc.), a river bank, a breakwater, a closure dam or an outlet sluice. On the mini level we look at dike protections, bank protections or bottom protections. The micro level, finally, consists of elements like stones, blocks etc. In this hierarchy, the title of this book indicates that it treats subjects on the third level. Level 1 should always play a role in the background, see e.g. section 1.1.1. Level 2 will be treated where and when adequate, while sometimes level 4 also plays a role e.g. when it comes to defining stone sizes. As a consequence of these levels, it can be said that the design of protections in a large project is usually more in the lower part of Figure 1-5, when it comes to the technical development of a plan.

1.1.3 Science or craftsmanship

Protections of the interface of land and water have been made for more than 1000 years. Science came to this field much later, as a matter of fact very recently. The second world war boosted the understanding of waves and coasts. In the Netherlands after 1953, the Delta project had an impact on the research into protection works. In the last decades, major contributions to the design practice have been made, thanks to new research facilities, like (large scale) wind wave flumes, (turbulent) flow measurement devices, numerical models etc. progress has been made in The scientific basis of our knowledge has progressed considerably, but even after 50 years, much of the knowledge of these matters is of an empirical nature. Most formulas in this book are also empirical, based on experiments or experience.

Working with these empirical relations requires insight, in order to prevent misconceived use. The idea underlying this book is to start with a theoretical approach of the phenomena, focussing on understanding them. In the design of protections, especially in the unusual cases, a mix of science and experience is required. Since undergraduates, by definition, lack the latter, a sound theoretical basis and insight into the phenomena is paramount. This book goes one step further than simply presenting empirical design relations; it aims to create a better understanding of these relations. Engineering is an applied science, which then, by definition, means that science is the basis but not the core. Creativity, experience and common sense are just as important.

Computer models play an increasing role in engineering. For a hydraulic engineer, however, a sheet of white paper and a pencil are still essential, especially in the preliminary stage of a design. A hand made sketch of a current or wave pattern is as valuable as the correct application of calculation rules. For both, a good insight into the physics of the processes involved is indispensable.

1.2 How to deal with protections

1.2.1 Protection against what?

Interfaces between land and water exist in all sizes and circumstances. Figure 1-6 gives an idea of typical values for the loading phenomena in various watersystems (of course it is always possible to find an example with different figures).

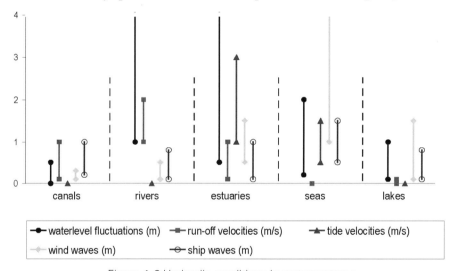

Figure 1-6 Hydraulic conditions in watersystems

This book treats the interface stability by looking at the phenomena instead of the watersystems. This is more exceptional than it seems, because most textbooks deal with shore protection or river training works or navigation canals etc. Much of the knowledge of these protection works is based on experience and experience is often gained in one of the mentioned fields, not in all of them. This is a pity because many of the phenomena involved are similar: ship waves and wind waves have different sources, but behave very much the same. The same holds for flow in a river, through a tidal closure or an outlet sluice, when it comes to protecting the bed or the bank. Moreover, in river bank protections, wind waves can sometimes play a role, which is often neglected in textbooks on river engineering. Therefore, an attempt is made to find the physical core of all these related problems.

One thing protections have in common, is that their function is to *withstand the energy of moving water*. Water in motion contains energy: currents, wind waves, ship movement, groundwater-flow etc, which can become available to transport material.

The energy comes from external sources, like wind or ships, and eventually ends up as heat by means of viscous friction. This is not an energy loss but an energy transfer, from kinetic energy, via turbulence, to heat. Turbulence plays an important role and will be discussed in more detail in the next chapter. For now it is sufficient to say that turbulence is related to the transformation of kinetic energy into heat. During this transfer, turbulence contributes to the attack of the interface.

Hydraulic engineering research is often empirical and fragmented. This leads to an avalanche of relations for each subject, while the connections remain unclear. One of the basic ideas of this book is to show similarities and differences between the various phenomena and therefore between the various formulas, in order to clarify the overall picture. Chapter 2 deals with open-channel flow, chapter 5 with porous flow (flow through pores of granular structures like soil or rock), chapter 7 with waves and chapter 9 with ships. These subjects can and will be treated separately, but there are more similarities than many textbooks reveal.

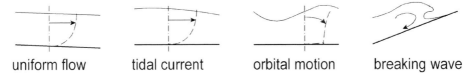

uniform flow   tidal current   orbital motion   breaking wave

Figure 1-7 Flow and wave situations

Uniform flow is the starting point for many hydraulic considerations, see Figure 1-7. The equilibrium between gravity and wall friction completely determines the flow. The boundary layer, connected with the wall friction, takes up the whole waterdepth, is turbulent and shows a logarithmic velocity profile. The velocity profile of tide waves (very long waves with typical periods of 12 hours and wave lengths of several hundreds of km's) only slightly differs from the uniform flow velocity profile. It is therefore justified, when designing a protection, to consider tidal currents as a succession of uniform flow situations with different velocities. For wind waves (typical periods of 5 – 10 s and wave lengths of 50 – 150 m), the situation is completely different with a non turbulent orbital motion and a thin turbulent boundary layer, although such a wave in very shallow water will again approach the situation with a tidal wave. Finally, a wave that breaks on a slope, leads to turbulence over the whole waterdepth.

hydraulic jump   bore

Figure 1-8 Jump and bore

A hydraulic jump and the roller of a broken wave (the bore) are very much the same. This can be seen when the jump is observed from a fixed position and the bore from a position that moves at wave celerity. The turbulence characteristics, caused by the friction between roller and flowing water, are also similar. Chapter 7 will show this in detail.

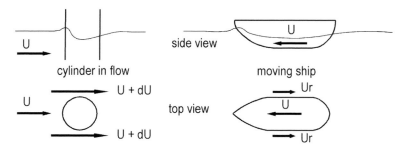

Figure 1-9 Flowing water versus moving object

The same similarity exists between a fixed object in flowing water and a ship sailing in still water. The water around the object accelerates, while around the ship a return current occurs, both leading to a water level depression.

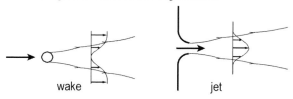

Figure 1-10 Mixing layers in wake and jet

Behind an object in flow or behind a ship, a wake occurs, where there is a velocity deficit compared with the environment. This velocity difference causes a so-called mixing layer where relatively slowly and quickly moving water mix which leads to a lot of turbulence. In a jet (an outflow in stagnant or slowly moving water) the same velocity differences (but now due to an excess velocity) occur, causing the same mixing layer and turbulence.

The last analogy in this chapter is between pipe flow and porous flow (see Figure 1-11) which is flow through a porous medium like sand or stones. In a straight pipe the (uniform) flow is determined by the wall friction. In an irregular pipe, uniform flow will never really occur, due to the irregularities in the cross section. Even with a constant discharge, accelerations and decelerations will always occur and, at sharp discontinuities, even flow separation with a mixing layer will take place. The flow between grains, when considered on a micro level, also show continuous accelerations and decelerations and the same basic equations describe this type of flow, including laminar and turbulent flow. In practice, however, the flow is integrated over many grains and pores, because it is not feasible and not necessary to have velocity information of every pore.

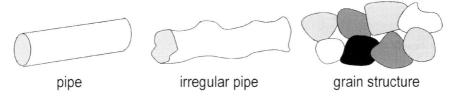

pipe   irregular pipe   grain structure

Figure 1-11 Pipe flow and porous flow

All of the above examples contain elements of three phenomena: *wall flow, mixing layer and oscillating flow (wave)* with turbulence playing a role in all three of them. Wall flow is present in uniform (pipe) flow, tidal flow and in the boundary layer of wind waves. A mixing layer is visible in hydraulic jumps and bores (between main flow and roller) and in wakes and jets. On a micro scale, porous flow includes both wall flow and mixing layers. It is a simplification to say that every flow situation can be reduced to these three basic phenomena or a combination of them but every situation contains at least one of these three features. It is therefore indispensable to be able to recognize and understand their elementary properties. This is what the chapters on loads are about: chapter 2 on flow, chapter 5 on porous flow and chapter 7 on waves.

Hydraulics and geotechnics

In general, hydraulic and soil-mechanical mechanisms determine the stability of a structure. Cause and effect can lie in both fields: failure of a protection can cause settlements of a structure, but vice versa is also possible. Figure 1-12 gives some examples.

In case (a) the sill under a water-retaining structure is a malfunctioning filter. Due to erosion, the structure will settle. As the maximum gradient inside the filter possibly occurs at the entrance side of the flow, the settlement can be against the head difference. In case (b) a canal is situated above groundwater-level. To prevent waterlosses, the bottom of the canal is coated with an impermeable protection. If the dike along the canal settles, due to insufficient strength of the subsoil, a fracture in the protection can occur and the canal water can drain into the subsoil. Case (a) looks like a soil mechanical problem but it has a hydraulic cause while in case (b) it is the other way around.

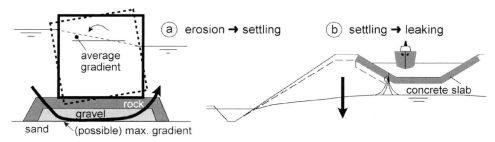

Figure 1-12 Cause and effect

1.2.2 Failure and design

The previous section already stressed that insight in phenomena is paramount for the design of a reliable interface protection. Neglectance of a relevant phenomenon can lead to a protection that causes more damage than it prevents or that shifts the problem to the neglected phenomenon, see Figure 1-13.

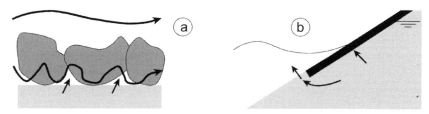

Figure 1-13 Ill-designed protections

In case (a), large rocks have been dumped on a sandy bottom which erodes because of currents. The rocks lead to a slightly lower velocity at the bottom, but to a considerable increase in turbulence and hence, maybe even an increase of erosion. Case (b) shows an asphalt-protection on a slope which would otherwise erode due to wave action. The protection now causes a difference between the water-tables inside and outside of the slope during low water. This head difference causes pressures on the protection which can result in lifting the protection layer. It also causes a concentrated groundwater-flow at the edge of the asphalt which leads to erosion at that spot.

Figure 1-14 shows the forces that act on a protected slope. A represents the loads from the water-side of the interface, the external load due to waves and currents. C is the load from inside due to a relatively high groundwater-potential in the soil-mass. B is the interaction between the external load and the inside of the structure. Although the external forces are usually rather violent and spectacular, many protections fail because of B or C.

The external forces A require a *strong* protection. This strength can be obtained by using large, heavy stones. The example in Figure 1-13 has shown that a protection should also be *sandtight* due to process B. To make the protection sandtight, some layer is needed between the top-layer and the subsoil e.g. a filter, a cloth or a foil. But if that layer is impermeable, C can become a threat. That means *permeability* can be required (unless there are other reasons to make the protection impermeable; in that case the protection has to be designed to withstand the possible pressures). Another way of increasing strength is ensuring coherence in the top-layer e.g. by using concrete or asphalt instead of dumped rocks. The protection can then become impermeable or stiff which can cause problems if settlements are expected. So *flexibility* is another factor to reckon with. Figure 1-14 gives an idea of the contradicting factors in design of a protection.

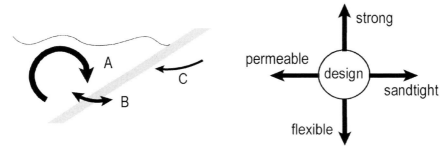

Figure 1-14 Contradicting demands

Failure mechanisms

In general, it is always necessary to keep the overall picture in mind. Figure 1-15 shows the relevant failure mechanisms for a revetment.

Even if these mechanisms are not completely open to computation, a mere qualitative understanding can help to prevent an unbalanced design. Insight is more important than having an accurate formula to compute dimensions of some part of the structure. *Protections seldom fail because of an underestimation of the loads of 10 %; most protections fail because a mechanism has been neglected*!

Sometimes, designers put most of their energy into the first failure mechanism, the instability of the protection layer, also including the filter action (mechanism B in Figure 1-14). But if the protection is too low, wave overtopping can destroy the revetment. Toe protection is often neglected or underestimated. Instability of the slope can be of a micro or macro nature, both connected with the slope angle; chapter 5 gives more details. Collision or agression is self evident, but hard to include in a design. A collision proof design will be unnecessarily expensive, unless the protection is situated at a notorious accident spot. A better approach can be to keep repair material in stock and to anticipate repair works in the maintenance programme.

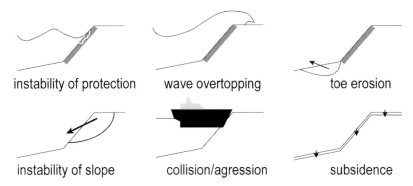

Figure 1-15 Failure mechanisms

The failure mechanisms of a structure can be combined in a fault tree (Figure 1-16), which gives the relations between the possible causes and the failure of the revetment on top of the tree. If you are able to assign probabilities to the events, it is possible to determine the total probability of failure of the structure and to find weak spots. But also without that quantitative information it is useful to draw a fault tree to get the overall picture. An experienced designer does so intuitively but even then, it is a useful tool.

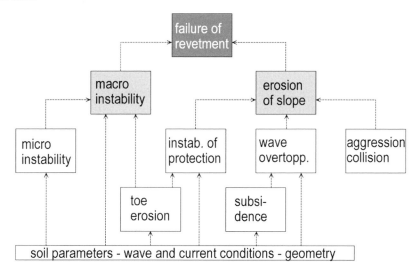

Figure 1-16 Fault tree

1.2.3 Load and strength

The core item in this book is the design of protections that can withstand the loads due to currents, water-level differences or waves. For structures consisting of relatively small elements (rocks, stones, blocks etc.) the definition of strength is somewhat ambiguous. A comparison with a steel structure is made to clarify this point (Figure 1-17).

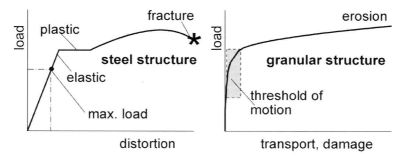

Figure 1-17 Load and strength for steel and granular structure

When steel is loaded, at first the distortion is elastic, obeying Hooke's law. At a certain point, without increase of the load, the distortion becomes plastic. After that, some strengthening occurs until the steel yields. The strength of steel is normally chosen well under the plastic limit, based on the statistics of the steel quality. The clear change of material behaviour serves as an indicator for the permissible strength

For a granular structure, things are less clear. When the load is small compared to the strength, nothing happens at all. At a certain load, some elements move and stop again after some time. Further increase of the load leads to more displacements, finally leading to complete erosion. Usually some damage, like the displacement of some stones on a slope, is not much of a problem, this also depends on the maintenance policy. A clear limit between acceptable and not-acceptable erosion is lacking and the threshold of motion has to be defined. This will be done for the various phenomena in the following chapters.

Another difference between steel and stones is that, for steel, both load and strength can be expressed in the same unit: Newton. For a protection this could be achieved by expressing the load on an individual stone in N and defining the strength as the weight of that stone (mass*$g = N$), but that is not very practical. It is customary to express the load in terms of the wave height or the current velocity. The strength is then indicated with a diameter or thickness, d, often as well as the relative density of the material ($\Delta = \rho_m - \rho_w/\rho_w$), which contributes to the strength. This leads to dimensionless parameters like $H/\Delta d$ or $u^2/\Delta gd$.

This can lead to confusion because, in hydraulic engineering, these dimensionless parameters are used both as *mobility* parameters and as *stability* parameters. The difference becomes clear when you consider the mobility parameter as an independent variable in a transport equation and the stability parameter as a dependent variable in a stability equation:

Transport, damage $= f\,(\textbf{mobility parameter}, \textit{geometry, etc.})$

Stability (parameter) $= f\,(accepted\ damage,\ geometry,\ etc.)$

The first type of equation includes many sediment transport equations, e.g. in this book the Paintal equation in chapter 3 or, in a modified form, in scour relations in chapter 4. Most relations in this book are of the second type, like those by Shields and Izbash in chapter 3 or those by Van der Meer and Hudson in chapter 8. It is good to be aware of the difference, as the use of these parameters in textbooks is not always consistent.

When used as a *mobility* parameter, a large value indicates more mobility (high load versus low strength). When used as a *stability* parameter, a larger value of $H/\Delta d$ indicates more stability (the same stone size can resist a larger wave or for the same wave, a smaller stone can be used). The *stability* parameter can be seen as a *critical value of the mobility* parameter, since the amount of acceptable damage or transport has been chosen. This may be confusing, but is essential in working with the different formulas.

The difference can be illustrated by the stability of stones on a slope in breaking waves (See Figure 1-18) based on the Van der Meer relations, see chapter 8. In the $H/\Delta d$ - damage plane, $H/\Delta d$ is a *mobility* parameter, in the $H/\Delta d - \alpha$ plane it is a *stability* parameter. For a certain slope angle, α, and given stone dimensions (Δd), a higher wave, hence a greater $H/\Delta d$, gives more mobility or more damage. For a certain acceptable damage, a given stone (Δd) can stand higher waves when the slope is gentler (smaller α). This has to do with gravity, which reduces the strength of a stone on a steep slope, but also with the different behaviour of breaking waves on different slope angles, resulting in different loads. This, again, illustrates that load and strength are not defined unambiguously the way they are for a steel structure, when both can be expressed in Newtons.

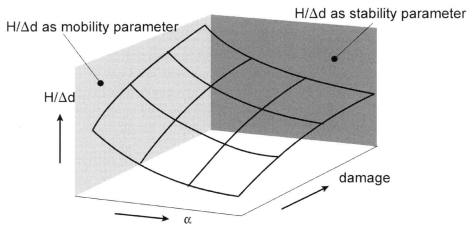

Figure 1-18 Mobility versus stability

Example 1-1

A beach coast with a wave height of 3 m during a storm and sand with a grain size of 0.2 mm gives a H/Δd of about 10000. During that storm, a lot of sand will be transported. The same wave height with a concrete caisson wall will give an H/Δd of less than 1 and no movement at all. So, a higher value of the mobility parameter indicates less stability.

The same coast is going to be protected with stones on a slope. From experiments it is found that the stability parameter for a slope 1:2 is about 2 (with hardly any movement of stones) and for a slope 1:4 is about 2.5 (with the same degree of damage). With the given wave height in a design storm of 3 m, this would lead to a stone size of 0.9 m for the 1:2 slope and 0.7 for the 1:4 slope. So, a higher value of the stability parameter indicates more stability.

Load and strength as design options

When the load exceeds the strength and measures have to be taken, there are two possible approaches: the strength can be increased or the load can be reduced. Figure 1-19 illustrates these possibilities. A bank, eroding due to wave action, can be protected by making a revetment (case A) or by constructing a wave reductor in front of the bank (case B). The latter can be chosen when the "natural" look of the bank has to be preserved, see also chapter 12.

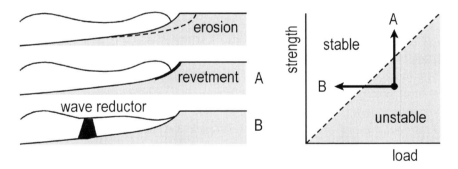

Figure 1-19 Strength increase or load reduction

Load and strength statistics

Loads in nature show a lot of variation. Waves depend on wind, velocities in a river depend on rainfall, so loads heavily depend on meteorology which has a random character. Probabilistic design methods, therefore, are important for protections. In a feasibility study it is often sufficient to work with a representative load. The choice of that load should be based on the relevant failure mechanisms (see Figure 1-15) and on the consequences of exceeding the load. Stability of the top layer of a bottom protection behind a sluice (e.g. see Figure 1-3) is mainly sensitive for exceptional loads and should therefore be based on an extreme event, while erosion behind the protection is also determined by everyday flow. The use of different exceedance frequencies results in different design values for the same loading phenomenon! More in general, the performance of a structure should be judged under various circumstances related to different limit states, see e.g. Vrijling et al, 1992. Two widely used limit states are:

Ultimate limit state (ULS): This limit state defines collapse or such deformation that the structure as a whole can no longer perform its main task. It is usually related to extreme load conditions. Related to the levels of section 1.1.2, it can mean e.g. the collapse of a dike (meso level). In the fault tree in Figure 1-16 the ULS is represented by the higher part of the tree.

Serviceability limit state (SLS): This limit state defines the required performance, e.g. the wave reduction by breakwaters in a harbour. In the context of this book it describes a state that needs to be maintained. Related to the levels of section 1.1.2 it means e.g. the damage of a dike protection (mini level).

Note: for the top layer of the protection this could be seen as the ULS to show that these definitions also depend on the level of detail. In Figure 1-16 the SLS is related to the bottom part of the tree.

The accepted probability of reaching both limit states is a function of the damage caused by exceeding that state. It is obvious that the chance of reaching the ULS should be much lower than reaching the SLS. Maintenance policy is closely related to these limit states. The strength of the structure *as a whole* can drop below the level that is needed under extreme conditions (see Figure 1-20). As long as these conditions do not occur, the ULS will not be reached. When deterioration goes on for a long time, the strength can become too small even for daily conditions and collapse will occur out of the blue. The extreme load in Figure 1-20 has some probability. When the strength is greater than the extreme load, the probability of reaching the ULS is considered acceptable. Without maintenance, the strength decreases and the failure probability increases until it reaches about 100 % when the strength becomes lower than the ever present loads. See chapter 10 for more detail.

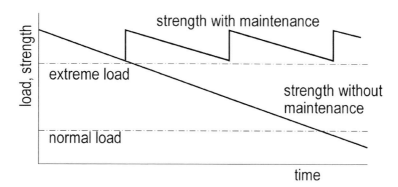

Figure 1-20 Strength as a function of time and maintenance

1.3 How to use this book

There are many textbooks on protection design. They are often aimed at professionals and deal with specific practical applications without treating the theoretical backgrounds. There are also many books on the theoretical backgrounds of flow and wave phenomena without practical application. This book aims to introduce protection design with a focus on the link between theory and practice. It is intended for use as a textbook in a graduate course on university level. The reader is supposed to be familiar with basic knowledge of hydraulics and soil mechanics; only the most important elements thereof will be treated in this book.

Some engineers are addicted to formulas and computing. Formulas are indispensible to calculate dimensions, but again, it is stressed that insight is often more important than numbers. There are many formulas in this book, and of course, they are meant to be used to calculate the dimensions of protections but there is also another way to look at them. Formulas are a very special form of language; they are the most concise way to express a phenomenon. By reading them carefully in this way, it is possible to gain some insight because they show the relations between different parameters, thus describing a phenomenon. The worst thing that can happen to a formula is to be learned by heart without being understood. Another accident that can happen with a formula is that it is considered algebra instead of physics. When doing so, cause and effect can be interchanged freely, sometimes with funny computational results.

The best way to read a formula is to start with the parameters. Do they seem logical for the process described, are any parameters missing? When a parameter's value doubles, what happens to the result and does that seem reasonable? What is the domain in which the formula is valid? Empirical relations are only valid in the range of experiments; theoretical formulae are often based on simplifications.

Another essential element in the book is made up of pictures. These too, present a concise language, either by means of "real life" pictures of some phenomenon or by means of graphs describing the relation between parameters. Text, formulae and pictures together tell the story of protecting the interface between soil and water. Interpretation of these three requires the ability to imagine what is happening to the water, the sediment and the structures. Keeping an open eye and mind when walking along a bank or coast, or in any other place where water moves, surely helps.

This introductory chapter tries to reveal the core of the whole subject and therefore also sometimes resembles a summary. The reader is advised to read it before and after studying this book. Much of what is not clear when reading this chapter the first time, might be recognized immediately when reading it again later. There is a saying: *"Understanding is nothing but getting used to"* which contains some truth.

The contents of the rest of the book can be divided into a more theoretical part (Chapter 2–9, an application of theoretical hydrodynamics and soil mechanics) and a part that deals with the applications of protections (Chapter 10-13). Chapter 2 to 9 contain the technical heart of the matter and have a logical composition. It starts with flow phenomena and related erosion and stability problems (2, 3 and 4). Porous flow is the next step with a small addition concerning geotechnical issues (5 and 6). Wind wave phenomena are treated in chapter 7 and related erosion and stability problems in chapter 8. Chapter 9 deals with all aspects of erosion and stability related to ships. This chapter refers to many of the previous chapters, since the ship-related phenomena contain both elements of flow and waves.

Chapter 10 is on dimensioning of protections and mainly deals with probabilistic methods. Chapter 11 contains examples of protections as made in several places in the world, the focus being on The Netherlands. Chapter 12 looks into environmental aspects, with a focus on nature-friendly protections. The construction of protection works is the subject of chapter 13.

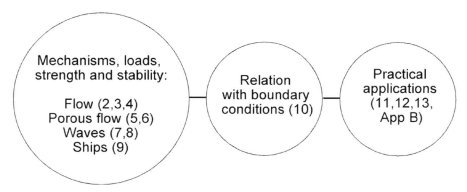

Figure 1-21 Main structure of this book

Consequently implementing the order of subjects appeared to be impossible. Filters are treated in chapter 6, including filters under wave loads, while waves are not treated until chapter 7. This has been done because a special section on filters in chapter 7 appeared to become too insignificant. The alternative of treating porous flow and filters after waves was not attractive, as porous flow plays a role in the stability of block revetments.

The main text in each chapter contains the basic message of this book with a one-page *summary* at the end of the chapter. *Intermezzos* sometimes clarify the main text or give some historic background. *Examples* are intended to illustrate the application of formulas. Some chapters have *appendices*. These are, by definition, no part of the main text. They serve as background information on subjects that are supposed to have been studied but that have possibly been forgotten somewhat. The same is valid for *Reminder I* which contains simple equations that you should know by heart already, but the reminder comes in handy when you do not. *Reminder II* contains some interesting details about the contents of this book. They may sometimes be overlooked easily in the avalanche of information, but can come in handy when one is confronted with a protection problem.

There are two general appendices: A and B. Appendix A gives information on materials, to be used in the many formulas in the book. Appendix B gives some elaborated example cases.

2 FLOW
Loads

Outflow Eastern Scheldt storm surge barrier (courtesy Rijkwaterstaat)

2.1 **Introduction**

When designing a protection it is necessary to have rather detailed information about the velocity field. For many projects, flow data is available from historical records or from a network model which calculates overall quantities, like discharges and hence, given the geometry of the situation, as average velocities: $\bar{u} = Q/A$.

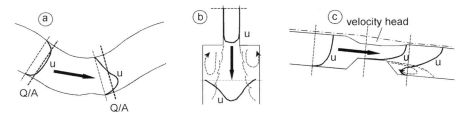

Figure 2-1 Velocity field in various situations

For the design of a bank protection in a river bend (see Figure 2-1a), the velocity near the bank must be known, which can be deducted from measurements in the river or in a scale model, from a numerical model or from a sketch, based on some understanding of the flow. In Figure 2-1b, it is inappropriate to work with a velocity averaged over the cross section downstream of the outflow, since the flow direction in some parts is opposite to the main flow direction due to separation between the main flow and the flow near the side-walls. Figure 2-1c shows a similar situation in a vertical cross section. Working with averaged velocity values ($\bar{u} = Q/A$), e.g. with Chezy's law for uniform flow: $\bar{u} = C\sqrt{(Ri)}$, would produce nearly the same value for the velocity upstream and downstream of the sill, since the geometry is the same. The figure, however, shows a completely different flow situation. Upstream, the velocity is well represented with a logarithmic profile, as can be expected in a (stationary) uniform flow. Downstream of the sill there is a flow separation with an eddy in which the flow direction near the bottom is opposite to the mainstream. The eddy can be seen as an ill-defined boundary for the flow, which also influences the turbulence level. In practice, the flow is always turbulent, but at the transition between main flow and eddies, the turbulence will be much greater and will persevere, far downstream from the transition. Protections of the interface between water and soil in these areas need to be relatively strong, because without protection, the erosion will be considerable.

Understanding the way water flows is paramount for every hydraulic engineer and a sketch of the velocity field should mark the start of every project. For such a sketch it is necessary to go from average values of the velocity to a local value and requires some understanding of the turbulence. In general one can say that it is necessary to have some insight in what is happening *inside* the water. Although turbulence is one of the most complex subjects in hydraulics, some basic facets of turbulence will be reviewed in the following section, focussing on phenomena rather than on formulas.

2.2 Turbulence

For the physical background of turbulence see appendix 2.8.2. This section discusses the characterization and importance of turbulence in hydraulic engineering. One of the most striking features of turbulent motion is that the velocity and pressure show irregular fluctuations, see Figure 2-2.

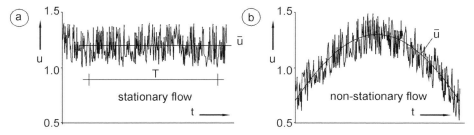

Figure 2-2 Velocity registration in turbulent flow

The abundance of definitions of turbulence indicate that the subject is complicated. The definition according to Hinze, 1975 is: *"Turbulent fluid motion is an irregular motion, but statistically distinct average values can be discerned and can be described by laws of probability"*. To do so, velocities are averaged over a certain period of time, "smoothing" out the turbulent fluctuations. The values of velocity and pressure can be written as:

$$u = \bar{u} + u' \quad v = \bar{v} + v' \quad w = \bar{w} + w' \quad p = \bar{p} + p' \tag{2.1}$$

in which ⁻ indicates the average value and ´ a measure of the fluctuations. The averaging period T (see Figure 2-2a), sometimes called the *turbulence period*, is chosen such that it is long enough to smooth out turbulence and short compared with the principal motion. Figure 2-2a is valid for a *stationary* flow, where stationary means that the *average* is constant in time. A turbulent signal in a non-stationary flow can be smoothed out, using a moving average, see Figure 2-2b.

The next step in obtaining *statistically distinct average values* is defining a measure of the intensity of the velocity fluctuations. This can not be done directly with the averages of the fluctuations, since these are 0 by definition, see Figure 2-2. Therefore the squares of the fluctuations are used and are averaged. The intensity is now defined as the square root of this average and again has the same dimension as a velocity: the root-mean-square value (r.m.s., which is equal to the standard deviation in a Gaussian distributed signal).

Turbulence can then be expressed in various ways, such as:

$$k = \frac{1}{2}\left(\overline{u'^2} + \overline{v'^2} + \overline{w'^2}\right), \; r_u = \frac{\sqrt{\overline{u'^2}}}{\overline{u}}, r_v = \frac{\sqrt{\overline{v'^2}}}{\overline{u}}, r_w = \frac{\sqrt{\overline{w'^2}}}{\overline{u}} \qquad (2.2)$$

k represents the total kinetic energy in a turbulent flow, r the relative fluctuation intensities of u, v and w, all of them compared with the main flow component, which is customary. When r is used without any index, it means r_u. One should always be aware of the parameter that is used to make the rms-value dimensionless. It is possible to relate the turbulent fluctuations (u') to an average value ($\overline{u}$) measured in a different location.

Reynolds stresses

Appendix 2.8.1, on basic equations, shows that, in a 2-dimensional situation, extra normal and shear stresses appear in the momentum equation, due to the use of an average velocity:

$$\rho\left(\frac{\partial \overline{u}}{\partial t} + \overline{u}\frac{\partial \overline{u}}{\partial x} + \overline{w}\frac{\partial \overline{u}}{\partial z}\right) = -\frac{\partial \overline{p}}{\partial x} + \mu\frac{\partial^2 \overline{u}}{\partial z^2} - \rho\left(\frac{\partial \overline{u'^2}}{\partial x} + \frac{\partial \overline{u'w'}}{\partial z}\right)$$

$$\qquad \text{inertia} \qquad\qquad \text{press.} \quad \text{visc.} \quad \text{Reynolds stresses} \qquad (2.3)$$

$$|{-}| \qquad |{-}{-}{-}{-}{-}{-}{-}{-}{-}|$$

$$\text{mean values} \qquad\qquad\qquad \text{turb. fluctuations}$$

These extra terms originate from the non-linear convective inertia terms and have the same dimension as stresses, but how logical is it to consider them as stresses? The following is a qualitative analogy with elementary mechanics (adapted from LeMéhauté, 1976).

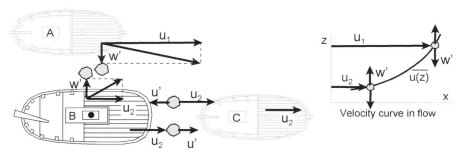

Figure 2-3 Exchange of momentum due to turbulence

Consider the quarrelling fishermen in Figure 2-3. A and B throw stones with mass m at each other with relative speed w'. When they hit the target, with the vectorially added velocities w' and u_1 or u_2 respectively, the stone from A accelerates B, because

$u_1 > u_2$ and, reversely, the stone from B decelerates A. The exchange of momentum is equal to $m(u_1-u_2)$, negative for A and positive for B.

Another analogy could be found in people walking in a very crowded street where those on the sidewalk walk slower than those in the middle of the street. When being pushed from the sidewalk, a pedestrian will be accelerated by the faster moving crowd and vice versa.

A simular situation exists in flowing water with a velocity gradient, see the velocity curve in Figure 2-3. Consider a mean flow with $\bar{u} = \bar{u}(z)$, and $d\bar{u}/dz > 0$. The particles that travel upwards arrive at a layer with a higher velocity $\bar{u}$. These particles (or lumps) preserve their original velocity causing a negative component u', thus decelerating the flow in x-direction. Conversely, the particles arriving from above give rise to a positive u', thus accelerating the flow. The flux of mass (per unit of area and per unit of time) between the layers is equal to $\rho\, w'$ (compare the masses of the stones thrown from the ships in Figure 2-3), hence the transferred momentum $= \rho w'(u_1-u_2)$ which looks similar to $\rho\, w'u'$ (with u' as a measure of u_1-u_2). The vertical flux of turbulent momentum per unit surface area is then equivalent to a stress:

$$M_z = \rho \cdot u' \cdot w' = \tau_{xz}\left(= \frac{\text{kg}}{\text{m}^3}\,\frac{\text{m}}{\text{s}}\,\frac{\text{m}}{\text{s}} = \frac{\text{kg m}}{\text{s}^2\text{m}^2} = \frac{\text{N}}{\text{m}^2} \right) \tag{2.4}$$

Similarly, one can see the analogy between the Reynolds normal stress ($\rho\, u'u'$) and stones thrown between B and C in the figure. Both stones result in a net force to the right between B and C. Due to this stress, the force acting on a motionless body in a current with a certain velocity can be 10% greater than the force acting on an identical body dragged with the same velocity through water at rest.

Flow resistance

In laminar flow the resistance, expressed as a shear stress, is proportional with the flow velocity. In turbulent flow, the quadratic terms in equation (2.3) become dominant and the relation between τ and u becomes quadratic, see Figure 2-4a:

$$\tau = c_f \rho u^2 \tag{2.5}$$

Figure 2-4b shows the velocity distribution in a pipe or channel for laminar and turbulent flow. The velocity is more homogeneous in a turbulent flow, as a result of the turbulent exchange of momentum.

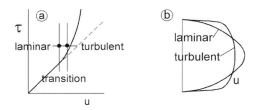

Figure 2-4 Resistance in laminar and turbulent flow

Intermezzo 2-1

The Reynolds stresses appear in equation (2.3) due to the use of average velocities, see appendix 2.8.1. But how can extra forces be created just because a mathematical procedure is followed? And who informs the water of these extra forces? Of course, averaging cannot create forces and water flows nicely, even without our equations.	When the flow is laminar, the fluctuations in equation (2.3) are equal to 0 and there are no Reynolds stresses. When the flow becomes turbulent, the fluctuations in Figure 2-2a create extra forces as was deducted using Figure 2-3. Since we use $\bar{u}$, both in laminar and turbulent flow, we would omit these extra forces without the Reynold stresses.

2.3 Wall flow

2.3.1 Uniform flow

The most elementary and one of the best known cases of wall flow is uniform flow in a channel. All of the phases of the energy cascade in section 2.8.2 are present in every cross-section of the flow. The slope causes a continuous transformation of potential energy, via kinetic energy in the main flow and in the turbulent eddies, into heat. An equilibrium exists between the bottom shear stress and the component of the fluid pressure on the slope, see Figure 2-5.

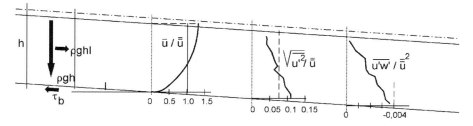

Figure 2-5 Uniform flow

The bottom shear stress is also related to the velocity, see equation (2.5). The velocity averaged over the height and in time is used (two overbars, in everyday use often reduced to one, or even none). This leads to:

$$\tau_b = \rho g h I = c_f \rho \, \bar{\bar{u}}^2 \, (= \rho u_*^2 = \rho \overline{u_b' w_b'}) \quad \Rightarrow \quad \bar{\bar{u}} = \frac{1}{\sqrt{c_f}} \sqrt{g h I} \qquad (2.6)$$

u_* is the so-called shear velocity, defined as: $u_* = \sqrt{\tau / \rho}$. c_f is a dimensionless friction coefficient. The use of h is only correct for an infinitely wide channel; otherwise the hydraulic radius R should be used.

In traditional hydraulics, empirical relations are used, such as:

$$\text{Chezy:} \quad \bar{\bar{u}} = C\sqrt{RI} \qquad \text{with: } C = \sqrt{\frac{g}{c_f}}$$

$$\text{Manning: } \bar{\bar{u}} = \frac{1}{n} R^{2/3} \sqrt{I} \qquad \text{with: } n = R^{1/6} \sqrt{\frac{c_f}{g}} \qquad (2.7)$$

C and n are not dimensionless; both definitions contain the acceleration of gravity, g. So gravity, which is responsible for the flow, is only implicitly present in these empirical equations! C is expressed in [m$^{1/2}$/s] and n in [s/m$^{1/3}$]. n is a roughness coefficient, while C is actually a "smoothness" coefficient (a large value of C means little roughness). From equations (2.6) and (2.7) together with the definition of u_* follows: $u_* = \bar{\bar{u}} \sqrt{g} / C$.

C, which is used in this book, can be related to the so-called equivalent sand roughness according to Nikuradse-Colebrook, for a hydraulically rough situation as follows:

$$C = \frac{\sqrt{g}}{\kappa} \ln \frac{12 R}{k_r} \approx 18 \log \frac{12 R}{k_r} \qquad (k_r \text{ is equivalent roughness}) \qquad (2.8)$$

For a smooth bed with grains, k_r usually equals several times the characteristic grain diameter (see chapter 3). For a moving bed, higher values are possible due to the formation of ripples or dunes.

Figure 2-5 also gives some measurements in a uniform flow. One bar over u means averaged over the turbulence period and the double bar means averaged over the turbulence period and over the waterdepth. For practical reasons, this notation will not be used consequently in the following sections. The vertical velocity profile is logarithmic with an average velocity at about 0.4 times the waterdepth from the bottom. The turbulent fluctuations can be approximated with:

$$\bar{r} = \frac{1}{\bar{\bar{u}} \, h} \int_0^h \sqrt{\overline{u'^2}(z)} \, dz = 1.2 \frac{\sqrt{g}}{C} \qquad \frac{\overline{u_b' w_b'}}{\bar{\bar{u}}^2} = \frac{g}{C^2} \qquad (2.9)$$

The expression for the depth-averaged fluctuation was derived from numerical computations by Hoffmans,1993, while the turbulent shear stress follows directly from equation (2.6) and (2.7). Now that we have simple equations that express the turbulent quantities, we can calculate the Reynolds-stresses. In uniform wall flow these are a function of the roughness only. This is not really surprising, as the wall roughness is the only source of turbulence in wall flow, which is also clearly visible in the measurements in Figure 2-5. C-values are normally in the range 40 to 60 $\sqrt{}$m/s, giving depth-averaged values of r of 0.06 to 0.1

Example 2-1

20 m³/s of water flows in a 10 m wide channel with vertical banks, a bed slope of 1/1000 and a roughness of 0.2 m. What is the depth, the velocity, the Chezy-value, the relative turbulence intensity and the relative turbulent shear stress?

Calculating the depth and velocity, using the Chezy equation, is an iterative process. First, a depth of 1 m is estimated (any estimate will work). The hydraulic radius then becomes: R = bh/(b+2h) = 10/22 = 0.83 m. The Chezy-value becomes 18log(12R/k$_r$) = 18log(12*0.83/0.2) = 30.6 $\sqrt{}$m/s. Using u = C$\sqrt{}$RI, u = 30.6$\sqrt{}$0.83*0.001 = 0.883 m/s follows. The discharge would then be: Q = bhu = 8.83 m³/s which is more than twice as little as the given discharge of 20 m³/s. A new estimate of h = 2 m, with the same procedure gives C = 34.8 $\sqrt{}$m/s, u = 1.32 m/s and Q = 26.3 m³/s, so, h = 2 m is too much and a lower value has to be chosen. The final result for Q = 20 m³/s is: h = 1.67 m, C = 33.8 $\sqrt{}$m/s and u = 1.2 m/s.

Note: with sloping banks the iteration involves more parameters.

The relative turbulence intensity is approached with equation 2.9: r = 1.2 $\sqrt{}$g/C = 0.11 and the relative turbulent shear stress is: g/C² = 8.6 10⁻³.

2.3.2 Non-uniform flow

In practice, flow is never uniform. Accelerations and decelerations influence the boundary layer and the turbulence in the flow. The boundary layer is defined as the region which is influenced by the presence of the wall, in contrast to the region outside the boundary layer. In stationary, uniform flow, the boundary layer is fully developed and takes up the entire water depth, leading to the logarithmic velocity distribution, see Figure 2-5. The following illustrations show wall flow with boundary layers that are not fully developed. When water suddenly flows along a wall, a boundary layer will start to grow, see e.g. Schlichting,1968. This is illustrated in Figure 2-6, showing the growth of a boundary layer when an infinitely thin plate is placed in a flow with $u = u_0$.

The shear stress along the plate slows down the flow and the exchange of momentum will lead to the growth of the boundary layer. This growth can be estimated roughly with $\delta(x) \approx 0.02\ x\ to\ 0.03\ x$, indicating that after a distance equal to 30 - 50 times the waterdepth, the flow will be fully developed and the boundary layer will take up the entire waterdepth. With the growth of δ, the shear stress decreases and eventually, c_f reaches the value expressed in equation (2.6).

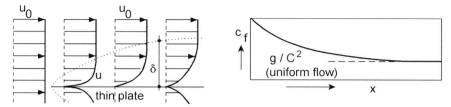

Figure 2-6 Growth of boundary layer

In the situation in Figure 2-6 there are no accelerations or decelerations on a macro level; the thin plate only creates a new boundary layer. With accelerations and decelerations, another situation arises. Acceleration is due to a pressure gradient in the flow direction and an opposite gradient leads to deceleration, both cause a change in the boundary layer thickness. The change in δ is roughly given by (see Booij, 1992):

$$\frac{d\delta}{dx} = \frac{-(4 \text{ to } 5)\delta}{u_0} \frac{d u_0}{dx} \tag{2.10}$$

indicating that acceleration causes reduction of the boundary layer thickness, while deceleration does the opposite. These changes of the boundary layer are visible in the velocity distribution, see Figure 2-7. In accelerating flow, the velocity profile becomes fuller, increasing $\partial u/\partial z$ and hence increasing the shear stress.

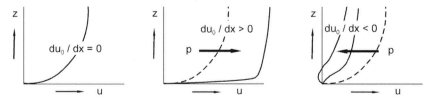

Figure 2-7 Influence of pressure gradient on velocity profile

The effect of acceleration on turbulence can be seen in Figure 2-8. In a windtunnel, turbulence is created by means of a grid. The fluctuations decrease downstream from the grid. In the contraction, the fluctuations in the flow direction (u') decrease even further, because the flow concentrates. The fluctuations perpendicular to the flow increase in the contraction. The total amount of turbulent kinetic energy, k, remains approximately constant. Due to the increased velocity in the contraction, the relative turbulence, r, using the *local* mean velocity, see equation (2.2), decreases.

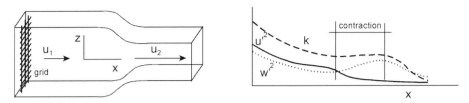

Figure 2-8 Turbulence in windtunnel contraction (from: Reynolds, 1977)

In decelerating flow the opposite happens: growth of the boundary layer with possibly flow separation, see Figure 2-7 and considerable increase in turbulence. The following section show several examples.

2.4 Free flow

2.4.1 Mixing layers

Whenever two bodies of fluid move along each other with different velocities, a mixing layer will grow between them, comparable with the growth of the boundary layer in wall flow. Figure 2-9 shows some characteristics of mixing layers (from Rajaratnam,1976). In this figure one layer flows and the other is stagnant for reasons of simplicity. This is not essential, any velocity difference causes a mixing layer. The stagnant fluid will accelerate, whereas the flowing mass will lose momentum. In the mixing layer the shear stress is intense, inducing turbulence.

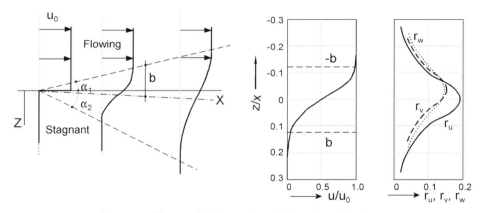

Figure 2-9 Flow, velocities and turbulence in mixing layer

The diverging angles of the mixing layer (see Figure 2-9) are $\alpha_1 \approx 5°$ and $\alpha_2 \approx 10°$, meaning that the flow penetrates into the stagnant area with a slope of 1:6. The line $u = u_0/2$ is taken as the x-axis, which deviates slightly in the direction of the stagnant zone. The width of the mixing layer, b, is defined as the distance between the points where the velocity is 0.99 u_0 and 0.5 u_0. The momentum equation shows that $b \propto x$

and from that we can assume similarity in the flow profiles, hence u/u_0 is a function of z/b only. The assumption of similarity has been confirmed in experiments, see Figure 2-9 which is based on measurements, see Rajaratnam,1976. u/u_0 has been plotted against z/x, which is, with $b \propto x$, equivalent to z/b.

The velocity distribution in the mixing layer can be described reasonably with:

$$u = u_0 \text{ for } z < -b \text{ and } u = u_0 \, e^{-0.693\left(\frac{z}{b}+1\right)^2} \text{ for } z \geq -b \text{ with } b \approx 0.12 \, x \quad (2.11)$$

Since we are interested in turbulent velocity fluctuations, which influence stability and erosion, some measurements of r are also presented in Figure 2-9. Note that the fluctuations in all directions have been made dimensionless by using the driving velocity of the shear layer: u_0. The maximum of u' occurs where the velocity gradient is maximum and where $u = 0.5 \, u_0$. Compared with the local velocity, the velocity fluctuations are twice as high. The total turbulent kinetic energy in the centre of the mixing layer, related to u_0, is approximately: $k/u_0^2 = (r_u^2 + r_v^2 + r_w^2)/2 \approx 1/2*(0.2^2 + 0.15^2 + 0.15^2) \approx 0.045$, see equation (2.2).

2.4.2 Jets

A jet flows into a large body of water with a surplus velocity compared with the ambient fluid. Many flow situations in hydraulic engineering show resemblance to jets, e.g. flow from a culvert or the flow behind a ship's propellor. The falling watermass in a plunging breaker (see chapter 7) also has jet-like features. A plane jet is a two-dimensional outflow from an orifice and a circular jet is 3-dimensional with axial symmetry. The jet attracts water from the ambient fluid, causing an increase the discharge in the flow direction with a constant momentum flux. The extremes are at one end an infinitely small orifice with an infinitely high velocity and at the other end an infinitely large flow area with an infinitely small velocity, but with the same momentum flux. Figure 2-10 shows some characteristics of jets.

Close to the orifice (the flow development region) turbulence penetrates into the core of the jet. This is the mixing layer as described in the previous section. The velocity in the centre-line is equal to u_0. After the flow has developed fully, the velocity decreases, the jet spreads and the dimensionless velocity profile no longer changes with x, see Figure 2-10.

Only results relevant to hydraulic engineering are presented. For a detailed description of jet phenomena, the reader is referred to Rajaratnam, 1976. The velocity distribution can be described as a Gaussian curve with only two parameters, u_m (u in the center of the jet) and b (a typical width, usually defined where $u = u_m/2$). Plots of u/u_m against z/b (for plane jets; for circular jets against R/b, R being the radial distance from the centre) are similar for all x (in the developed region!, see

Figure 2-10). For plane jets: $u_m \propto 1 / \sqrt{x}$ and $b \propto x$ can be derived and for circular jets: $u_m \propto 1/x$ and $b \propto x$. Numerical values for these proportionalities must be determined by experiments. For free jets the following proportionalities were found, see also Figure 2-10:

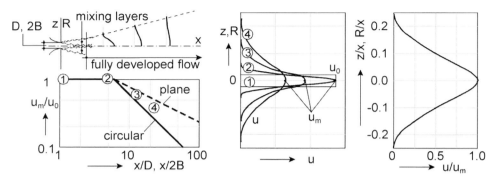

Figure 2-10 Flow and velocities in jets

$$\text{Plane jets: } u_m = \frac{3.5\, u_0}{\sqrt{x/B}} \quad b = 0.1\, x \quad u = u_m\, e^{\left(-0.693\left(\frac{z}{b}\right)^2\right)}$$

$$\text{(2.12)}$$

$$\text{Circular jets: } u_m = \frac{6.3\, u_0}{x/D} \quad b = 0.1\, x \quad u = u_m\, e^{\left(-0.693\left(\frac{R}{b}\right)^2\right)}$$

These expressions are only valid in the region of fully developed flow, which starts at about $x = 12B$ for plane jets (B is half the width of the orifice) and $x = 6D$ for circular jets (D is the radius of the orifice). In the flow development region the velocity can be approximated with u_0.

Since we are again interested in turbulent fluctuations, Figure 2-11 shows some measurements for a circular jet: Figure 2-11a in the centre-line and Figure 2-11b from the centre to the edges of the jet. Note that the fluctuations in all directions are relative to the velocity in x-direction, namely with u_m, which is the maximum velocity in the centre-line, which decreases proportionally with x. The relative fluctuations in the centre-line become constant with increasing x, reaching a value of about 30% in the centre-line for r_u. For R or $z = b \approx 0.1x$ the velocity is, by definition, $u_m/2$. The fluctuation there $(R/x = 0.1x/x = 0.1)$ is about 25% relative to u_m, so, relative to the local velocity, it is about 50%.

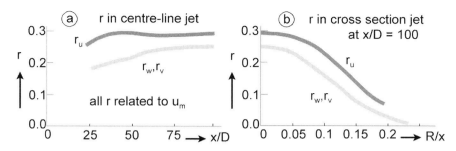

Figure 2-11 Turbulent fluctiations in circular jet

Example 2-2

A circular jet flows with 6 m/s from a nozzle with a diameter of 1.5 m. What is the velocity at a location 15 m downstream from the nozzle and 1 m from the axis of the jet? From equation 2.12 the velocity in the jet axis 15 m downstream follows: $u_m = 6.3u_0/(x/D)$ = 6.3*6/10 = 3.78 m/s. The velocity 1 m from the axis is: $u = u_m exp(-0.693(R/0.1x)^2 = 2.78$ m/s.

2.5 Combination of wall flow and free flow

2.5.1 Flow separation

Most flow situations are combinations or alternations of wall flow and free flow. Examples are flow along sills, abutments, groynes or bridge piers but also constrictions and expansions in a river or canal. For the archetypes of geometrical variations (vertical and horizontal constrictions and expansions, detached bodies), acceleration, deceleration and turbulence are discussed and illustrated with experimental results. Again, the focus is on phenomena rather than on formulas.

In these situations both boundary layers and mixing layers are present. Flow separation can be seen as the transition from wall flow to free flow. In the case of a sharp edge (see Figure 2-12a) it is easy to see that the flow has to separate, since no separation would mean that a streamline would make a right angle with a radius $R = 0$, leading to an infinite centripetal force ($F_c = u^2/R$).

So, the water cannot follow such a sharp curve and goes straight on. Here, the relation between (convective) inertia and viscosity plays a role. With (very) low *Re*-numbers, the flow can make the turn. Viscosity can then be seen as the "tension strength" of a fluid. When there is no fixed separation point, the situation is like in Figure 2-12b. From B onwards, the velocity decreases causing growth of the boundary layer up till point D (see also Figure 2-7 and equation (2.10)) and eventually flow separation. This point of separation depends on the shape, *Re*-number and roughness of the body surface and can not be predicted accurately for an arbitrary body. Flow retardation and separation usually increase turbulence and hence the load caused by the flow.

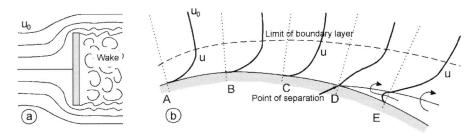

Figure 2-12 Flow separation around blunt and round body

2.5.2 Vertical expansion (backward-facing step)

A sudden increase in depth in the flow direction is often indicated as a backward facing step. The uniform wall flow before the step becomes free with a mixing layer immediately downstream of the step and eventually becomes uniform wall flow again with an increased depth. Figure 2-13 shows some measurements near a backward-facing step, mainly from Nakagawa/Nezu,1987. Figure 2-13a shows the flow situation with some characteristic areas. Between the mixing layer and the bottom, a recirculation zone exists, which is absent in a completely free mixing layer. The location where the main flow "touches down" is the reattachment point. From there a new wall boundary layer grows and the flow tends to reach a new equilibrium situation. Figure 2-13b, c and d show the vertical distributions of $\bar{u}/\bar{u}_1$, r_u (also related to $\bar{u}_1$) and $u'w'/\bar{u}_1^2$, respectively. $\bar{u}_1$ is the maximum (time averaged) velocity on top of the step. The mixing layer and the tendency to revert to uniform wall flow are clearly visible. The velocity distribution and the values of r_u correspond quite well with equation (2.11) and Figure 2-9. The reattachment point lies 5-7 times the stepheight downstream of the step. The area near and slightly downstream of the reattachment point, appears to be the critical area of attack on a bottom protection, see also chapter 3. It is clear that 10 times the stepheight downstream of the step, the flow is still far from uniform.

Figure 2-13e shows the deviations from hydrostatic pressure ($\rho g(h\text{-}z)$) made dimensionless with the pressure accompanying the velocity on top of the step ($\Delta p/2\rho u_1^2$). Near the reattachment point, where the flow hits the bottom, the pressure is high. In the recirculation zone the pressure is low due to the entrainment of water into the mainstream. This pressure pattern causes flow from the reattachment point in a direction opposite to that of the mainstream.

For calculations of stone stability and scour it is important to have an idea of the degree of recovery of the flow as a function of the distance downstream of the step. Figure 2-13f is derived from Hoffmans,1993 and shows the transition of the turbulence in the mixing layer to that of the new wall boundary layer. The value of k/u_1^2 downstream of the step is 0.045 as found from Figure 2-9, section 2.4, while the final turbulence in the boundary layer is calculated according to an expression similar to equation (2.9).

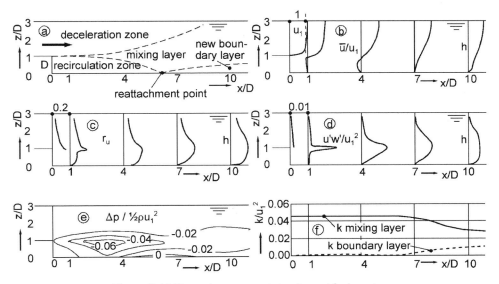

Figure 2-13 Flow phenomena in backward-facing step

For locations downstream of the reattachment point, Hoffmans proposes to calculate the relative depth averaged turbulence, r as follows:

$$r(x) = \sqrt{0.5\, k_0 \left[1 - \frac{D}{h}\right]^{-2} \left[\frac{x}{\lambda} + 1\right]^{-1.08} + 1.45\, \frac{g}{C^2}}$$
(2.13)

in which λ is a relaxation length ($\lambda \approx 6.67\, h$). k_0 is the relative turbulent energy in the mixing layer (≈ 0.045). The factor 0.5 accounts for the substitution of k_0 by the depth-averaged r. The velocity on top of the step is substituted by the velocity downstream with a factor ($1\text{-}D/h$), derived from the continuity equation. x is the *distance downstream of the reattachment point* ($\approx 6D$ downstream of the step). The second term represents the equilibrium value of r in uniform wall flow.

2.5.3 *Vertical constriction and expansion (sill)*

A sill is a good example of a vertical constriction. At Delft University of Technology, research on the flow over a sill has been carried out (Blom,1993). Computations were done and experiments were carried out in a flume. Figure 2-14 shows measurements obtained with the experiments. The slopes of the sill were such that no real flow separation occurred.

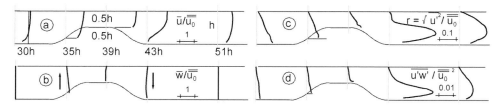

Figure 2-14 Flow characteristics around sill (from Blom, 1993)

In Figure 2-14a we see the changes of the velocity: a logarithmic profile before the sill, accelerating flow and a more rectangular profile on top, decelerating flow and almost separating flow beyond the sill and a slow recovery to a logarithmic profile in uniform flow again, see also Figure 2-7. The vertical velocities are negligible, except near the sill (Figure 2-14b).

The turbulent fluctuations r_u, related to u_0 (the average velocity upstream, Q/A, Figure 2-14c), are rather small before the sill and remain more or less the same on the sill, leading to a lower local r-value (the average velocity doubles). Beyond the sill the turbulence increases considerably, due to the deceleration and the energy transfer to turbulence (this internal process is the so-called Carnot energy-loss in basic hydraulics). The r-value beyond the sill is about 0.3. This is slightly lower than in a mixing layer, where it would be about 0.35 - 0.4 (~ 0.2 related to the velocity on top of the sill, twice as much related to the average downstream velocity). So, even without flow separation, this flow pattern closely resembles the pattern of a backward facing step and deceleration always causes a considerable increase in turbulence.

In the acceleration zone, the shear stress tends to increase, which is visible in Figure 2-14d. In the deceleration zone, the shear stress is great with the maximum velocity gradient and much smaller at the bottom. It is good to bear in mind that the shear stress behind a constriction does not determine the load; the Reynolds normal stresses can be more important. Beyond the sill, the flow slowly recovers to uniform flow.

2.5.4 *Horizontal expansion*

The situation is similar to the backward facing step of section 2.5.2, but now the flow is constricted in a horizontal direction. At DUT this situation was investigated (see e.g. Tukker, 1998 and Zuurveld, 1998); Figure 2-15 shows some results.

A mixing layer originates beyond the sudden expansion, but the growth is now limited compared with the situation in section 2.4.1 due to bottom friction (Figure 2-15a). Both near the bottom and the surface, the velocity spreads into the stagnant zone, again with recirculation beyond the expansion (Figure 2-15b). The turbulent fluctuations show large differences between the bottom and the surface (Figure 2-15c): the effect of the mixing layer is only visible at the surface, at the bottom the friction dominates. Figure 2-15d shows the peak velocities near the bottom. These peak velocities ($\bar{u} + 3\sigma = [1 + 3r]*\bar{u}$) are often held responsible for stability and erosion. It appears that the highest peak velocities near the bottom occur in the mainstream and not in the mixing layer.

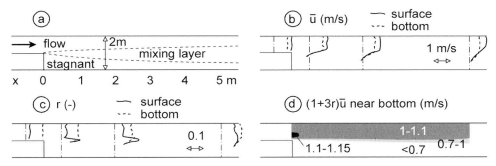

Figure 2-15 Flow characteristics in horizontal expansion (Zuurveld,1998)

2.5.5 Horizontal constriction and expansion (groyne)

Abutments and groynes are clear examples of horizontal constrictions. Natural variations in the width and protruding points in a river show the same phenomena. For a study on scour and grain stability at Delft University of Technology, 1/6 of the width of a flume was blocked and the flow pattern for this situation was investigated, see Ariëns, 1993.

Figure 2-16 shows measured values of the velocity (vertically averaged and averaged over the turbulence period, see equation (2.1)), $\bar{u}$, the relative turbulence (related to the *local* value of $\bar{u}$), r and the (absolute) value of the peak velocity, $(1 + 3r)\bar{u}$. The acceleration and deceleration are clearly visible in the figure showing the average velocity. Acceleration again causes a decrease in the *relative* turbulence from 8% to less than 5%, with again an increase in the deceleration zone. In the mixing layer, the relative turbulence reaches high values. (The mixing layer is now curved, because the flow is forced from the side of the flume to the centre.)

The diagram representing peak velocities contains a large zone with high values. Now, in contrast with the expansion in section 2.5.4, the area with the highest peak velocities coincides largely with the mixing layer. This is, however, due to the velocity pattern associated with the acceleration around the groyne.

2.5.6 Detached bodies

Examples of detached bodies are bridge or jetty piers, nautical structures and offshore platforms. An elementary shape of a detached body, which has been studied intensively, is a cylinder. As this shape is important to the engineering practice, it is useful to pay attention to the complicated flow characteristics around a cylinder. To a large extent, the flow pattern and the forces in the flow are determined by flow separation around the cylinder but as the separation depends on the roughness of the surface and the *Re*-number (now defined as *uD/v* where *D* is the diameter of the cylinder), it is still very hard to predict. Moreover, the flow around a cylinder is truly 3-dimensional.

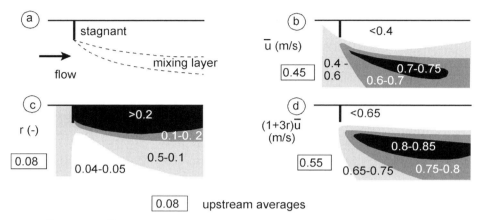

Figure 2-16 Flow characteristics for a horizontal constriction (from Ariens, 1993)

Potential flow (see e.g. LeMehauté,1976) is a classical approach to flow around a cylinder. It neglects some of the above-mentioned complications but is still useful as a starter. Figure 2-17a and c show streamlines and velocities (relative to the upstream average) according to potential flow. Figure 2-17 b and d show measurements near the bottom. A comparison of a with b shows the merits and the limitations of the potential approach. In the upstream area and in the acceleration area, roughly up to the centre-line of the cylinder, the similarity is quite good. In the deceleration area, the flow separates and the whole pattern changes. In the wake of the cylinder, a recirculation occurs like we have seen before in the case of an expansion. The potential theory predicts a doubling of the velocity alongside the cylinder. The measurements near the bottom show the same, albeit with lower absolute values near the bottom.

Figure 2-17e shows the water surface according to potential flow and as measured. Along the central streamline, the water in the vicinity of the cylinder decelerates in front of the body, accelerates at the sides and decelerates again at the back. The associated waterlevel, rises, falls and rises again (see points 1-2-3) with the local velocity head $(u^2/2g)$. The conversion of potential energy into kinetic energy on its way from 1 to 2 can be considered a "ride downhill", from 2 to 3 it is "uphill" into an area with increasing pressure. In a perfect fluid there is no friction and the pressures in 1 and 3 are equal (pressure recovery), hence the net force on the body is zero.

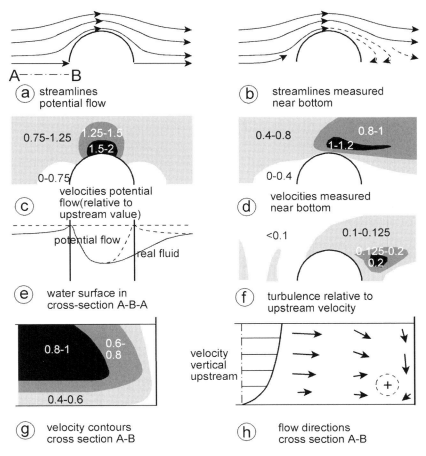

Figure 2-17 Flow details around cylinder (from Melville/Raudkivi)

The viscosity of a real fluid, however, changes the picture completely. The friction in the boundary layer stops the uphill motion between 2 and 3 causing flow separation from the body. The flow separation and wake result in an extra drag force on the body (in addition to the friction along the body).

Figure 2-17 illustrates some more interesting details of the flow around a cylinder (from Melville/Raudkivi,1977). Figure 2-17f shows the turbulence intensity ($r = rms(u')/u_0$) near the bottom. In the acceleration area on the sides of the cylinder the turbulence intensity diminishes, as can be seen in the diagram. In the mixing layer between the wake behind the cylinder and the main flow, the maximum intensity, around 0.2, is found.

Figure 2-17g and h show the vertical cross-section at the centre-line of the cylinder. Figure 2-17g gives the contourlines of the velocity in the direction of the main flow. The deceleration is clearly visible. The rise in pressure that comes with this decrease of velocity causes another interesting phenomenon. The pressure rises with $\Delta p = $

$1/2\rho\, u^2$. The velocity near the water surface is higher than near the bottom, hence Δp, with a constant piezometric level across the waterdepth, results in a pressure gradient downwards. This causes a vertical jet (Figure 2-17h) which plays an important role in the scour around a cylinder, see chapter 4. Theoretically, the vertical velocity in the jet can reach the same value as the approach velocity but measurements show less than half that value. At the foot of the pier, a circulation is visible. In chapter 4 we will see that this is the beginning of the so-called horse-shoe vortex. Some authors claim that this vortex is always present around a pier, others say that it originates in the scour hole in front of the pier and that without scour there is no vortex. More information on flow around cylindrical piers is found in e.g. Melville/Raudkivi, 1977, Raudkivi, 1985, Zdravkovich, 1997 and Ahmed/ Rajaratnam, 1998.

2.6 Load reduction

Sometimes it can be desirable to reduce the flow as a means of protection or to avoid the need for heavy protections. Figure 2-18 shows three ways to reduce the flow, two at a river bank and one at an outflow.

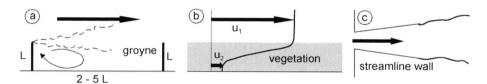

Figure 2-18 Possible reduction of flow induced loads

Groynes keep the main flow at a distance from the banks. In rivers, they are usually part of a complete river training scheme, not just to reduce the velocity, but also to maintain a stable channel with sufficient depth for navigation. In this case the focus is on flow reduction. At the tip of a groyne, the flow separates and an eddy is formed (see Figure 2-18a). To keep the main flow away from the bank, the distance between two groynes should be less than 5 times their length. A more strict demand is the eddy pattern. For a proper guidance of the main flow, there should be one eddy between two groynes which requires a distance between groynes of about twice the groyne length. The velocity in the eddy is about 1/3 of the velocity in the main current, so the reduction is about 2/3. For more information see Jansen, 1979. Groynes, however, are very costly and cause their own problems, e.g. scouring holes at the tip. The cost-benefit relation should be considered in all cases.

Instead of using groynes, flow near the bank can be reduced in a diffuse way by increasing the roughness. This can be done by making the bottom more rough or by placing resistance elements in the flow. The latter is the case when vegetation, e.g.

reed or trees, is planted along the bank. It is possible to express the resistance of a roughened bed with a Chezy-coefficient. A first approximation of the velocity reduction is found by assuming that the waterlevel slope is equal for the main flow and the bank, see Figure 2-18b:

$$u = C \sqrt{h I} \rightarrow \frac{u_2}{u_1} = \frac{C_2}{C_1} \sqrt{\frac{h_2}{h_1}}$$ (2.14)

In chapter 12 examples of roughness values for various types of vegetation will be given.

When the turbulence caused by a mixing layer has to be avoided or postponed till a point where the velocity is lower has been reached, walls are often streamlined with slopes of 1 : 8 or even less to prevent flow separation, see Figure 2-18c. This is often done with outflows at culverts and similar structures. The fact that the flow does not separate within the streamlined walls does not mean that there is no increase in the turbulence intensity. Deceleration always causes more turbulence, so, the main advantage is the reduction of the mean flow velocity.

2.7 Summary

To design a protection, or to judge whether a protection is necessary, detailed information on the velocity field is needed to determine the loads. This can be obtained from measurements, a flow model or a sketch, based on overall discharge data and insight into the flow phenomena. In fact, every preliminary design should start with a simple sketch on a sheet of paper.

The same is true for turbulence in the velocity field, albeit that the determination of turbulence is much more difficult and uncertain. The turbulent velocity fluctuations are a result of velocity differences, either between a flow and a wall (wall turbulence) or between two fluid bodies (free turbulence). Turbulence is always coupled with loss of kinetic energy, so, especially in deceleration areas great turbulence can be expected. In accelerating flow the relative turbulence is less, but the shear stress increases. The turbulent fluctuations cause so-called Reynolds stresses: a normal stress due to the velocity fluctuations in one direction and a shear stress due to the correlation between the fluctuations in two directions.

For uniform (wall) flow, relatively simple, partly empirical relations are available for the average velocity (like $\bar{u} = C\sqrt{RI}$) and accompanying expressions for wall roughness and turbulence which originates at the wall. The boundary layer in uniform flow covers the whole waterdepth. Mixing layers play an essential part in free turbulence and occur in many situations in hydraulic engineering. Another example of free turbulence can be found in jet flow, which is also frequently encountered in flow situations near hydraulic structures. A typical value for wall *turbulence is r (= r.m.s. u'/ū̄) ≈ 0.1 and for free turbulence r_{max} ≈ 0.2-0.3. These values also depend on where ū̄ is defined.*

Flow situations around bodies, like (bridge) piers, through vertical constrictions, like sills and through horizontal constrictions, like abutments and groynes, all show acceleration followed by deceleration with the accompanying turbulence. The acceleration leads to higher velocities than in uniform flow situations, causing a greater load on any interface in the neighbourhood. The same is true in the deceleration zone where the increase of the Reynolds stresses is caused by an increase of the turbulent fluctuations. Section 2.5 gives several examples of combinations of turbulent wall flow and free flow. You are advised to study these thoroughly and, above all, to imagine what is happening. This will be all the more fruitful when coupled with observations of flow phenomena everywhere around you in rivers, ditches or in your sink.

Finally, some examples of load reduction are presented in section 2.6.

2.8 Appendices

2.8.1 Basic equations

The basic equations for fluid motion (conservation of momentum and mass) will be used for demonstration purposes only and mathematical purity is sacrificed in search of simplicity. For more detail, see e.g. Schlichting (1968) or Le Méhauté (1976).

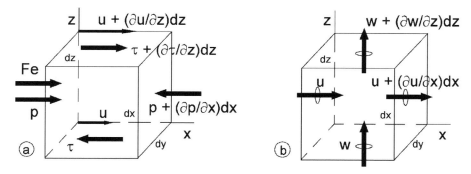

Figure 2-19 Forces and flow with regard to dxdydz

The momentum equation is in fact Newton's second law: $F = ma$. For a 2-dimensional flow the resulting force in the x-direction is (see Figure 2-19a):

$$F_x = -\frac{\partial p}{\partial x} \, dx \, (dy \, dz) + \frac{\partial \tau}{\partial z} \, dz \, (dx \, dy) + Fe(x) \tag{2.15}$$

The second term on the right-hand side of equation (2.15) represents the rate of change of viscous shear stress perpendicular to the flow (in the z-direction there is a shear stress of equal magnitude). The viscous shear is an internal stress caused by the transfer of molecular momentum. For a Newtonian fluid: $\tau = \mu\, \partial u/\partial z$, so, $\partial\tau/\partial z = \mu(\partial^2 u/\partial z^2)$, in which μ is the dynamic viscosity. The external force, *Fe*, can result from gravity, which is supposed to work in *z*-direction, so, in the *x*-direction, external forces are neglected. The acceleration in *x*-direction can be written as:

$$a_x = \frac{Du}{Dt} = \frac{\partial u}{\partial t} + \frac{\partial u}{\partial x}\frac{dx}{dt} + \frac{\partial u}{\partial z}\frac{dz}{dt} = \frac{\partial u}{\partial t} + u\frac{\partial u}{\partial x} + w\frac{\partial u}{\partial z} \quad (\frac{dx}{dt} = u\,,\ \frac{dz}{dt} = w) \quad (2.16)$$

With $m = \rho\,(dxdydz)$ this leads to:

$$m\,a_x = \rho\,(dx\,dy\,dz)\left(\frac{\partial u}{\partial t} + u\frac{\partial u}{\partial x} + w\frac{\partial u}{\partial z}\right) =$$

$$F_x = -\frac{\partial p}{\partial x}\,dx\ (dy\,dz) + \mu\,\frac{\partial^2 u}{\partial z^2}\,dz\,(dx\,dy) \tag{2.17}$$

Dividing by $(dx\,dy\,dz)$ finally gives the (simplified) *Navier-Stokes equation* for the *x*-direction, which is valid for both laminar and turbulent flow:

$$\rho\left(\frac{\partial u}{\partial t} + u\frac{\partial u}{\partial x} + w\frac{\partial u}{\partial z}\right) = -\frac{\partial p}{\partial x} + \mu\frac{\partial^2 u}{\partial z^2}$$

$$\text{local} \qquad \text{convective} \qquad\qquad \text{pressure} \quad \text{viscous} \tag{2.18}$$

$$\text{inertia} \qquad \text{inertia} \qquad\qquad \text{gradient} \quad \text{shear}$$

The **continuity equation** (conservation of mass) for a non-compressible fluid with constant density states that the net flow through a fluid element without a free surface must be zero (see Figure 2-19b):

$$\frac{\partial u}{\partial x} + \frac{\partial w}{\partial z} = 0 \tag{2.19}$$

For turbulent flow, the values of velocity and pressure in the equations can be split up. For the velocity this gives: $u = \bar{u} + u'$, in which $\bar{u}$ indicates the average value and u' the turbulent fluctuation. In order to work with average values, equation (2.18) can be averaged over the turbulence period (see Figure 2-2). The linear terms, like $\partial u/\partial t$ become:

$$\overline{\frac{\partial u}{\partial t}} = \frac{1}{T}\int_0^T \frac{\partial u}{\partial t}\,dt = \frac{\partial}{\partial t}\,\frac{1}{T}\int_0^T (\bar{u} + u')\,dt =$$

$$\frac{\partial}{\partial t}\,\frac{1}{T}\int_0^T \bar{u}\,dt + \frac{\partial}{\partial t}\,\frac{1}{T}\int_0^T u'\,dt = \frac{\partial\bar{u}}{\partial t} + 0 = \frac{\partial\bar{u}}{\partial t} \qquad (2.20)$$

since:

$$\bar{u} = \frac{1}{T}\int_0^T u\,dt \quad and \quad \bar{u}' = \frac{1}{T}\int_0^T u'\,dt = 0 \qquad (2.21)$$

The same holds for w and p, the linear terms for laminar and turbulent flow are expressed in the same type of function. This is not true, however, for the quadratic terms:

$$\overline{u^2} = \frac{1}{T}\int_0^T u^2\,dt = \frac{1}{T}\int_0^T \left(\bar{u}^2 + 2\cdot\bar{u}\cdot u' + u'^2\right)\,dt = \bar{u}^2 + \overline{u'^2}$$

$$since: \quad \frac{1}{T}\int_0^T 2\cdot\bar{u}\cdot u'\,dt = 2\,\bar{u}\,\frac{1}{T}\int_0^T u'\,dt = 2\cdot\bar{u}\cdot 0 = 0 \qquad (2.22)$$

$$and: \quad \frac{1}{T}\int_0^T u'^2\,dt = \overline{u'^2}\,(\,mean\ value\ of\ u'^2 \neq 0\ !)$$

Similarly:

$$u\,w = (\bar{u} + u')\cdot(\bar{w} + w') = \bar{u}\,\bar{w} + u'\bar{w} + w'\bar{u} + u'w' \quad hence: \quad \overline{uw} = \bar{u}\,\bar{w} + \overline{u'w'}$$
$$(2.23)$$

and equation (2.18) becomes (note: the viscous term is of the second order but linear):

$$\rho\left(\frac{\partial\bar{u}}{\partial t} + \bar{u}\,\frac{\partial\bar{u}}{\partial x} + \bar{w}\,\frac{\partial\bar{u}}{\partial z} + \overline{u'\,\frac{\partial u'}{\partial x}} + \overline{w'\,\frac{\partial u'}{\partial z}}\right) = -\frac{\partial\bar{p}}{\partial x} + \mu\,\frac{\partial^2\bar{u}}{\partial z^2}$$

| local | conv.inertia | conv.inertia | press. | visc. | (2.24) |
| inertia | by mean vel. | by fluct.vel. | grad. | shear | |

With the continuity equation, this can be rewritten to obtain the so-called *Reynolds equation*:

$$\rho \left(\frac{\partial \overline{u}}{\partial t} + \overline{u}\,\frac{\partial \overline{u}}{\partial x} + \overline{w}\,\frac{\partial \overline{u}}{\partial z} \right) = -\frac{\partial \overline{p}}{\partial x} + \mu\,\frac{\partial^2 \overline{u}}{\partial z^2} - \rho \left(\frac{\partial \overline{u'^2}}{\partial x} + \frac{\partial \overline{u'w'}}{\partial z} \right)$$

$$\qquad\qquad\text{inertia}\qquad\qquad\qquad\text{press.}\quad\text{visc.}\qquad\text{Reynolds - stresses}\qquad (2.25)$$

|- -| |- - - - - - - - - - -

mean values   turb. fluct.

The Reynolds-stresses are a consequence of working with time-averaged values. The turbulent shear stress ($u'w'$) is usually much larger than the viscous shear stress, while the turbulent normal stress (u^2) can become important in situations with flow separation, see the examples in section 2.5.

2.8.2 Why turbulence?

Everybody who has been on an aeroplane in rough weather, or on a bike in gusty wind, has some idea of turbulence. And everybody who looks carefully at the fluid motion in a river, or even in a straight flume in a laboratory, will notice the turbulent character of the flow. Turbulent means the opposite of laminar. Laminar flow is smooth and orderly, while turbulent flow is chaotic and fluctuates, even when the flow as a whole looks quiet.

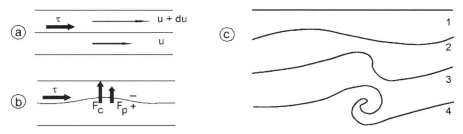

Figure 2-20 (In)stability of laminar flow

Figure 2-20 (from LeMéhauté,1976) shows the origin of turbulence. Between two fluid layers with different velocities, e.g. the outflow in Figure 2-1b, one always finds friction. A small disturbance of the flow, regardless of its cause, will induce forces that increase the disturbance. The curve in the flow line results in a centrifugal force (F_c in Figure 2-20b). The same curve causes a decrease of the velocity in the lower layer and an increase in the upper layer (continuity!) resulting in a pressure difference (F_p in Figure 2-20b), which follows from the Bernoulli equation (piezometric head + velocity head = constant). The pressure difference and the centrifugal force work in the same direction and tend to increase the instability. The result is an increasing undulation, the so-called Kelvin-Helmholtz instability, see Figure 2-20c, which shows successive stages of this instability. In the end, packets of fluid travel in every direction: the turbulent eddies.

The centrifugal force $(= \rho\, u^2/R)$ and the pressure force (proportional to the difference in velocity, hence to $\rho\, u(\partial u/\partial x)$, see section 2.8.1) are destabilizing forces. They are related to the non-linear convective inertia terms in the basic equations and are expressed in the same units. The viscous damping (represented by $\mu(\partial^2 u/\partial z^2)$, see also section 2.8.1), has a stabilizing effect as an increase in the path length in the undulation causes an increase in viscous friction. All forces can be expressed with a characteristic *velocity difference*, U, over a characteristic length, L. The ratio between destabilizing and stabilizing forces then leads to the well-known Reynolds-number:

$$Re = \frac{\rho \cdot U^2 / L}{\mu \cdot U / L^2} = \frac{U \cdot L}{\upsilon} \tag{2.26}$$

where μ is the so-called dynamic viscosity and υ the kinematic viscosity ($\upsilon = \mu / \rho$). υ is a property of the fluid and is independent of geometry or flow velocity. So, there will always be a value for U and L, for which the flow will be naturally unstable.

The chosen value L in the Re-number, depends on the geometry of the flow. In river flow the depth is a logical choice; for the flow around a cylinder, the diameter is a clear choice. For the velocity difference U, the velocity itself is the most appropriate value (the difference between the flow and the river bottom or the stagnant zone behind the cylinder). Typical values for the transition between laminar and turbulent flow are $Re = 1000\text{-}2000$. For normal circumstances, with $\upsilon \approx 10^{-6}$ m²/s, this means that the flow is already turbulent in a flume with a depth of 0.1 m, and a velocity of a few centimeters per second! In the hydraulic engineering practice, flow will always be turbulent.

From the foregoing we have learned that turbulence is caused by a velocity gradient perpendicular to the main flow direction. This gradient can exist between a wall and the flow or between two flow zones. In the first case we speak of *wall turbulence* and in the second case of **free** *turbulence*, see Figure 2-21 a and b. Wall turbulence exists where water flows along a bottom, a bank or a body. Free turbulence occurs where two masses of water are forced to move along each other with different velocities, like the outflow of a jet in a stagnant ambient fluid or the wake behind a body, in a hydraulic jump or a breaking wave.

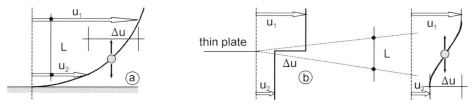

Figure 2-21 Wall turbulence and free turbulence

The energy cascade

We have seen that turbulence is induced by a velocity difference, that is too large for the molecular viscosity which can otherwise prevent the formation of turbulent eddies. Another way to look at turbulence is as follows. A rubber ball, falling from some height, gains kinetic energy due to the loss of potential energy. This kinetic energy is transferred into heat, by means of friction between the ball and the ground and internal friction inside the ball due to deformation, untill it finally comes to a halt on the ground. Usually, a ball has to bounce several times before the energy transfer is completed. Something similar goes on in decelerating flow. The kinetic energy of the mean motion is transferred into heat by the viscous shear stress: $\tau = \rho\ \upsilon(\partial u/\partial z)$. υ has a fixed value, coupled with the physical properties of the fluid. Maximum values of $\partial \bar{u}/\partial z$ in the flowing water are too small to transfer all of the kinetic energy into heat at once. Turbulence is a means of increasing the viscosity (hence the so-called turbulent viscosity) and can be interpreted as the process that allows the kinetic energy of the main flow to dissipate, leaving the system as heat due to viscous friction.

Another analogy is dissolving sugar in a cup of tea. The molecular diffusion (the mechanism that makes the sugar dissolve) and the gradient of the sugar concentration in a cup at rest have limited values and it can take some time before all of the sugar is dissolved. Stirring is an effective way of increasing the surface with a gradient such that diffusion can take place. Turbulence can then be seen as the preprocessing in the treatment of "waste" kinetic energy.

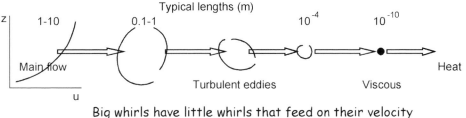

Big whirls have little whirls that feed on their velocity
and little whirls have smaller whirls and so on to viscosity
(L.F.Richardson)

Figure 2-22 Energy cascade in turbulent motion

Figure 2-22 shows the transfer process with some typical lengths. Only in the smallest eddies, the velocity gradient is large enough for the viscous friction to play an effective role in the transfer into heat.

Illustration: Energy dissipation in hydraulic jump

To demonstrate some of the phenomena discussed in the preceding, the flow in a hydraulic jump will be reviewed in detail. A hydraulic jump can be seen as the ultimate energy dissipator. From the one-dimensional momentum equation one can derive:

$$\frac{h_2}{h_1} = \frac{1}{2}\left(\sqrt{1 + 8\,Fr_1^2} - 1\right) \quad (Fr_1 = u_1 / \sqrt{gh_1}) \tag{2.27}$$

Figure 2-23a shows the profile and the main characteristics as defined in basic hydraulics. Since we are interested in what is happening inside the flow, we take a look "under the hood" of the hydraulic jump. Figure 2-23b to f show the results of measurements in a hydraulic jump with a *Fr*-value of 4, as performed by Rouse,1958.

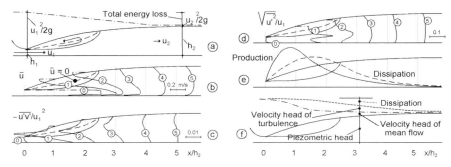

Figure 2-23 Flow and turbulence in hydraulic jump (Fr = 4, from Rouse, 1958)

Figure 2-23b shows the velocity in the jump, $\bar{u}$, averaged over the turbulence period. The spread of the velocity is clearly visible, with a reversal of the flow in the roller. Figure 2-23c shows the turbulent shear stress. Where the gradient of $\bar{u}$ is maximal, the largest values of $u'w'$ can be found. Immediately after the roller starts, the turbulent shear stress starts to grow quickly. The turbulent fluctuations in the flow direction are shown in Figure 2-23d. The fluctuations also grow where the velocity gradient is large, but they are then diffused through the water and the location of the maximum turbulence shifts. Turbulent fluctuations in the vertical direction are slightly smaller than those in the flow direction, but they show the same pattern. Figure 2-23e gives the production and dissipation of turbulence, while Figure 2-23f shows the energy content of the flow. The figures show that turbulence is indeed an intermediate phase in the transformation of kinetic energy of the main flow into heat by means of viscous shear stress. Velocity gradients are important in the transfer of energy from the main motion to the turbulent eddies.

Finally, since all motion ends in heat, it is interesting to compute the raise in temperature in a hydraulic jump. For a jump with $Fr = 4$, $h_1 = 1$ m and $u_1 = 12.5$ m/s, given that 4000 J is needed to create a temperature rise of 1 °C in 1 kg water, the total raise in temperature is less than 0.01 °C. So, turbulence may cause a lot of excitement in engineering circles, it leaves the water cold.

3 FLOW
Stability

Moved bed protection in Eastern Scheldt model (courtesy Delft Hydraulics)

3.1 Introduction

The focus in this chapter will be on the stability of loose non-cohesive grains, like rock and gravel. In addition, section 3.5 presents some stability considerations for coherent material. Natural rock is a very important construction material for protections. It is almost omnipresent in the world and relatively easy to obtain. The specific mass of natural rock normally lies between 2500 and 2700 kg/m³ while extremes of more than 3000 kg/m³ can be found. This is favourable for the stability and is difficult to match by artificial material like concrete. Rocks come in all sizes, varying from millimeters up to one or two meters, corresponding with a stone mass of around 10000 kg. Only for very heavily attacked structures, like breakwaters in deep water with high waves, this will be insufficient. Thus, as a construction material for hydraulic engineering, rock is still indispensable. Sand and gravel also consist of non-cohesive, but much smaller and rounder grains. They are not such stable construction materials, but to determine erosion, it is necessary to know the same limits of stability.

In the following sections, the stability of stones in all kinds of flow situations will be discussed, starting again with uniform flow and a (nearly) horizontal bottom. At Delft University of Technology, several studies of stability in flow have been made in recent years aiming at the translation of theoretical concepts into practical design rules. Some of them will be mentioned when appropriate.

3.2 Uniform flow - horizontal bed

3.2.1 Basic equations

Despite many research activities and an avalanche of reports, the knowledge about the behaviour of grains exposed to currents, is still very empirical. In an attempt to understand the stability of loose grains, it is necessary to know which forces make a stone move. We start with turbulent flow and a (nearly) horizontal bed.

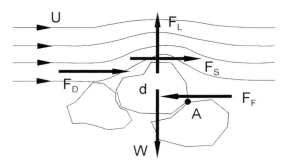

Figure 3-1 Forces on a grain in flow

It is easy to imagine that there is a critical value of the flow velocity at which a grain is no longer in equilibrium and starts to move. It is not so easy to indicate exactly how the flow induces movement and how the grains resist it. Figure 3-1 shows the forces acting on a grain. A drag force by the flow on a stone is easy to imagine. The concept of a shear stress for a single stone is somewhat more troublesome; it seems more appropriate for a certain area of a flat bed with relatively small grains. A lift force is mainly caused by the curvature of the flow around a grain. The various flow forces on the grain can be expressed as follows:

$$
\left.
\begin{aligned}
\text{Drag force: } & F_D = \text{\textperthousand } C_D \ \rho_w \ u^2 A_D \\
\text{Shear force: } & F_S = \text{\textperthousand } C_F \ \rho_w \ u^2 A_S \\
\text{Lift force: } & F_L = \text{\textperthousand } C_L \ \rho_w \ u^2 A_L
\end{aligned}
\right\} \quad F \propto \rho_w u^2 d^2 \qquad (3.1)
$$

in which C_i are coefficients of proportionality and A_i are the exposed surface areas. So, all forces are proportional to the square of the velocity (defined somewhere in the vicinity of the grain) and since the grain surfaces are proportional to the square of a representative size d (to be defined later on), the resultant load can be expressed in one term where the constant of proportionality remains unknown.

The same can be done for the equilibrium. Here again, several resisting forces and mechanisms can be present. The lift force is balanced directly by the (submerged!) weight. Shear and drag, are balanced, either by the moment around A or by the friction force (which is like the shear a somewhat peculiar notion for a single grain). It does not matter whether the horizontal, vertical or moment equilibrium is considered, only one proportionality between load and strength remains:

$$
\left.
\begin{aligned}
\Sigma H = 0: \quad & F_{D,S} = f \, x \, W = F_F \\
\Sigma V = 0: \quad & F_L = W \\
\Sigma M = 0: \quad & F_{D,S} \cdot O(d) = W \cdot O(d)
\end{aligned}
\right\} \quad \rho_w \ u_c^2 \ d^2 \ \propto \ (\rho_s - \rho_w) \, g d^3 \qquad (3.2)
$$

The velocity used is the critical velocity: u_c since we are dealing here with a stability parameter (see chapter 1). This leads to a dimensionless relation between load and strength:

$$
u_c^2 \ \propto \ \left(\frac{\rho_s - \rho_w}{\rho_w} \right) g \, d = \Delta \, g \, d \quad \rightarrow \quad u_c^2 = K \, \Delta \, g \, d \qquad (3.3)
$$

All formulae on grain stability come down to this proportionality. The constant in this relation has to be found experimentally and that is where the problem starts. There are numerous formulae but there are two names you cannot avoid when dealing with the stability of stones in flowing water: Izbash and Shields.

Izbash

Izbash,1930, expressed relation (3.3) as (with equivalent forms, used in other sources):

$$u_c = 1.2 \sqrt{2\,\Delta g\, d} \quad \text{or} \quad \frac{u_c}{\sqrt{\Delta g\, d}} = 1.7 \quad \text{or} \quad \Delta d = 0.7\,\frac{u_c^2}{2\,g} \tag{3.4}$$

There is no influence of depth in this formula; in fact Izbash did not define the place of the velocity, neither is it very clear how the diameter is defined. It is presented as a tool for first approximation in cases where a velocity near the bottom is known but the relation with the waterdepth is not clear, like a jet entering a body of water. In other cases the Shields approach is recommended.

Shields

Probably the best-known formula for uniform flow is the one by Shields from 1936. Shields gives a relation between a dimensionless shear stress and the so-called particle Reynolds-number:

$$\psi_c = \frac{\tau_c}{(\rho_s - \rho_w)g\,d} = \frac{u_{*c}^2}{\Delta g\,d} = f(\text{Re}_*) = f\!\left(\frac{u_{*c}\,d}{\upsilon}\right) \tag{3.5}$$

Shields chose the shear stress as the active force. We know that this is not necessarily the right choice, but in practice the differences are not so important. Note that the grain diameter and the shear velocity appear on both sides of the formula. ψ_c is usually called the Shields parameter and is a stability parameter which is defined using a critical value of the (shear) velocity. When the diameter present in a bed and the actual velocity, instead of the critical velocity, are used, ψ is a mobility parameter. *Note: the subscript c is not applied consequently. It is always necessary to be aware of the use of ψ!* The particle Reynolds-number indicates whether the grain protrudes into the turbulent boundary layer or stays witin the viscous sub-layer. It differs from the normal Reynolds-number and says nothing about the flow characteristics of the flow as a whole (which is usually turbulent).

Figure 3-2 shows two presentations of the Shields relation. Figure 3-2a is the classical form. For high Re_*-numbers (large grains, larger than the viscous sub-layer, turbulent flow around the grain), ψ_c is no longer dependent on Re_* and becomes constant with a value of about 0.055. The lower ψ_c-values of intermediate Re_*-numbers (indicating a lower stability) are sometimes explained with the idea that the shear stress dominates the drag forces, giving a larger arm for the moment, leading to a less favourable situation. For very low Re_*-numbers (very small grains), ψ_c increases linearly with Re_*^{-1} compensating for the square of the velocity in equation (3.5) which is no longer valid in the viscous sub-layer.

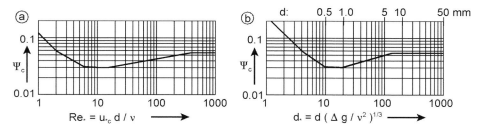

Figure 3-2 Critical shear stress according to Shields - van Rijn

Intermezzo 3-1

Many attempts have been made to determine the constant of proportionality in equation (3.3) analytically. For instance by considering the horizontal equilibrium and equating the possible friction force to the tangent of the angle of repose of the material: $\Sigma F_H = W.\tan\phi$. With $\tan\phi \approx 1$ for rocks this leads to values of the critical velocity which are much too high, even

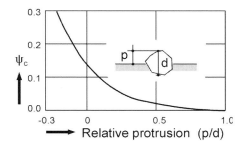

when the turbulent peak velocities are applied instead of the average value. The same is true when the moment equilibrium around point A in Figure 3-1 is considered. This is further complicated by the fact that the protrusion of an individual stone highly determines the protection by the surrounding stones and, hence, the stability. The figure shows the influence of protrusion on the value of ψ_c as found in experiments (Breusers/Raudkivi, 1991, see also Wiberg/Smith, 1987). The differences in protrusion of grains in a bed and, more in general, the differences between the size and shape in a natural material, make an analytical approach of stone stability a dead end. In fact, the whole concept of a critical velocity is a doubtful notion. The next section will deal with that problem.

Figure 3-2b gives the same stability relation, but now Re_* is replaced by a dimensionless particle diameter, d_*, as used e.g. by Van Rijn,1984 in his presentation of the Shields relation, which has the advantage that it avoids iteration on u_{*c}. The elimination of u_{*c} from the particle Re-number is possible because every grain diameter has a corresponding value of u_{*c}. d_* can be seen as the ratio of the submerged weight of a grain to the viscous forces. It can be derived that ψ_c becomes a constant for d = 6-7 mm, depending on ρ and ν. This is important, since it indicates that for stones in a normal bed protection, ψ_c can be used as a constant without having to iterate with the diameter in the Shields curve. Another important conclusion is that no scale effects will be present in tests when the stone size exceeds 6-7 mm. The use of Figure 3-2b is always recommended as iteration with u_{*c} is never necessary. For $v = 1.33*10^{-6}$ m²/s and $\rho = 2650$ kg/m³, values of the grain size in mm are indicated on the graph.

Example 3-1

What is u_{*c} for sand ($\rho = 2650$ kg/m³) with $d = 2$ mm?

In the "classical" Shields-curve (Figure 3-2a) u_{*c} appears in both axes, so iteration is necessary. Suppose you don't have the faintest idea how much u_{*c} is and you make a wild guess, say 1 m/s. Re$_*$ then becomes: $1*0.002/1.33*10^{-6} = 1500 \rightarrow \psi_c = 0.055 \rightarrow u_{*c}$ $= \sqrt{(1.65*9.81*0.002*0.055)} = 0.042$ m/s$\rightarrow$ Re$_* = 63 \rightarrow u_{*c} = 0.036$ m/s. This is also the final value. Using Figure 3-2b, you would have found directly $d_* = 0.002*(1.65*9.81/(1.33*10^{-6})^2)^{1/3} = 42 \rightarrow \psi_c = 0.04 \rightarrow u_{*c} = 0.036$ m/s.

3.2.2 Threshold of motion

In the previous section, a critical velocity was mentioned in relation to stability. The idea of a critical value of the velocity, however, is a complicated affair as you can see when an experiment is carried out in a flume. With a low velocity one or more stones will move a little. They are probably stones with a high protrusion or an otherwise unfavourable position. Having found a new position, these stone will not move anymore. So, this can not be seen as the threshold of motion for the whole bed. After increasing the flow velocity, here and there some stones will move a certain distance. A further increase of the velocity will eventually lead to transport of stones everywhere. The situation is then clearly beyond the threshold of motion. So, how can this threshold be defined?

In fact, no such thing as a critical velocity exists! Due to the irregularities of natural stones, the position, the protrusion and hence the exposure and the stability of every stone in the bed is different. The flow itself is turbulent, which means that peak pressures and velocities can be much higher than the average value. In theory there is no limit. Figure 3-3a shows the probability distribution of occurring and critical shear stress for a given situation. The average shear stress in the flow is lower than the average critical shear stress. Due to the deviations from the mean value of the load and the strength, movement of stones can still occur. It can be seen that there is no critical value below which movement is impossible.

The threshold of movement is a subjective matter when judged in an experiment. Figure 3-3b shows the Shields graph of Figure 3-2a together with the results of an investigation into incipient motion (DHL, 1969). 7 transport stages were discerned:

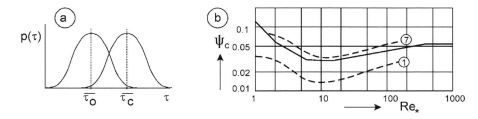

Figure 3-3 Load and strength distribution and extrapolation of transport

0. no movement at all
1. occasional movement at some locations
2. frequent movement at some locations
3. frequent movement at several locations
4. frequent movement at many locations
5. frequent movement at all locations
6. continuous movement at all locations
7. general transport of the grains.

It appeared that the Shields criterion fits stage 6 rather well. Shields found his values by extrapolating a measured transport of material to zero. Figure 3-4a gives an idea of the transport situation and an explanation of the result as obtained by Shields.

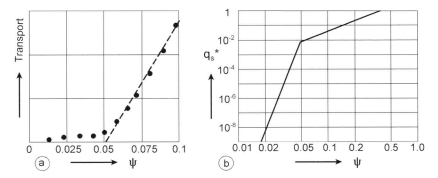

Figure 3-4 Threshold of motion: extrapolation of transport to zero (a) and Paintal (b)

Paintal,1971, performed similar tests (for rather coarse material) to Shields' tests, but did not extrapolate his transport measurements to 0. Instead, he defined the transport per m width, q_s, see Figure 3-4b:

$$
\left.\begin{array}{ll}
q_s^* = 6.56 \cdot 10^{18}\, \psi^{16} & (\text{for } \psi < 0.05) \\
q_s^* = 13\, \psi^{2.5} & (\text{for } \psi > 0.05)
\end{array}\right\}
\quad \text{with } q_s^* = \frac{q_s}{\sqrt{\Delta g\, d^3}}
\qquad (3.6)
$$

Note that ψ is now used as a mobility parameter in a transport equation. This relation gives a transport for every non zero shear stress. But it decreases with the 16th power of the shear stress, which means with the 32nd power of the velocity. Most sediment-transport formulae have a power relation of 3 to 5 with the velocity, which is in line with the relation by Paintal for $\psi > 0.05$. So, the Shields criterion for ψ_c could be seen as the start of sediment transport or erosion. Both the description of stage 6 by DHL and the kink in the Paintal relation seem to justify this. Moreover, Shields performed his research to find the beginning of sediment transport. Interesting reflections on Shields and the threshold of motion can be found in Buffington/Montgomery, 1997, and Buffington, 1999.

Example 3-2

A value of 0.03 for ψ_c is considered a safe choice for the threshold of motion. For stones with a characteristic diameter of 0.4 m this gives $q_s = 6.56*10^{18}*\psi^{16}*\sqrt{(\Delta g d^3)} \approx 3*10^6$ $m^3/m/s$. This is equivalent with $86400*q_s/d^3 \approx 4$ stones per day per m width. The design velocity for an apron is usually a value which occurs only in exceptional cases e.g. with a chance of 1% per year. This makes such a transport quantity acceptable. Moreover, this is the transport per m width, not per m^2. This example also shows that there is always a chance of some damage and inspection and maintenance is necessary for every structure.

De Boer,1998, did similar tests to Paintal's for rip-rap aprons with low values of ψ. He found much lower transports than Paintal and he also found that the transport stopped completely after some hours for ψ-values lower than 0.06. Above 0.06 the transport went on continuously. The transport process stopped because moving stones found a new location where they were stable enough for the present load. Especially small stones were trapped by the bed. De Boer also varied the roughness of the bed and found that the transport on a densely packed bed, which is relatively smooth, was larger than on a loose bed.

Although there are some doubts regarding the absolute values of the Paintal transport formula, the message is obvious: there is no clear threshold of motion and there will always be some transport of material. The choice of ψ_c depends on the amount of transport that is acceptable, hence ψ_c can also be regarded as a damage number.

3.2.3 Stone dimensions

For stability and erosion, the size of a stone or sand grain is an important parameter. There is, however, a lot of ambiguity when it comes to defining the size of a single stone or grain, let alone characterizing a shipload of material. This has to do with the shape of natural materials which can resemble spheres, discs, cubes or cylinders with rounded or very angular surface, making the definition of one size complicated. Another complication is the way the dimensions of a grain are determined in practice. For small grains it is customary to use sieves, which is impossible for large stones. A large rock can be weighed which is, in turn, impractical for small grains. One can imagine that in stability, the weight of a grain is the dominating factor. This has led to the use of the so-called nominal diameter in hydraulic engineering during the last decades. This is simply the side of a cube with the same volume as the stone considered:

$$d_n = \sqrt[3]{V} = \sqrt[3]{M / \rho} \tag{3.7}$$

A third complication comes from the fact that natural grains all differ in size, the grading of the material. Sand contains smaller and larger grains and stone from a quarry is sorted into classes. This can be represented by a sieve curve or any other distribution curve, see also Appendix A. To characterize such a class with one parameter, usually the median value is applied. d_{50} or d_{n50} indicates that 50 % of the weight of all grains is smaller or larger than that value. Since the large grains contribute more to the total weight, the number of smaller grains will be much greater than the number of larger grains.

Boutovski,1998, studied the influence of shape and grading on the stability of stones in flow. For stones with narrow or wide grading ($d_{85}/d_{15} = 1.4 - 3.1$) he found no difference in stability when d_{n50} was used as the characteristic dimension. For a mixture containing 80% stones with a disk or cylinder shape, the same stability was found as for "normal" stones, provided again that d_{n50} was used. *The use of d_{n50} has become common practice for natural protection material. For small grains, like sand, the sieve diameter remains the standard, indicated as d_{50}. d is normally 20 %* larger than d_n. Appendix A gives more detail.

In more dated research it is often not very clear how the applied diameter is defined, leading to some uncertainties when interpreting the results.

3.2.4 Waterdepth

With the uniform flow relations of chapter 2, equation (3.5) can be rewritten in the same format as the Izbash formula: u_* (the shear velocity) $= \bar{u} \sqrt{g}/C$, where $\bar{u}$ is the velocity averaged over the vertical or the cross section, see chapter 2. This leads to:

$$\frac{\overline{u_C}}{\sqrt{\Delta g \, d_{n50}}} = \frac{C\sqrt{\psi_c}}{\sqrt{g}} \tag{3.8}$$

Note that, via C, the waterdepth now enters the relation implicitly as, given a bottom roughness, C increases with the waterdepth: $C = 18 \log(12 \, R/k_r)$. k_r is the so-called equivalent sand roughness. Unfortunately, this introduces another uncertainty in the idea of a threshold of motion. k_r depends on the gradation and on the stacking of the stones, which is in turn determined by the way the bed is constructed. Section 3.2.5 will give some more information. In this section we will simply use $k_r = 2_{dn50}$. With $k_r = 2d_{n50}$ equation (3.8) becomes: $u_c/\sqrt{(\Delta g d_{n50})} = 5.75 \cdot \log(6\ h/d_{n50}) \cdot \sqrt{\psi_c}$.

In Figure 3-5 this relation has been drawn for several values of ψ_c and compared with Izbash, equation (3.4). Shields gives a greater permissible velocity at greater depths. This has also been found in experiments, making Shields "superior" to Izbash. The following remarks, however, need to be considered.

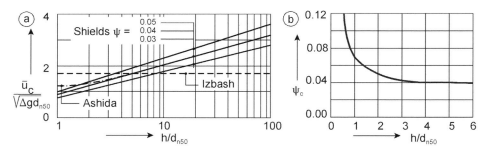

Figure 3-5 Influence of waterdepth on critical velocity

The favourable influence of the waterdepth on the critical velocity cannot go on infinitely, since the logarithmic velocity profile in natural streams with large depths is only found in the lower half of the depth. $h/d_{n50} = 100$ can be seen as a safe upper limit.

For small waterdepths, somewhere around $h/d_{n50} = 5$, Izbash gives higher permissible velocities than Shields. For $h/d_{n50} < 5$, Ashida,1973, found a considerable increase in ψ_c, see Figure 3-5b. This more favourable value of ψ_c can possibly be attributed to the fact that the whole concept of a shear stress on the grains is no longer valid and the flow exerts only a drag force on the grains. As we have seen in section 3.2.1, drag and shear are not treated separately in the empirical relations. Another possible explanation comes from the turbulent eddies: at waterdepths in the order of magnitude of the stone diameter, d, the maximum size of these eddies are in the same order of magnitude as d. For larger waterdepths, the eddy dimensions can and will grow, leading to a heavier attack on the bottom.

3.2.5 Practical application

Normally in a design situation, a stone class to withstand a certain extreme value for the velocity has to be found. Equation (3.8) has been reworked for that purpose:

$$\frac{\overline{u}_c}{\sqrt{\Delta g\, d_{n50}}} = \frac{C\sqrt{\psi_c}}{\sqrt{g}} \quad \rightarrow \quad d_{n50} = \frac{\overline{u}_c^{\,2}}{\psi_c \Delta C^2} \tag{3.9}$$

This formula is the basis for all stability relations in flow situations. In the coming sections additions will be given for flow situations other than uniform flow with a horizontal bed. In many of these situations, the load on protection elements is very complex, but for a first estimate the Shields relation with correction parameters can be applied. It has been decided to reserve the symbol ψ_c for the stability parameter in uniform flow on a horizontal bottom and to include deviations in extra parameters.

However, some decisions remain to be made on how to deal with parameters like roughness and threshold of motion. We start with the roughness. For a bed that is made under favourable conditions, e.g. on a dry and visible subsoil, $k_r \approx 2d_{n50}$ seems achievable. Lammers,1997, and Boutovski,1998, found for a flat bed in a flume experiment: $k_r \approx 6d_{n50}$ (without side-wall corrections). Van Rijn, 1986, proposed $k_r \approx 3d_{90}$ which is about 4-5 d_{n50}. When the stones are dumped with barges, the irregularities, and hence the bottom roughness, can be much larger. On the other hand, when stones move in that case, the irregularities will be flattened again. This also has to do with the choice of ψ_c.

In section 3.2.2 we saw that there will always be some movement when a certain value of ψ_c has been chosen. Below a level of 0.06, De Boer,1998, found transport stopped after some time. Lammers,1997, also found something similar. Figure 3-6a shows the results of an experiment with a very rough bed (artificial ripples of loose stones) which was loaded twice with an increasing flow velocity. The first time, the roughness decreased from $18d_{n50}$ to $6d_{n50}$ (which was also the value found for a flat bed) because the transport of stones has lead to a flatter bed. The numbers next to the curve represent the transport stages from section 3.2.2. After this, the flow was stopped and was increased again. Now the roughness kept the flat-bed value, but much more importantly, the transport stage numbers were much lower. This was due to strengthening of the bed. Stones in an initially unfavourable position, found better locations and the bed as a whole became more stable.

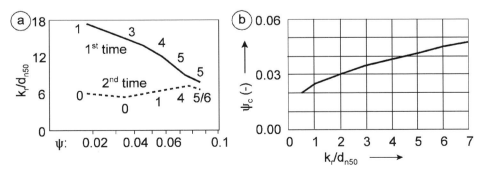

Figure 3-6 Roughness and the threshold of motion (Figure 3-6a from Lammers,1997)

The choice of ψ_c remains subjective as long as there is no quantitative probabilistic information on optimal costs of construction and maintenance. The definition of

Serviceability Limit State and Ultimate Limit State is therefore ambiguous. For the time being, given all the uncertainties, a practical choice has to be made for ψ_c and for k_r.

The result in Example 3-3 would also have been found with a ψ_c-value of 0.045 and a roughness of 6 times d_{n50} or with $\psi_c = 0.04$ and $k_r = 4.5d_{n50}$. Equation (3.9) shows that different combinations of values of ψ_c and k_r can give the same diameter. Figure 3-6b shows the relation between k_r/d_{n50} and ψ_c for the computed diameter in Example 3-3. Given the outcome of the research by Lammers,1997, a choice for a high roughness value and a high ψ_c-value seems logical. The result, however, is not very sensitive to the choice as long as a reasonable combination of the two values is used. *$\psi_c = 0.03$ and $k_r = 2d_{n50}$ is therefore a practical choice.*

Example 3-3

> *Given a (design) velocity of 4 m/s, a waterdepth of 10 m and $\Delta = 1.65$, what stone size (d_{n50}) is necessary to withstand this (uniform) flow?*
>
> *We take a ψ_c-value of 0.03 and a roughness of 2 times d_{n50}. Since the roughness in equation (3.9) depends on the still unknown diameter, we have to iterate. So we have to start with a guess of either a value for d_{n50} or for C. We start with C=50 √m/s (any guess is good). From this we compute $d_{n50} = 4^2/0.03*1.65*50^2 = 0.129$ m. From there: C = 18 log(12*10/2*0.129) = 48 √m/s. From equation (3.9) again: $d_{n50} = 4^2/0.03*1.65*48^2 = 0.14$ m. And so on. Finally we find $d_{n50} = 0.146$ m. Using common sense, this iteration will converge in 3 or 4 calculations. From appendix A we find that a stone size of 80/200 mm will do.*

3.3 Sloping bed

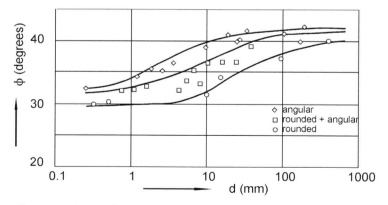

Figure 3-7 Angle of repose for non-cohesive materials (Simons, 1957)

The relations in section 3.2 are valid for stones on a horizontal bed. A sloping bed will decrease the strength. When the slope is equal to the angle of repose, ϕ, a stone is already on the threshold of motion and any load will induce movement. Figure 3-7 gives an idea of the value of ϕ for various materials. When the bed is between horizontal and the angle of repose, the diameter calculated with equation (3.8) needs a correction, as is explained in this section.

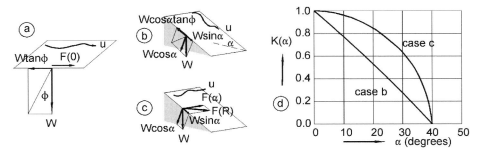

Figure 3-8 Influence of slope on stability

To determine the influence of a slope on the stability, it is assumed that for a stone on a flat bed the critical flow force, or the strength, can be expressed as $F(0) = W \cdot \tan\phi$, see Figure 3-8a. This implies that the friction factor of a stone on a horizontal bed is equal to the tangent of the angle of repose. As we have seen in Intermezzo 3-1, this is not correct but the slope correction factor we are going to derive will not be used absolutely but relative to the horizontal situation.

The reduction of strength can be divided into two components. There is a destabilizing component of the weight along the slope ($W \sin\alpha$), while the effective weight perpendicular to the slope becomes less ($W \cos\alpha$). In case of a slope in the direction of the flow (Figure 3-8b) the strength then becomes: $F(\alpha_{//}) = W \cos\alpha \tan\phi - W \sin\alpha$. (Note: when the flow is reverse, the situation becomes more stable and $F(\alpha_{//}) = W\cos\alpha \tan\phi + W \sin\alpha$). The relation between $F(\alpha)$ and $F(0)$ now gives the reduction factor to be applied on the strength:

$$K(\alpha_{//}) = \frac{F(\alpha_{//})}{F(0)} = \frac{W\cos\alpha \tan\phi - W\sin\alpha}{W\tan\phi} = \frac{\sin\phi\cos\alpha - \cos\phi\sin\alpha}{\sin\phi} = \frac{\sin(\phi - \alpha)}{\sin\phi}$$

(3.10)

For a side slope with angle α, the same procedure can be followed, see Figure 3-8c, but now the components have to be added vectorially: $F(R)^2 = F(\alpha_\perp)^2 + (W\sin\alpha)^2 = (W\cos\alpha \tan\phi)^2$. This leads to a strength: $F(\alpha_\perp)^2 = (W\cos\alpha \tan\phi)^2 - (W\sin\alpha)^2$ and the reduction factor becomes:

$$K(\alpha) = \frac{F(\alpha)}{F(0)} = \sqrt{\frac{\cos^2\alpha \tan^2\phi - \sin^2\alpha}{\tan^2\phi}} = \cos\alpha \sqrt{1 - \frac{\tan^2\alpha}{\tan^2\phi}} = \sqrt{1 - \frac{\sin^2\alpha}{\sin^2\phi}} \quad (3.11)$$

In Figure 3-8d, both of the reduction factors $K(\alpha_{//})$, for a slope in flow direction, and $K(\alpha_{\perp})$, for a slope perpendicular to the flow, have been drawn for $\phi = 40°$. It is clear that a slope in the flow direction is much less favourable.

Note: $K(\alpha)$ is a correction factor for the diameter and has to be applied in the denominator of equation (3.9). $K(\alpha)$ represents a reduction of strength, hence $K(\alpha) < 1$.

Both equations have been checked in laboratory experiments. Chiew/Parker,1994, gave the results for a slope in flow direction, equation (3.10). The results are quite satisfactory for slopes not too close to the angle of repose. Slopes with the flow in an upward direction, making the stones more stable, were also investigated with the same result. DHL,1970 gives some results for laboratory experiments with gravel on slopes 1:3 and 1:4 perpendicular to the flow. Here too, the results were satisfactory. Wijgerse,2000, investigated how to deal with equation (3.11) in practice. The choice to be made is where on the slope, the velocity and the depth have to be used. At the toe, the waterdepth is maximal which is favourable but the velocity decreases from the toe to the waterline, so it is not clear which combination of velocity and depth is dominant. The result of this study was that the situation at the toe is normative.

3.4 Non-uniform flow

Relations like the one given by Shields are valid for (stationary), uniform flow. When the flow is not uniform, the load will be higher locally. For large hydraulic structures this can be investigated with a scale model and in the future possibly with advanced computational models in which the turbulent velocities can be determined at every location. For relatively small structures and for preliminary designs, the higher load can be expressed in a factor for the velocity, K_v, in stability relations.

In accelerating flow we have seen an increase in shear stress and a decrease of (relative) turbulence; in decelerating flow it is the other way round (see chapter 2). For these phenomena in vertical and horizontal constrictions, only the results of empirical research are available. With the information on fluid motion, as outlined in chapter 2, these results will now be discussed.

3.4.1 Acceleration

Vertical constriction – Overflow

Figure 3-9a shows the various zones that can be distinguished when a flow passes a vertical constriction. In the uniform flow zone, relations like the one by Shields are valid. For stones on top of a sill, one might expect a more unfavourable relation, because of the increased shear stress due to the acceleration. In experiments has been found that the first damage on a sill occurs at the downstream crest. There, the velocity is maximal and the waterdepth and strength minimal (because of the downward slope in flow direction). When a single stone on that crest has a protruding position, it will move at almost any velocity, reshaping the crest somewhat. This is usually not considered as damage. Figure 3-9b gives experimental results, compared with the Shields formula for the same values of ψ_c of Figure 3-5, using the *velocity on top of the sill* (which is of course larger than in the upstream cross-sections). The experimental flume tests were done in the framework of the Deltaworks, see DHL,1985. All measured critical velocities lie above the line for ψ_c = 0.04 indicating that the relation between critical velocity and stone diameter is no less favourable than for a situation with uniform flow.

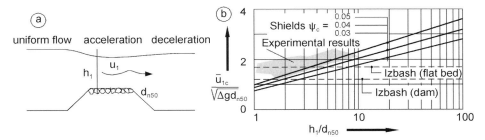

Figure 3-9 Investigation of stability on top of a sill (DHL, 1985)

This can be explained by the fact that the peak velocities are lower than in uniform flow. $u + 3\sigma = (1 + 3r)u$ while r is lower in the acceleration zone than in uniform flow.

Beside the earlier mentioned equation by Izbash for a flat bed, Izbash, 1970, gives a second equation for (protruding) stones on top of a dam. The Izbash line for dams in Figure 3-9 can be seen as a conservative value of the stability.

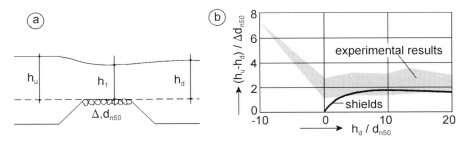

Figure 3-10 Stability on top of sill and head difference (DHL, 1985)

When a sill becomes relatively high, especially when critical flow can occur, the whole flow-pattern becomes completely different. The flow lines over the sill are strongly curved and at the downstream crest the waterflow through the stones becomes important. In that case working with velocities is useless. In DHL, 1985, many experimental results were re-analyzed and for higher dams the head difference across the sill appeared to be a satisfactory parameter, see Figure 3-10b. Moreover, the head difference, $(h_u - h_d)$, can be more easily defined and measured for high dams.

As a comparison, Shields (equation (3.9)), has been reworked in the format of Figure 3-10, with $k_r = 2d_{n50}$ and $\psi_c = 0.03$, using the Bernoulli equation with an expression for the velocity on top of the sill and a discharge coefficient, μ, taken from the same DHL-experiments:

$$u_1^2 = \mu^2 \, 2g \, (h_u - h_d) = (0.5 + 0.04 \, \frac{h_d}{d_{n50}}) \, 2 \, g(h_u - h_d) \qquad (3.12)$$

This relation has been drawn in Figure 3-10b, using h_d and u from equation (3.12). For relatively high waterdepths downstream, Shields is satisfactory and on the safe side as we had found earlier. For low waterdepths downstream, Figure 3-10b should be used, where the stability parameter is now defined as $(h_u - h_d)/\Delta d_{n50}$. For high waterdepths downstream, the stability simply becomes: $(h_u - h_d)/\Delta d_{n50} \approx 2$, while for low waterdepths the value increases greatly, see Figure 3-10b. $h_d/d_{n50} \approx 5$ can be seen as a transition point, which also appears to be the transition between critical and sub-critical flow.

Vertical constriction - Underflow

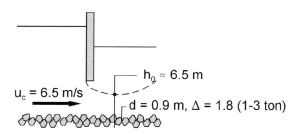

Figure 3-11 Stability with flow under weir

Figure 3-11 shows flow under a weir as investigated for the Eastern Scheldt storm surge barrier. In the flow contraction under the weir, the stability relation (3.9) matches when $\psi_c = 0.03$ is applied and the contracted jet height is used as the waterdepth. So, here too, no extra coefficients are necessary to calculate the stability.

Horizontal constriction

A groyne or a dam causes an accelerated flow around the obstruction. Moreover, stones on the groyne head lie on a slope perpendicular to the flow, for which a correction factor has to be applied, see equation (3.11). In DHL, 1986 the stability in that situation was analyzed, based on experiments on horizontal closure of a river with stones. The stability can be approached with Shields and a slope correction (equation (3.9) and (3.11):

$$\frac{\overline{u_{gap}}}{\sqrt{\Delta g d_{n50}}} = C\sqrt{\frac{\psi_c}{g}} \sqrt[4]{1 - \frac{\sin^2 \alpha}{\sin^2 \phi}} = 4.5\log\left(\frac{3h}{d_{n50}}\right)\sqrt{\psi_c} \qquad (3.13)$$

The damage appears to occur at half depth, so C becomes 18 log $(12 \cdot \frac{1}{2}h)/2d_{n50} = 18$ log $(3h/d_{n50})$. For closure works in a river and with $\alpha = 30°$ and $\phi = 40°$ (normal values for a rockfill dam), the stability parameter reduces to the right hand side of equation (3.13). Figure 3-12 compares equation (3.13) and experimental measurements. The result is similar to Figure 3-9: the situation is no less favourable than a uniform flow situation (using the velocity in the gap).

In a nutshell: in an accelerated flow, the stability of loose stones is no less than in a uniform flow, **provided the local velocity is used in the Shields relation.** *Or, when working with a velocity factor, K_v, as indicated in the first paragraph of section 3.4, the design value with regard to the* **local velocity** *in an accelerating flow is 1.*

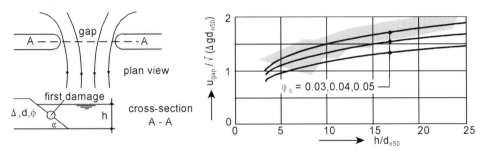

Figure 3-12 Stability of stones on head of dam or groyne

3.4.2 Deceleration

In decelerating flow the flow situation differs completely from that in uniform flow or in accelerating flow. Flow separation and a mixing layer often occur and in all cases the turbulence level is much higher due to the overall loss of kinetic energy. Now the velocity coefficient, with regard to the local velocity, is expected to be higher than 1. K_v is defined as:

$$K_v = \frac{u_c \text{ uniform flow}}{u_c \text{ with load increase}} \qquad (3.14)$$

in which: $\overline{u}_{cu}$ is the vertically averaged critical velocity in uniform flow and $\overline{u}_{cs}$ is the velocity in the case with a structure. $\overline{u}_{cs}$ can be vertically averaged at some location or averaged in a cross-section, see Intermezzo 3-2.

Intermezzo 3-2

In an empty flume (uniform flow, left hand side of figure) for some stone size we find a critical velocity, $\bar{u}_{cu}$ = 0.78 m/s. With an abutment in the same flume, which blocks 30% of the width, we find a critical velocity averaged over the cross section of the flume of 0.40 m/s (1 in right hand side of figure). Hence, K_v = 0.78/0.40 = 1.95. It is also possible to use the average velocity in the cross section at the abutment (2) where the velocity is 1/0.7 times higher. This leads to K_v = 0.78/(0.4/0.7) = 1.37. Under critical flow conditions the maximum velocity along the abutment (3) was measured to be 0.72 m/s, giving K_v = 0.78/0.72 = 1.08. In uniform flow there is hardly a problem to define the critical velocity but when a structure is present, a choice has to be made. A value for K_v can never be seen apart from its definition, which becomes clear from the example above. Only when a local (maximum) velocity is used, K_v will represent the extra turbulence. When using the average upstream velocity, u (= Q/A), K_v will be "polluted" by the variations in the velocity field. For 2-dimensional flow situations, this difficulty does not exist, see the section on vertical constriction.

Note: The critical velocity in the denominator of equation (3.14) *is lower than the critical value in uniform flow (the numerator). This means that, with some obstacle, stones begin to move at a lower velocity in the flume. It does not mean that the strength of the bed is lower. The bed remains the same, but the load increases. Language can be a pitfall!*

It can be expected that the extremes in the velocity field are responsible for incipient motion. These extremes can be represented by e.g. $\bar{u} + 3\sigma \approx (1 + 3r) \bar{u}$, so, we would expect a relation between K_v from equation (3.14) and the turbulence level:

$$(1+3r_{cu})\bar{u}_{cu} = (1+3r_{cs})\bar{u}_{cs} \longrightarrow K_v = \frac{\bar{u}_{cu}}{\bar{u}_{cs}} = \frac{1+3r_{cs}}{1+3r_{cu}} \tag{3.15}$$

in which r_{cs} is the vertically averaged turbulence intensity at some location (see Intermezzo 3-2) and r_{cu} the intensity in uniform flow.

Note: The use of vertically averaged values is a first-order approximation. The velocity and turbulence profile will influence the attack on the bottom.

In the following examples we will determine K_v from equation (3.14) and check equation (3.15). When, in future, sophisticated flow models become available and the turbulent velocities are known at every location, it will be possible to design bottom protections without scale models. Now, equations like (3.15) serve mainly to get some insight into the processes. For the time being, we will have to work with rough estimates of the velocity coefficient for simple structures and preliminary designs and a scale model to check calculations for complicated structures.

Vertical constriction – Overflow

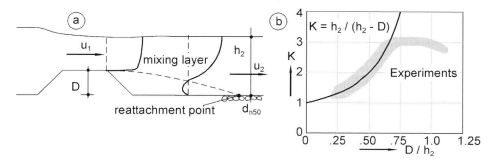

Figure 3-13 Stability downstream of a sill

Figure 3-13a shows the situation behind a sill with flow separation and a mixing layer. We focus on the area near the reattachement point, where the maximum flow attack can be expected. We could try to find an expression for the velocity and the turbulence at the reattachment point, using the relations for a mixing layer as described in chapter 2. A even more crude approach is to assume that the total energy of the flow from the top of the sill remains constant up to the reattachment point and that the stability there can be described with the velocity on top of the sill. From continuity we find:

$$u_1(h_2 - D) = u_2 h_2 \;\rightarrow\; u_1 = \frac{h_2}{h_2 - D}\, u_2 \;\rightarrow\; K \approx \frac{h_2}{h_2 - D} \tag{3.16}$$

where K_v is related to the local velocity (vertically averaged) at the reattachment point, u_2. Figure 3-13b shows a surprising agreement (given the crudeness of the assumption) between this equation and experiments (see DHL, 1985) for sills with a height up to 65% of the water depth. The load increases till the situation of critical flow over the sill is reached, which is normally the case when $D/h_2 \approx 0.75$. When D/h_2 increases further, the total energy loss also increases, but the "extra" loss, compared with subcritical flow, is absorbed at the downstream slope of the sill. This approach is only valid for the heaviest attacked point beyond the sill, near the reattachment point. Wang/Fontijn,1993, found, based on measurements of forces on an element downstream of a backward facing step, that the heaviest attack took place 10 to 20 step heights downstream. De Gunst, 1999, also investigated stability downstream of a backward facing step with $D/h_2 = 0.34$ for which $K_v = 1.5$ according to equation (3.16). De Gunst found $K_v \approx 1.2$ with equation (3.14), which is in the lower region of the experimental results of Figure 3-13b (a sill possibly gives a less favourable flow situation than a backward facing step). With turbulence measurements and equation (3.15), De Gunst found $K_v \approx 1.2$, so the attack on the bottom is in line with the idea of a peak velocity. (The definition of K_v is unambiguous in case of a vertical constriction)

Further downstream, the situation tends to uniform flow again and K_v can be computed using the turbulence decay relation as given in section 2.5.2.

Vertical constriction - Underflow

For deceleration in underflow situations no data on stone stability is available. Therefore we look at the stability downstream of a hydraulic jump, for which RWS, 1987 gives some examples based on research by Kumin, see Figure 3-14.

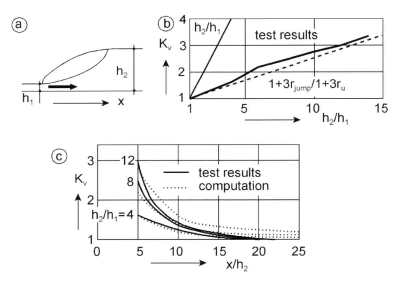

Figure 3-14 Stone stability downstream of a hydraulic jump

Applying the same approach as used for an overflow, we would find from the continuity equation: $K_v = u_1 / u_2 = h_2 / h_1$. Compared with the test results, this is much too high, see Figure 3-14b, while in the overflow situation of Figure 3-13 the similarity was surprisingly good. The explanation is to be found in the location of the energy dissipation. In the case of a sill this takes place predominantly at the bottom, in the hydraulic jump mainly at the surface in the roller of the jump. This appears to be favourable for the attack on the bottom. Using some data concerning turbulence in hydraulic jumps, K_v according to equation (3.15) is also represented in Figure 3-14b. This similarity is quite reasonable.

The tests by Kumin also included determining K_v-values for locations further downstream of the jump, see Figure 3-14c. Computations using the turbulence decrease expression from section 2.5.2 together with the divergence in a plane jet ($u \propto 1/\sqrt{x}$) produce the dotted lines in Figure 3-14c. The result is encouraging. The velocity reduction with the root of the distance appears to be dominant in the process. With $d \propto u^2$, this means that the necessary stone diameter will be proportional to the distance from the outflow. Research for the Eastern Scheldt storm surge barrier indicated approximately the same for the diameter of the stones in the top layer of the bottom protection, which is several hundreds of meters long.

Horizontal constriction

Results of systematic investigations into stability in decelerations behind horizontal constrictions are not available. Zuurveld,1998, investigated stone stability in a mixing layer beyond a sudden widening, see also section 2.5. Figure 3-15 a and c show that incipient motion occurs in the area where the peak velocities are maximal. This is not the case in the center-line of the mixing layer but in the mainstream. K_v, with regard to the outflow velocity, ≈ 1. With regard to the average velocity downstream the value becomes 2, which is in line with a vertical constriction (Figure 3-13).

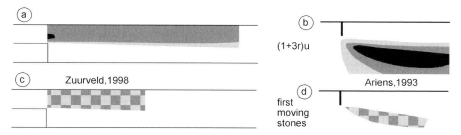

Figure 3-15 Peak velocities and incipient motion in horizontal constriction

Ariens, 1993, investigated stone stability behind groynes, see Figure 3-15 b and d. Here too, the first movement of stones is found in the region with the highest (peak) velocities. The maximum velocity, $(1 + 3r)\bar{u}$, at the threshold of motion was $(1 + 3*0.053)0.72 = 0.83$ m/s. In the empty flume, without obstacles, incipient motion was found at an average velocity of 0.69 m/s and a, vertically averaged, turbulence intensity, $r = 0.077$. This leads to a maximum velocity $(1 + 3r)\bar{u} = 0.85$ m/s, so, here again the peak velocities seem responsible for the movement. The erosion pattern might give the impression that incipient motion occurs in the mixing layer, but this is deceitful! Due to the geometry the flow accelerates around the groyne and compared with the local velocity K_v is not much above 1. So, the flow pattern greatly determines the stability highlighting again the importance of insight into this pattern.

3.4.3 Practical applications

In practice, K_v-values are used, based on experiments and applying equation (3.14). Ariens,1993, and Van Breugel/Ten Hove,1995, have established values which can be useful for a preliminary design. Table 3-1 gives K_v-factors in accordance with the various definitions of $\bar{u}$ which will be clarified herafter.

Table 3-1 K_v-factors for various types of structures

Structure	Shape	K_{v0}	K_{vG}	K_{vM}
Groyne	Rect-angular	b_0*K_{vG}/b_G	1.3 - 1.7	1.1 - 1.2
	Trape-zoidal	$b_0*K_v b_G$	1.2	1
Abutment	Rect-Angular	b_0*K_v/b_G	1.3 - 1.7	1.2
	Round	b_0*K_v/b_G	1.2 - 1.3	1.2
	Stream Lined	b_0*K_v/b_G	1 - 1.1	1 - 1.1
Pier	Round	b_0*K_v/b_G ⊗ $2*K_v$	1.2 - 1.4 ⊗	1 - 1.1
	Rect-Angular	b_0*K_v/b_G ⊗ $2*K_v$	1.4 - 1.6 ⊗	1.2 - 1.3
Outflow	Abruptly	--	1	--
	Stream Lined	--	0.9	--
Sill	Top	Section 3.4.1	Section 3.4.1	Section 3.4.1
	Down Stream	Figure 3-13	Figure 3-13	Figure 3-13

⊗ *For many piers in a river the first expression for K_v is appropriate. The second is valid for a detached pier in an infinitely wide flow, where K_G is not defined.*

Which one of these three is used depends on the availability of velocity data and on the nature of the flow situation. Here again: beware of the following pitfalls. For a horizontal constriction, $\bar{u}_G$ and $\bar{u}_0$ are related by the continuity equation. When the horizontal constriction is small, $\bar{u}_G$ and $\bar{u}_0$ will be almost the same. This is no problem when we consider the situation in a river or a canal, but in an infinitely wide flow this will lead to nonsense. Figure 3-16b shows the situation with a single pier in a sea. The average velocity upstream and the maximum velocity can be defined, but there is no "gap".

Figure 3-16 Definition of velocities in relation to K_v

Concerning Table 3-1, the following remarks can be made:
1. When the maximum velocity can be determined or estimated, the use of K_{vM} is preferable.
2. In general it can be said that streamlining pays, but the cost-effectiveness has to be determined for each case separately.
3. In an outflow, there is no extra influence of turbulence. The maximum velocity inside the aperture determines the stability.

All parameters of the previous sections are now combined with equation (3.9) into:

$$d = \frac{K_v^2 \, {}^*\overline{u}_c^2}{K_s \psi_c \Delta C^2} \tag{3.17}$$

in which K_s is the slope correction factor as defined in section 3.3. Equation (3.17) is a design formula for practical applications. One could think of combining K_v and $\overline{u}$ directly, K_v being a load increase factor. But, in that case K_v would interfere in the iteration with the roughness via C, which would result in a too conservative value for d_{n50}. The origin of K_v from equation (3.14) implies that tests with and without structures were carried out with the same waterdepths, hence with the same value for h/d. So, first, d_{n50} has to be determined without applying K_v.

Example 3-4

In the narrowest cross-section of a river (depth = 5 m) with a round abutment, the design velocity is 3.5 m/s. What stone class is necessary in front of the abutment?
First we compute the diameter in uniform flow with u = 3.5 m/s, see Example 3-3. The result is d_{n50} = 0.128 m. From Table 3-1 we find K_v = 1.3. So, the diameter becomes: d_{n50} = 0.128.1.3² = 0.22 m. From Appendix A we find stone class 10/60 kg.

3.5 Coherent material

The word coherent is used here for all materials that show some coherence in contrast to the loose stones of the previous sections. This contrast is not so black and white. A placed block that is not connected with, but getting support from its neighbours, forms the transition between the two categories. For various types of material, some data will be given. This is based on very few investigations, so this information should be handled with care.

Gabions

Gabions consist of loose stones, packed together into larger elements, see appendix A. Gabions can be useful when available stones are not large enough for the flow to be expected or they may be used because they are easy to place. Due to the pores between the stones, a gabion has a smaller relative density than solid rock:

$$\Delta = \frac{(1-n)\ (\rho_S - \rho_W)}{\rho_W} \qquad (3.18)$$

The porosity, n, usually lies around 40%. Gabions are more stable than a single stone, as the porosity makes the load of the flow less effective. The coefficient of a single gabion using the Izbash approach is about 2 times higher than the coefficient of loose stones, so the final gain in stability coefficient is about $2 \times \sqrt{(1-n)} \approx 1.5$.

Clay soils

Ven te Chow,1959, gives some data on the critical velocity for various clay types without vegetation. The results can be presented as follows:

Table 3-2 Critical velocity clay

	Porosity		
	20 %	40 %	60 %
Critical velocity (m/s)	1.5 - 1.8	1.0 - 1.2	0.1 - 0.4

Vegetation

Chapter 12 gives some information on flow resistance of grass.

Placed blocks

Square or rectangular blocks can be piled up, forming a vertical wall. So, the angle of repose is 90°, leading to an infinite resistance (tan 90° = ∞). Theoretically this is correct; blocks that are well-placed and interlocked cannot turn and are jammed, see Figure 3-17a. The only way to lift a block is by pressure from below, or by suction from above. In practice blocks will not interlock ideally. In Izbash-form, critical values of $u/\sqrt{(\Delta gd)}$ of 1.5 to 2 times higher than for loose stones were found (RWS, 1990).

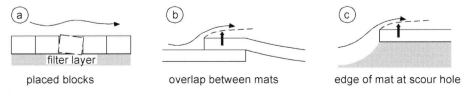

| placed blocks | overlap between mats | edge of mat at scour hole |

Figure 3-17 Flow situations with coherent materials

Mats

Mats can consist of a geotextile with blocks attached to it or they can be any other mattress type protection, see appendix A. The critical velocity for the mat as a whole is almost infinite. The vulnerable parts are the transitions, e.g the overlap between two mats or the edge of the protected area in a scour hole, see Figure 3-17b and c. (in unidirectional flow, these situations do not necessarily occur when the mats are placed like roof tiles; in tidal flow there is no choice). Due to the curved streamlines, or even flow separation, there will be a relatively low pressure at the upper side. The result can be lifting of the mat, making the difference in pressure even greater (part of the protection standing as a vertical wall in the stream gives a considerable drag). The magnitude of this lifting force also depends on the permeability of the mat; gabion type mats can be considered permeable, concrete blocks on a sandtight geotextile impermeable. With the stability expressed in an Izbash-type of formula, the following values of $u_c/\sqrt{(\Delta gd)}$ were found:

Table 3-3 Stability of mats

	Permeable	Impermeable
Overlap between mats	2	1.4
Mat at edge of scour hole	1.4	1.1
Loose stones on flat bed	1.7	

The reversed flow in a scour hole is the most unfavourable situation, while with a permeable mattress, at an overlap, the value is still higher than the Izbash value for a stone on a bed.

3.6 Summary

This chapter has treated the stability in flow. The base relation for non-cohesive material like rock is the Shields relation with correction factors for non-horizontal bottoms and flow situations that deviate from uniform flow (equation (3.17):

$$d = \frac{K_v^2 * \overline{u}_c^2}{K_s \psi_c \Delta C^2}$$

The following remarks concerning the use of the parameters in this formula can be made:

$\overline{u}_c$ *the critical, depth-averaged, velocity in **uniform flow** for incipient motion.*

ψ_c *the threshold-of-motion parameter, indicating the degree of transport near incipient motion. $\psi_c = 0.06$ can be seen as the start of sediment transport and* *can be used in erosion computations. For bed protections, where only minor losses of material are acceptable, a lower value, e.g. $\psi_c = 0.03$ is recommended.*

C *the Chezy-coefficient, indicating the roughness. $k_r = 2d_{n50}$ can serve as a first estimate of the equivalent roughness (together with $\psi_c = 0.03$, see section 3.2.5.*

K_s *is a strength-reduction parameter for stones on a slope, see equations (3.10) and (3.11).*

K_v *the velocity/turbulence factor, indicating a load deviating from uniform flow. For various flow situations, Table 3-1 gives values for K_v. This factor should be applied as a correction after the diameter in uniform flow has been determined. Otherwise K_v will influence the iteration of d and C (which also contains d via k).*

In accelerating flow, $K_v = 1$ can always be used, provided the local velocity is used in the equation.

Finally, some information on stability of coherent material like mats, gabions and clay is presented.

4 FLOW
Erosion

Scour in a horizontal constriction (courtesy Delft Hydraulics)

4.1 Introduction

Before even thinking of protection works, one should have an idea of the possible erosion. Erosion (and sedimentation) occur everywhere in nature where the bottom consists of fine sediments. This book does not cover morphological processes, it is limited to local scour in the vicinity of structures. This scour may be caused by the structure itself, like a bridge pier in a river, or it may be the result of a natural process which the structure is meant to prevent or limit, like a protected bank in a meandering river. It must be stressed that awareness and knowledge of morphological processes is very important for engineering in natural systems. Bridges have been built across rivers that have later chosen another path, leaving the bridge as a monument to lousy engineering.

4.1.1 Scour as sediment transport

Scour is a special case of sediment transport and occurs when the local transport exceeds the supply from upstream. The difference in transport can be due to a difference either in velocity or in turbulence or both. The general expression for the conservation of mass for sediment reads:

$$\frac{\partial z_b}{\partial t} + \frac{\partial S}{\partial x} = 0 \qquad\qquad (4.1)$$

in which z_b is the position of the bed and S the total sediment transport per unit width. Three possible situations for local scour are, see Figure 4-1:

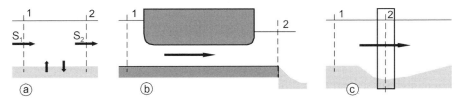

Figure 4-1 General picture local erosion

a) $S_2 = S_1 > 0$. This is a dynamic equilibrium situation. Sediment can be picked up and can settle again, but there is no net change of the position of the bottom.

b) $S_2 > S_1 = 0$. There is no sediment supply from upstream, while there is sediment transport downstream. This case is known as *clear-water scour*.

c) $S_2 > S_1 > 0$. There is sediment supply from upstream but the sediment transport downstream is larger. This case is known as *live-bed scour*.

Clear-water scour (case b) can result either from a lack of transport capacity or from a lack of erodable material upstream. Figure 4-1b shows an example of the first case, in which cross section 1 is in a lake, upstream from a dam, and cross section 2 is downstream of an outlet structure. The flow in 1 is too weak to transport material. As shown in Figure 4-1c it is also possible that the velocity in cross section 1 is below the critical value, while the disturbance due to the pier causes transport in 2. A bottom protection of infinite length upstream is an example of the second case. In cross section 1 the *transport capacity* is equal to the transport capacity in 2, but the *transport* in 1 is 0, because the flow cannot pick up sediment. So, an infinitely long bottom protection does nor prevent scour. Clear-water scour stops when a depth is reached such that the velocity drops below the critical value.

In *live-bed scour* (case c) the upstream flow in Figure 4-1c is strong enough to transport sediment but the structure causes accelerations and decelerations with turbulence, leading to an increase of the transport capacity. The scour stops when the eroding capacity in 2 equals the supply from 1. The eroded sediment will settle downstream from cross section 2, so on a larger scale case c can be a part of case a.

Sediment transport is a function of velocity and sediment properties and is often presented as (see also chapter 1 and 3 on mobility and stability parameters):

$$S = f(\psi - \psi_c) \quad \text{or} \quad S = f(\psi) \tag{4.2}$$

The first type of expression includes a threshold of motion. An example is the well-known formula by Meyer-Peter-Müller. The second type neglects ψ_c. Examples are Engelund-Hansen and Paintal. In cases where $\psi >> \psi_c$, suspended sediment load plays an important role in scouring. In uniform and stationary conditions, there is a dynamic equilibrium in the vertical direction (for more detail, see e.g. Graf, 1971):

$$w_s \bar{c} + \upsilon_s \frac{\partial \bar{c}}{\partial z} = 0 \tag{4.3}$$

w_s is the fall velocity of the sediment, c the time-averaged concentration and υ_s the turbulent diffusivity. The first term in equation (4.3) stands for the settling of the sediment and the second term represents the stirring up. w_s depends on the sediment, υ_s depends on the flow situation and is related to the turbulence in the flow.

It can be expected that scouring and a high turbulence intensity are somehow correlated. Figure 4-2 shows the scour in one of the channels of the Eastern Scheldt storm surge barrier during construction (from DHL, 1988). The flow in the channel separates twice: around the left abutment of the channel and around the (temporary) edge of the placed concrete beams. The turbulence in the mixing layers acts as a whirlwind and results in a relatively deep scour hole.

In chapter 3 we saw that the location with the heaviest attack on the bottom is not necessarily found at the mixing layer. In scouring this is possibly different. When suspension becomes important, equation (4.3) indicates that turbulence also becomes important. Moreover, Zuurveld,1998, and Booij, 1998, found that the ratio between the size of a grain and a vortex determines the stability. In a horizontal expansion when $\psi \approx \psi_c$, the vortices in the mainflow induce larger forces on the grains than those in the mixing layer. When $\psi \gg \psi_c$, this picture can change completely.

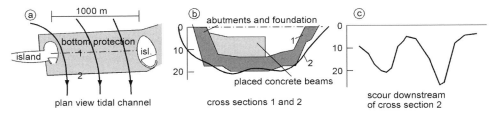

Figure 4-2 Example of scour during construction of Eastern Scheldt storm surge barrier

Being a special case of sediment transport, it seems logical to describe scour with formulas like (4.2) with an additional factor for turbulence, as found in stability relations. However, practically all scour relations are empirical expressions for the scour depth. In the following sections some of those expressions will be presented for different cases.

4.1.2 The scour process

Figure 4-3 shows some characteristics of scour. Initially, the bottom is still flat, the velocity profile is not yet influenced by the scour hole and the flow is still nearly uniform. Sediment is then suspended in the water, see equation (4.3), and is carried along with the main flow. The upstream slope is formed, β increases and then becomes constant.

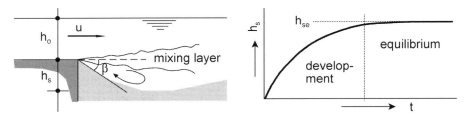

Figure 4-3 General definition scour hole and development in time

Because of this slope, the flow separates, causing a mixing layer in the scour hole and a recirculation zone in which water flows against the main flow direction. This also indicates that the scouring process is influenced by the scour hole itself, which is a complication compared with the stability phenomena discussed in chapter 3 *("scour causes scour")*. Sediment is picked up from the lower parts of the slope, the velocity in the scour hole reaches values close to the critical value and the process slows down. Material is no longer moved from the upstream slope. The final scour depth is reached asymptotically.

For practical purposes, the development and the equilibrium depth are important. When the erosion process is fast, only the equilibrium depth is of interest. When the process is slower, the development in time can also become important. The importance of the development in time is reflected in the maintenance policy, which can be to cover the scour hole when the depth becomes dangerously high. Another situation in which the development in time is important, is the development of scour during construction; different construction phases give different scour intensities.

4.2 Scour without protection

Protection against erosion is not always necessary, as scour as such is not the problem. Only if the stability of a structure is endangered, a protective apron will be needed. Insight into the degree of scour without protection is important to the designer, in order to be able to decide whether measures are to be taken. Scour without protection usually develops quickly, so only the equilibrium depth (h_{se}, see Figure 4-3) is of interest to the designer.

4.2.1 Scour in jets and culverts

There are many types of jets, circular and plane, horizontal and vertical, plunging and submerged, see Breusers/Raudkivi, 1991. Here, the focus is on submerged, horizontal jets, see Figure 4-4a. The difference between a (circular) jet and a culvert is the waterdepth downstream (the tailwater depth). When the tailwater depth is smaller than the diameter D, it is called a culvert although it is not crystal clear where the transition between a jet and a culvert lies.

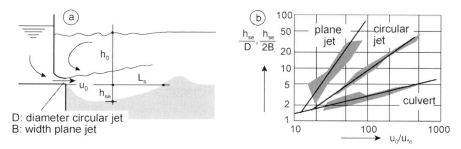

Figure 4-4 Scour in horizontal jets and culverts

For scour in horizontal jets, the approach by Breusers/Raudkivi,1991, is adopted. Experimental results suggest that the length-depth ratio of a jet scour hole, L_s/h_{se} see Figure 4-4a, is constant ($\approx$ 5 - 7). Assuming that L_s is determined by the location where the jet's u_m (see chapter 2) is equal to the critical velocity of the bottom material, we find (applying the jet velocity relations of chapter 2):

Circular jets: $u_c = u_m \propto \dfrac{D\,u_0}{x} \propto \dfrac{D\,u_0}{L_s} \propto \dfrac{D\,u_0}{h_s} \quad \rightarrow \quad \dfrac{h_{se}}{D} \propto \dfrac{u_0}{u_c} \propto \dfrac{u_0}{u_{*c}}$

Plane jets: $u_c = u_m \propto \dfrac{u_0\,\sqrt{B}}{\sqrt{x}} \propto \dfrac{u_0\,\sqrt{B}}{\sqrt{h_s}} \quad \rightarrow \quad \dfrac{h_{se}}{B} \propto \left(\dfrac{u_0}{u_{*c}}\right)^2$ (4.4)

u_{*c} is used instead of u_c, being an unambiguously defined critical velocity when the waterdepth is not clearly defined. These proportionalities, combined with experimental results, finally lead to, see also Figure 4-4b:

plane jet: $\dfrac{h_{se}}{2B} = 0.008 \left(\dfrac{u_0}{u_{*c}}\right)^2$ circular jet: $\dfrac{h_{se}}{D} = 0.08 \dfrac{u_0}{u_{*c}}$ (4.5)

The scour with a plane jet is much higher than with a circular jet, due to the horizontal divergence of the flow in a circular jet. For circular jets, the width of the scour hole is approximately half its length.

Experiments with *culverts* have given much smaller scouring depths than with circular jets. For culverts, Breusers/Raudkivi,1991, present the scour depth as (see also Figure 4-4b):

$$\dfrac{h_{se}}{D} = 0.65 \left(\dfrac{u_0}{u_{*c}}\right)^{0.33}$$ (4.6)

An explanation can possibly be found in the energy loss. To reach the same outflow velocity, u_0, one expects the necessary head difference for a jet to be greater than for a culvert, due to the "drowning" of the flow in the large tailwater depth. The loss of energy (or better the conversion from potential energy into kinetic energy and hence turbulence) is equal to $\rho g(h_u - h_d)Q$, so with equal u_0, hence equal Q, the amount of turbulent energy is larger for a jet, resulting in more scour. Another simple way of looking at this difference is that a greater discharge ($\propto \bar{u} \cdot h_0$) is able to transport more sediment. As said before, the transition between jets and culverts will not be so clear with varying tailwater depths, so equations (4.5) and (4.6) only serve as a first indication.

4.2.2 Scour around detached bodies

Piers are the most common examples of detached bodies and are important in many fields of hydraulic engineering. Figure 4-5a shows a clear picture (from Breusers/Raudkivi, 1991) of the flow around a cylinder. In chapter 2 we already saw that there is a downflow in front of the cylinder, which acts more or less as a vertical

jet, due to the difference in pressure, which is in turn cause∕
difference in horizontal velocity. This jet, together with the acce
sides of the pier (with $u_{max} \approx \bar{u}_0$), initiates erosion. Once a.
appeared, the same circulating current as found in the scour hole in ∟
appears. This eddy is carried along with the flow, leading to a horseshoe-shapeu
vortex. This vortex, together with the accelerated flow and wake vortices, transports
the sediment further downstream. Behind the pier, the transport capacity decreases
again and a bar is formed.

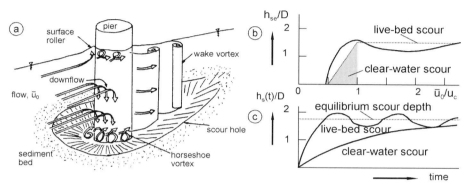

Figure 4-5 Scour around a cylinder (from Breusers / Raudkivi, 1991)

Figure 4-5b shows the equilibrium depth (made dimensionless by dividing by the
pier diameter) with increasing velocity. When $\bar{u}_0 < u_c$, scour already occurs due to
the above mentioned aggravation of the load around the pier, starting at $\bar{u}_0 \approx 0.5u_c$
(clear- water scour). The scour reaches a maximum when $\bar{u}_0 \approx u_c$. When $\bar{u}_0 > u_c$ it
can decrease again slightly, because then the whole bed moves (live-bed scour). This
is the transition between clear-water and live-bed scour. A second peak can occur at
the transition between ripples/dunes and anti-dunes where the bed is flat. In that case,
the flow does not need to spend energy to overcome the form drag of the dunes and
can devote itself completely to sediment transport. The dotted line indicates the way
the velocity factor is used in the final scour formula (equation (4.7).

Figure 4-5c, finally, shows the different development of clear-water and live-bed
scour. Live-bed scour reaches an equilibrium much faster than clear-water scour.
Moreover, the depth of the scour hole fluctuates due to the migration of ripples and
dunes on the bed. Usually, the scour process is so fast that only the equilibrium depth
is of interest in hydraulic engineering.

The complex three-dimensional flow pattern around a pier rules out an analytical
approach for scour depths; only experimental results are available. For slender piers
(waterdepth/diameter ratio > 2-3), the scour depth is found to be proportional to the
pier diameter with a constant of proportionality ~ 1.5-2, see Breusers et al., 1977.
The increase of velocity around the pier is not a function of the diameter (potential

flow gives doubling of the approach velocity for any diameter), but this proportionality can possibly be explained by the downflow in front of the pier and the vortices in the wake. The downflow is expected to increase when a broader object halts the flow, the vortices behind the pier grow as the wake behind the cylinder widens. It can be expected that this growth cannot go on when the diameter of the pier approaches or exceeds the waterdepth. The downflow will be limited when the pier acts as a complete "wall" and the size of the vortices will be limited by the waterdepth. Think of a "pier" with a diameter of 100 m in a waterdepth of 10 m (e.g. an offshore oil storage device) or, even larger, an island. A direct relation between scour depth and diameter does not seem logical with such large diameters; the flow process around one edge of the structure will hardly be related to what happens around the other edge. A relation with the waterdepth seems more appropriate, which has also been found in experiments.

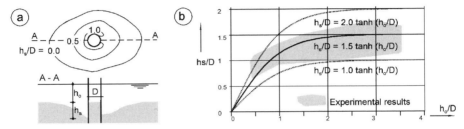

Figure 4-6 Scour around cylinder as function of waterdepth and diameter

Figure 4-6 shows the range of experimental results as given by Breusers et al, 1977. A mathematical description of the whole range of situations sketched above can be given by means of a hyperbolic tangent of the dimensionless parameters h_s/D and h_0/D. For large values of h_0/D (slender piles in deep water), $\tanh(h_0/D)$ becomes 1, giving $h_s \propto D$. For small values of h_0/D (large bodies in shallow water), $\tanh(h_0/D)$ becomes h_0/D, giving $h_s \propto h_0$. A constant of proportionality of 2 is recommended by Breusers et al, 1977, for design purposes.

Note: The tanh-function has no physical basis. It is just a function with convenient mathematical properties to describe the above-mentioned relations.

Figure 4-6 is valid for cylindrical piers. For other shapes, Table 4.1 gives values of a shape factor. Streamlining is favourable against scour, but when the flow direction deviates from the main axis of the pier, the effect can become negative, see Figure 4-7. For a first design the following encapsulating formula is proposed:

$$\frac{h_s}{D} = 2\, K_S\, K_\alpha\, K_u\, \tanh\left(\frac{h_0}{D}\right) \tag{4.7}$$

in which:

K_S = shape factor, see Table 4.1

K_α = angle of attack factor, see Figure 4-7

K_u = velocity factor, see Figure 4-5b: $K_u = 0$ for $u/u_c < 0.5$, $K_u = 1$ for $u/u_c > 1$ and K_u = $(2u/u_c – 1)$ for $0.5 < u/u_c < 1$

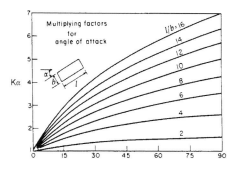

Figure 4-7 Multiplying factor for piers not aligned with flow

Table 4.1 Shape factors for various pier shapes

Pier shape	l/b	K_S
Cylinder	-	1.0
Rectangular	1	1.2
	3	1.1
	5	1.0
Elliptic	2	0.85
	3	0.8
	5	0.6

Note:

- floating debris or ice can enlarge the effective width of a pier or pile
- for a bridge with many piers, extra scour may occur due to the flow constriction
- wave action around the pier will hardly increase the scour depth (max. ~ 10%)

Example 4-1

A cylindrical pier with a diameter of 5 m is located in a river, 5 m deep, with a gravel bed, $d_{50} = 5$ mm. The flow velocity is 1 m/s. What is the expected scour depth?
The dimensionless diameter, d, in the Shields-Van Rijn graph, see figure 3.2b, is: 0.005*((1.65*9.81)/(1.33*10-6)2)1/3 = 105. This gives $\psi = 0.05$ (**Note**: scour refers to sediment transport, not to damage to a bottom protection, so, the original Shields values should be used). With an assumed roughness of twice the median grain diameter we find C = 18log(12*5/0.01) = 68 √m/s. From this we find a critical velocity: u_c = 68*√(1.65*0.005*0.05) = 1.38 m/s. The velocity coefficient in equation 4.7 then becomes: 2*1/1.38–1=0.45. The scour depth becomes: 5*0.45*tanh(5/5) = 1.7 m.*

4.2.3 Scour around attached bodies and in constrictions

Scour around constrictions will be treated by considering various elements: abutments, gradual constrictions and combinations as is the case with groynes.

Abutments

The first approach to abutment scour is simply to consider the abutment as a half pier. Figure 4-8 shows the relation between the scour depth, the abutment width (*D*, the protrusion into the water) and the waterdepth. This is the same pattern as can be seen in Figure 4-6, although D is now half the diameter, since the wall can be seen as the central axis of a pier. This makes abutment scour less than pier scour which is possibly because the wall suppresses the downflow and the vortices behind the structure. Streamlining is now relevant, because the angle of attack is less important. Note that the width of the flow is implicitly assumed infinite; narrowing of the flow width by the abutment does not play a role in these experiments.

Table 4.2 Shape factors for abutments

Abutment shape	K_S (to be used in equation (4.7)
Rectangular ("Blunt")	1.0
Cylindrical	0.75 - 1.0
Streamlined	0.5 – 0.75

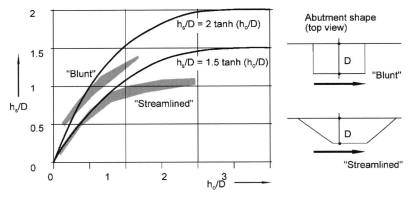

Figure 4-8 Scour around abutments (adapted from Breusers / Raudkivi, 1991)

Figure 4-9 shows the flow pattern and the erosion after the construction of the first Zeebrugge breakwater, which can be seen as an enormous abutment ($D \approx 1500$m). The velocity has increased with a factor, a little less than 2, which is in line with what could be expected according to potential flow. The scour depth is about 17.5-7 = 10 m, about 1.5 times the waterdepth, which is in accordance with equation (4.7) and Table 4.2. So, for a first approach, these equations are useful, despite the physical shortcomings.

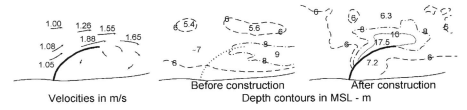

Figure 4-9 Flow velocities and scour around breakwater Zeebrugge

Gradual constrictions

Narrowing a river stretch leads to general deepening of the bed. The depth inside the constriction can be calculated using continuity equations for water and sediment. Both the discharge, Q, and the sediment transport, S, (see Figure 4-10) remain constant in the equilibrium stage.

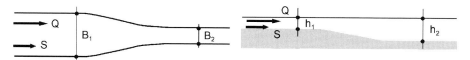

Figure 4-10 Erosion in gradual constriction

Assuming a simple relation between sediment transport and velocity: $S = k.u^m$, we find:

$$
\left.
\begin{aligned}
Q &= B_1 \, u_1 \, h_1 = B_2 \, u_2 \, h_2 \;\rightarrow\; u_2 = u_1 \, \frac{B_1 \, h_1}{B_2 \, h_2} \\[2mm]
S &= B_1 \, k \, u_1^m = B_2 \, k \, u_2^m
\end{aligned}
\right\}
\;\rightarrow\;
\frac{B_1^{m-1}}{B_2^{m-1}} = \frac{h_2^m}{h_1^m}
\;\rightarrow\;
\frac{h_2}{h_1} = \left(\frac{B_1}{B_2} \right)^{\frac{m-1}{m}}
$$

(4.8)

m usually lies between 4 and 5, so the power in equation (4.8) lies between 0.75 and 0.8.

Groynes

Groynes can be seen as blunt abutments, but the narrowing of the river bed also plays a role. The constricted river width therefore appears in the relation for scour depth. Breusers/Raudkivi, 1991, recommend as a first approximation for a vertical groyne perpendicular in a straight channel (see Figure 4-11):

$$
h_0 + h_{se} = 2.2 \left(\frac{Q}{B - b} \right)^{2/3}
$$

(4.9)

Note that the constant 2.2 in this equation is not dimensionless!

When the head of the groyne is not vertical, the separation of the flow behind the groyne does not take place in one vertical plane, which has a favourable influence on the erosion. With a slope of 45°, the depth is about 15% less. With very gentle sloping heads, reductions of 50% seem possible. When the groynes are situated at the concave side in a river bend, the depth can increase 10 - 50%, depending on the bend radius. For more detail, see Breusers/Raudkivi, 1991.

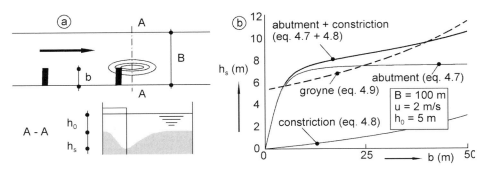

Figure 4-11 Scour around groyne

Figure 4-11 also shows a comparison of equation (4.9) with equations (4.7) (applied to abutments) and (4.8). Abutment and constriction scour are simply added together. The similarity for this case is rather good, but for other riverwidths or waterdepths, the results can deviate considerably. The comparison is only presented here to show that there are "family ties" between the empirical relations and that it should be possible in the future to replace them by one approach, dealing with the whole phenomenon.

4.3 Scour with bed protection

A bed protection around a bridge pier can be made to cover the whole area of increased load leading to almost negligible scour. With protection downstream of a jet or a culvert, the additional load vanishes slowly and the main effects are that the flow becomes calmer and the scour development moderates. Many investigations on this subject were carried out for the design of closure dams in the framework of the Deltaproject in the Netherlands. Development in time was an important factor there, as equilibrium depths for every building stage were never reached, thanks to the progress of the closure. But in other situations it can also be good to know the scour as a function of time (the development phase in Figure 4-3). It can play an important role in maintenance policies, as relatively little is known about the slopes of scouring holes and the soil-mechanical stability. It is therefore important not to be taken by surprise by unforeseen rapid developments.

4.3.1 Scour development in time

The research for the Deltaproject aimed mainly at establishing reliable scale relations to interpret model results, see the comparison in Intermezzo 4-1. The derived relations, however, can also be used for prediction purposes in a preliminary design. Several hundreds of tests have been performed and the experimentally established relations have gained some general validity. From a dimensional analysis and many experiments, the following expression was finally developed for clear-water scour behind a bed protection:

$$h_s(t) = \frac{\left(\alpha\,\bar{u} - \bar{u}_c\right)^{1.7} h_0^{0.2}}{10\,\Delta^{0.7}}\; t^{0.4} \tag{4.10}$$

in which $h_s(t)$ is the maximum depth in the scour hole as a function of time, while h_0 is the original waterdepth; $\bar{u}$ is the vertically averaged velocity at the end of the protection. t is the time in *hours* in this empirical relation. The constant 10 is not dimensionless. α represents, among other things, turbulence and increases the effective velocity. This is the "dust-bin" in the scouring formula.

From equation (4.10) we see that:

- the scour depth grows in time with a power 0.4
- there is some influence of the specific weight of the sediment (power 0.7) and very little of the waterdepth (power 0.2)
- the difference between the actual velocity multiplied by α and the critical velocity is dominant, due to the power 1.7

Note 1   Equation (4.10) is valid for the development phase of the scouring process; $\alpha\bar{u} - \bar{u}_c$ should therefore not be too small ($> \approx 0.1$ m/s).

Note 2   The power 0.4 for time in equation (4.10) is strictly speaking for 2-dimensional flow situations. For 3-dimensional flow (horizontal constriction, outflow from a discharge sluice etc.) the power should be higher in the beginning and slightly smaller later on. As a first approximation 0.4 will be used for all time dependent scour.

4.3.2   Factor α

For preliminary design purposes, it is necessary to have some insight into the relation between geometry and α. α is an amplification factor for the velocity, which expresses the disturbance in the flow, hence it is expected to be related to the turbulent fluctuations in the flow. But, similar to the velocity factor K_v in the stability relations for stones, see chapter 3, much depends on where and how the velocity is defined, see also Jorissen/Vrijling, 1989.

Figure 4-12a shows two ways to define α and $\bar{u}$ in equation (4.10). When $\bar{u}$ is defined as Q/A, the influence of a locally increased velocity must be included in α, thus "polluting" this coefficient with the velocity distribution. When $\bar{u}$ is defined locally, α only represents the amplification factor due to turbulence. The product $\alpha.\bar{u}$ is of course the same in both cases.

For $\bar{u} = Q/A$, Figure 4-12b shows α as a function of the vertical constriction for 2-dimensional and 3-dimensional scouring (without and with horizontal constriction respectively). This data has resulted from systematic scour investigations for the

Deltaproject. The α's differ considerably, largely due to the velocity distributions. Using the local velocity, Figure 4-12c gives the relation between α_L and the vertical constriction. The difference is much less, as the velocity distribution does not influence α. It is to be expected, therefore, that a correlation between α and the turbulence of the flow will only be found when u_L and α_L are used. Figure 4-12d shows the relation between α_L and the turbulent fluctuations, expressed as r (depth-averaged), which reads:

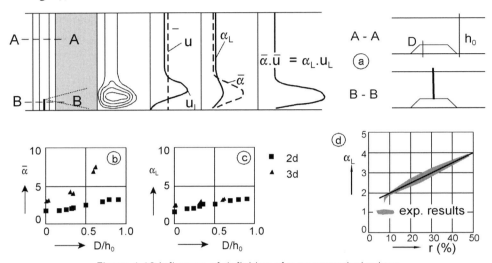

Figure 4-12 Influence of definition of α on numerical values

$$\alpha_L = 1.5 + 5r \quad \text{for} \quad \alpha_L > 1.8 \tag{4.11}$$

So indeed, a relation between α_L and r exists, indicating that turbulence together with a local velocity, accounts for time-dependent scour.

Intermezzo 4-1 Comparison model test and prototype

Equation (4.10) has been based on model-tests for the Deltaproject. α represents turbulence in the flow and is related to the geometry of the flow situation. It is often derived from a scale model in which h_0, $\bar{u}$, u_c and Δ are known and in which h_s is measured as a function of time. α can then be calculated from equation (4.10). With this α, equation (4.10) is used again to calculate the scour in prototype circumstances. The scale model is only used to obtain the geometry related α.

Comparisons of prototype data were very difficult due to the ever-changing situation during the construction of the Delta-dams. To increase confidence, a special test was performed using a real outlet sluice and its scale model, both equipped with an extra disturbing dam for extra scour. Figure 4-13 shows the result: conformity in development is reasonable. The resemblance between the scour holes in the model and in reality was good.

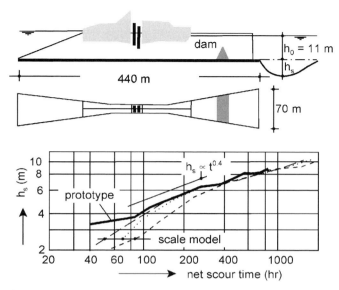

Figure 4-13 Comparison model-prototype scour development (de Grauw/Pilarczyk, 1981)

For a preliminary design, Figure 4-14 gives values for α based on local velocities in situations with vertical and horizontal constrictions. For a project of some importance, experiments using a scale model will be necessary. In the case of an outlet structure that discharges into a large body of water, there are no clear vertical and/or horizontal constriction dimensions. In that case the scour can be approximated to the α-value for a vertical vortex-street which lies around 4 for a bottom protection that is $10\,h_0$ long.

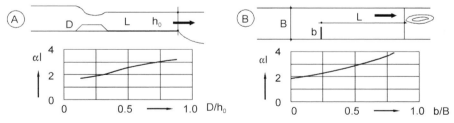

Figure 4-14 Values for α_L as a function of vertical and horizontal constriction for $L/h_0 =10$

4.3.3 Protection length and roughness

It is to be expected that a longer bottom protection will lead to lower α-values, due to dissipation of turbulence, see chapter 2. Figure 4-15 shows the influence of the length of the bottom protection on α, based on the results of systematic scour research. For $L/h_0 > 5$ this influence can be described using the following relation:

$$\alpha(L/h_0) = 1.5 + (1.57\,\alpha_{10} - 2.35)\,e^{\left(-0.045\,L/h_0\right)} \tag{4.12}$$

where α_{10} is the α-value for $L/h_0 = 10$. Note that $\alpha_{min} \approx 1.5$ for an infinitely long bottom protection. So, indeed, as said in section 4.1.1, such a protection does not prevent scour!

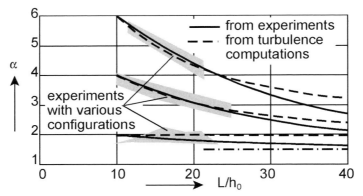

Figure 4-15 Relation between α, turbulence and length of bottom protection

The roughness of the bottom protection is also of influence. With a smooth protection, the velocities near the bed are high and cause more scour. The influence of the roughness on α is given by Booij/Hoffmans,1993:

$$\alpha = 1.5 + 5\,r_0\,f_c \quad \text{with} \quad f_c = \frac{C}{40} \quad (f_c = 1 \text{ for } C \le 40) \tag{4.13}$$

Smooth (e.g. asphalt) bottom-protections lead to 20-30% higher α-values than rough protections. Very rough protections induce extra turbulence, compensating the effect of a lower bottom velocity, hence the minimum value for $f_c = 1$.

In Figure 4-15, a comparison is also made between the experimental results (equation (4.12)) and the dissipation of turbulence due to a rough bottom protection ($C = 40$). The comparison is made for equal α-values at $L/h_0 = 10$. As can be seen the correspondence is quite good, for short protections, while for long protections, the empirical results give no clue as no data are available. The minimum α-value in equation (4.13) is 2 with $C = 40$ ($r = 1.2\ \sqrt{g}/C$, see chapter 2), while the experiments produced a mimimum value for $\alpha = 1.5$. Here again, the experimental data do not prove of which value is correct.

4.3.4 *Varying conditions*

For scour in a flow with gradually varying intensity, equation (4.10) can be used, replacing $(\alpha\bar{u} - \bar{u}_c)^{1.7}$ by:

$$\frac{1}{T}\int_0^T \left(\alpha\,\bar{u} - \bar{u}_c\right)^{1.7} dt \qquad\qquad (4.14)$$

in which T is the time used in the averaging process. In a tidal situation, the flow direction turns. and e.g. an outlet sluice, which discharges into the sea during low water, works only part of the tidal period. The scour at one side of a structure should then be computed using only the flow in one direction, see Figure 4-16a.

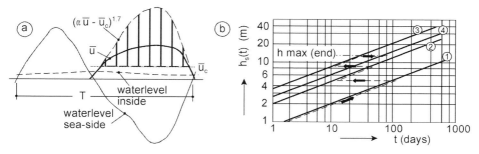

Figure 4-16 Scour in tidal flow and during constriction

When scour has to be determined for the construction period of a structure, Figure 4-16b shows the procedure. For various representative building stages, each one lasting Δt_i, the various scour-time curves need to be determined. The first curve is followed during Δt_1. At the end of this period the line continues horizontally to the second curve (the scour hole according to the second curve starts with the end value of the first period) and so on till the end. This procedure can be used as the similarity of the shapes of scour holes is an important finding of the systematic scour research.

4.3.5 Equilibrium scour

Clear-water scour

A simple approach to find the equilibrium depth in the case of clear-water scour with bed protection is to say that the equilibrium depth is reached when the velocity in the hole, multiplied by α equals u_c. This will overestimate the equilibrium depth, because α is defined at the edge of the protection and not in the hole itself. A reduction of $\alpha\,u$ with a factor of about 0.5 seems reasonable.

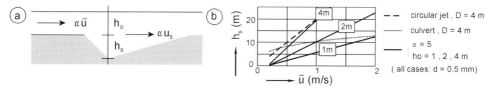

Figure 4-17 Equilibrium scour compared with equations for jets and culverts

With the continuity equation this leads to, see Figure 4-17a:

$$\left.\begin{array}{l} u_c = 0.5\,\alpha\,\bar{u}_s \\[2mm] \bar{u}_s\left(h_0 + h_{se}\right) = \bar{u}\,h_0 \end{array}\right\} \rightarrow \frac{h_{se}}{h_0} = \frac{0.5\,\alpha\,\bar{u} - \bar{u}_c}{\bar{u}_c} \tag{4.15}$$

Figure 4-17b shows the equilibrium depth computed with equation (4.15) compared with the depth in circular jets and culverts computed with equations (4.5) and (4.6) respectively. The comparison is only qualitative again to show the family ties between empirical relations. $\alpha = 5$ for a situation with a protection length of 0, very roughly estimated from Figure 4-15 (starting from $\alpha = 4$ for $L = 10h_0$). h_0 is 1, 2 and 4 m. The results show an increase in scour depth with increasing tailwater depth, which is the same trend shown by the jet and culvert relations, see section 4.2.1. The jet formula seems appropriate when the waterdepth equals or exceeds the outflow diameter.

Example 4-2

At the outflow of a culvert an apron serves as bottom protection. At the end of the protection the waterdepth is 3 m, the velocity is 1.2 m/s and the bottom consists of sand with $d_{50} = 0.3$ mm. The coefficient $\alpha = 2.5$ is used as an example in this situation. What is the scour depth after 10 days of effective use?
From chapter 3 we find $u_c = 0.4$ (see also Example 4-1) Equation 4.10 gives: $h_s = (2.5*1.2 - 0.4)^{1.7}*3^{0.2})*(24*10^{0.4}/(10*1.65^{0.7})) = 4$ m. This means that there seems no real danger of collapse at the edge of the apron within a short period of coming into use. The final depth (without supply of sediment) is derived from equation 4.15: $h_{se} = (0.5*2.5*1.2 - 0.4)*3/0.4 = 8.3$ m. So, within the first 10 days of use, the scour reaches almost half the final depth.

Live-bed scour

In the case of live-bed scour with bed protection, the equilibrium depth will be reached when the sediment outflow from the hole becomes equal to the sediment supply from upstream, see section 4.1. This depth will be smaller than follows from equation (4.15), which has been developed for clear-water scour. The scouring formula, equation (4.10), represents the clear-water scour and the sediment supply turns it into a formula for live-bed scour. Using equation (4.10) and representing the hole as a triangle, see Figure 4-18a, the volume of the hole per m width, without sediment supply, is:

$$\begin{aligned} I &= \frac{1}{2}\left(\cot\beta_1 + \cot\beta_2\right)h_s^2 \\[2mm] &= \left[.005\left(\cot\beta_1 + \cot\beta_2\right)\varDelta^{-1.4}\,h_0^{0.4}\left(\alpha\bar{u} - \bar{u}_c\right)^{3.4}\right]t^{0.8} = K\,t^{0.8} \end{aligned} \tag{4.16}$$

where K is short for the expression between brackets.

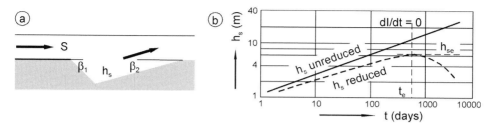

Figure 4-18 Scour reduction by sediment supply

With sediment supply this becomes:

$$I_{red} = K\, t^{0.8} - S \cdot t \quad \rightarrow \quad h_{s\,red} = \sqrt{\frac{I_{red}}{0.5 \cdot \left(\cot \beta_1 + \cot \beta_2\right)}} \tag{4.17}$$

An equilibrium is reached when, see Figure 4-18:

$$\frac{dI}{dt} = 0 \rightarrow 0.8\, K\, t^{-0.2} = S \rightarrow t_e = \left(\frac{0.8\,K}{S}\right)^5 \rightarrow h_{se} = \sqrt{\frac{K\, t_e^{0.8} - S \cdot t_e}{\frac{1}{2}\left(\cot \beta_1 + \cot \beta_2\right)}} \tag{4.18}$$

There are, however, many uncertainties in this procedure: $\cot \beta_{1,2}$ have to be estimated (e.g. 4 and 40), S has to be calculated and t_e is very sensitive to variations of K and S. Therefore, in closures, it is preferable to use a value of t_e measured during an early building stage, see Figure 4-16. Better results can be expected in the future when the scouring can be computed as sediment transport, taking the local velocities and turbulence into account. The sediment supply from upstream is then the input at the upstream boundary of such a model.

4.3.6 Stability of protection

Scour itself is not a problem as long as it does not undermine the considered structure. Figure 4-19a represents a situation with a stable scour hole that does not cause any problem. In Figure 4-19b, the scour depth, in combination with the upstream slope of the scour hole, has become too large and the upstream slope has slid. The bed protection has been damaged, but the structure is still intact. After this has been observed, repair is still possible, e.g. by dumping stones on the slope and on the damaged protection. If the damage is not observed in time or if it is neglected, the quality of the protection decides what happens next. If the protection is coherent, e.g. because of a strong geotextile, and/or mats with sufficient overlaps have been used, the consequences can be rather insignificant. But if the protection consists of loose stones, the scouring process will go on, with even more vigour because of the shorter protection length. The final result can be like the situation shown in Figure 4-19c. In that case, the protection is no longer effective and the structure will collapse.

Figure 4-19 Instability of bottom protection

These examples clearly illustrate that *the main function of a bottom protection, is not to minimize scour (see also Figure 4-15), but to keep the scour hole far away from the structure that needs protection*. To judge whether a protection is long enough, it is necessary to know the upstream slope β, (in Figure 4-18, β_1 is the upstream slope, but the subscript can now be left out), the depth of the hole, the stability parameters of the slope and the process that will take place once the slope has become unstable.

The slope angle β

From the systematic scour research, Hoffmans, 1993, derived the relation:

$$\beta = \arcsin\left[3 \cdot 10^{-4} \frac{u_0^2}{\Delta g d_{50}} + \left(0.11 + 0.75 r_0\right)f_c\right]$$

$$(f_c = \frac{C}{40}, f_c = 1 \text{ for } C \le 40)$$

(4.19)

indicating that velocity, turbulence, a smooth protection and fine sand, cause steep slopes.

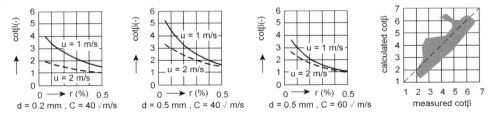

Figure 4-20 Slope of scour hole vs u, r, d and C (after Hoffmans, 1993)

Figure 4-20 shows the results of equation (4.19) for some likely values of the relative turbulence, the velocity, the roughness of the protection and the grain diameter. For the diameters used here (sand of 0.2 and 0.5 mm) the angle of repose, ϕ, lies around 30°, hence cot $\phi \approx 2$. The graphs indicate that steeper slopes are very much possible. The model experiments showed no slopes steeper than 1:2, but in reality steeper slopes and failures of these slopes have been observed. Figure 4-20 also indicates large differences between calculated and measured values, so, a steep slope calculated using equation (4.19) should be seen as a warning. It is also important to measure the scour as a function of time to be able to deal with maintenance, such as dumping protective material on the slope.

Stability and slides

When the slope becomes too steep with a certain height, it is no longer able to withstand the gravitational forces and it will slide. Soil will move side- and downwards and will result in a flatter slope. In sand, it can roughly be assumed that unprotected slopes will slide when $\cot \beta \leq 2$ with $h_s > 5\text{-}10$ m. Chapter 5 will discuss stability of slopes in more detail; now the focus is on the process after stability is lost.

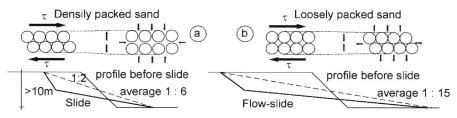

Figure 4-21 Slides and flow-slides

Sliding occurs after a slope has lost its stability, see Figure 4-21a. The grains roll over each other till the friction is sufficient to stop the sliding. Due to the inertia of the moving soil mass, the final slope will be gentler than the angle of repose ϕ. 1:6 can serve as an indication of an average slope for densely packed sand.

When the soil consists of loosely packed sand, the situation is worse. Figure 4-21 shows the difference between loosely and densely packed sand. When a shear stress is exerted on loose sand, the grains tend to a denser packing, producing an excess pressure on the pore water and thus forcing it out of the pores, see Figure 4-21b. This excess pore pressure in loosely packed sand decreases the contact forces between the grains (the effective stress, sse chapter 5) and hence, leads to a complete reduction of the shear strength. The soil becomes, temporarily, a thick fluid. This phenomenon is called *liquefaction*. The excess pore pressure can propagate as an elastic shock wave through the sand, turning a large area into *quicksand*. The result is a much flatter slope after failure than would result from a normal slide. Slopes of 1:25 have been reported; an average value of 1:15 is usually a reasinable estimate, see Figure 4-21b. With densely packed sand the reverse happens, water is sucked into the pores, leading to an increase of the effective stress, see Figure 4-21a.

4.4 Summary

In general, local scour is a special case of sediment transport. The local velocity and the turbulence pattern determine the local erosion. When this erosion takes place in a flow in which sediment is transported, even without local disturbances, we speak of live-bed scour. When the sediment transport is due to the local disturbance only, it is clear-water scour. Scour develops in time, reaching an equilibrium depth when the flow forces can no longer take away sediment (clear-water scour) or when the inflow and outflow of sediment become equal (live-bed scour).

In literature, two main approaches to scour can be distinguished: one is to study the final equilibrium scour and the other approach is to study development in time. The choice depends on the interest of the designer and the process itself. When the equilibrium is reached very quickly, it is of no use to know the development in time and the design can be based on the final scour depth. When the process is slower, it can be economical to study the development in time and to tune the design to a certain maintenance policy. When scour is important during construction, it may be necessary to know the scour as a function of time, since the geometry of the flow, and hence the scour, will change continuously.

Without protection, the scour will reach an equilibrium faster than with protection. Empirical relations for scour in jets and around bodies are presented which can be used for preliminary design purposes. An attempt has been made to give maximum physical reasoning behind the empirical relations, although the physical basis of some of these equations is meager or virtually non-existent. Important relations are (4.7) and (4.8).

In cases with protection, the scour as a function of time is coupled as much as possible with turbulent flow phenomena. Equation (4.10) is the basis of this approach, in which a paramount role is played by the coefficient α, which can be seen as a magnifying factor for the velocity. α is related to the relative turbulence of the flow, provided the local velocity is used. The estimate of equilibrium scour in the case with protection and the comparison with some formulas without protection, make the story as complete as possible.

*Finally, considerations are given concerning the stability of scour holes and bottom protections in which soil properties like the angle of repose and the packing of the grains play a role. **It is stressed that the scour itself is not the problem, but its threat to the stability of a structure.** The main function of protections behind outflows is to keep the scour hole as far away as possible from the structure that needs to be protected. Preventing scour is virtually impossible. Only in a flow field with considerable local variations, like the field round a pier, it is possible to cover the area with increased flow load, thus reducing scour depths.*

5 POROUS FLOW
General

Slide of dike slope (Zealand), courtesy GeoDelft

5.1 Introduction

Porous flow is the expression used for flow through a granular medium, like sand, pebbles or stones. The loads due to porous flow often come from the soil side of the interface soil-water, see Figure 5-1. As in open channel flow, *pressures* and *velocities* are related and both can be considered as loads.

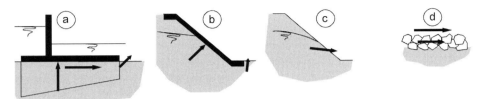

Figure 5-1 Examples of loads due to porous flow

Under the impervious structure in Figure 5-1a there is a resulting water pressure against the bottom downstream, due to the head difference across the structure. To counterbalance this pressure, the bottom of the structure should have enough *weight.* The head difference also causes a flow under the structure, which, again, may cause erosion at the downstream end, against which a *filter* may be necessary. If the bottom of the structure is relatively short, the outflow at the end can cause *piping*, (where erosion creates a "pipe"). The same phenomena occur with an impervious layer on a slope, e.g. at a river bank or a dike (Figure 5-1b). Due to the slope, not only is weight necessary to counterbalance the excess pressure, but *friction* between the layer and the slope also plays a role. At the toe, the outflow can, again, necessitate a filter to prevent erosion. When the slope is not closed and consists of loose grains, a head difference will cause *seepage* and possibly erosion (Figure 5-1c). A *milder slope* may be necessary to prevent erosion. Finally, the flow *over* a bottom protection consisting of large stones, which are stable themselves, will also cause flow *through* the protection. This may, again, cause erosion at the interface underneath the protection, for which a *granular filter* or a *geotextile* could be the right solution.

To create stable situations, a good insight into the porous flow phenomena and loads is paramount. The following sections discuss the basic equations for porous flow and some of their applications. It will be shown that, similar to open channel flow, porous flow can be either laminar or turbulent. In ground masses consisting of clay or sand, the flow is always laminar, leading to linear relations between pressure and velocity, making computations relatively easy. In coarse material the flow is usually turbulent, which is still less accessible for computation. Coarse materials are used in filters, sills, breakwaters, etc. For these applications, the design still depends heavily on empirical rules. Much research in this field remains to be done.

5.2 Basic equations

5.2.1 General

The starting point is again the Navier-Stokes equation, written in the Reynolds-form for the *x*-direction (see chapter 2):

$$\frac{\partial u}{\partial t} + u\frac{\partial u}{\partial x} + w\frac{\partial u}{\partial z} = -\frac{1}{\rho}\frac{\partial p}{\partial x} + \upsilon\frac{\partial^2 u}{\partial z^2} - \frac{\partial \overline{u'^2}}{\partial x} - \frac{\partial \overline{u'w'}}{\partial z} \tag{5.1}$$

With this equation the flow in a porous medium can be calculated, provided that every stone and every pore is taken into account, which is practically impossible and not necessary. An averaging procedure will be followed to reach practical solutions. The first step in this procedure is to define an average velocity, the filter-velocity:

$$u_f = \frac{1}{A}\iint_A u\, dA = n\cdot u \quad \left(n = \frac{V_P}{V_T}\right) \tag{5.2}$$

in which *n* is the porosity and *u* the real velocity in the pores. The porosity, *n*, is usually defined as the pore volume, V_P, divided by the total volume, V_T. This is not necessarily a measure for the permeability (compare with Emmenthal cheese: many pores, but not permeable), but for a normal grain-structure, *n* can be used also as a permeability parameter.

Note: The averaging has to include enough pores to represent an average flow in a process which is of a random nature. On the other hand, the area d*A* must be small compared with the mean motion (LeMehaute,1976), in other words: a large flow pattern through relatively small grains. Wave action in the armour layer of a breakwater with elements of several meters, cannot be described adequately by porous flow formulae.

The next step is to combine terms in equation (5.1), see also Figure 5-2. The term ∂*u*/∂*z* has a physical meaning in a particular pore, where there is a velocity gradient from the centre of the pore to the grain (no slip at the boundary), but with an average of many pores it becomes meaningless. It is common practice to combine all square inertia and turbulence terms in one quadratic friction term and to replace the (linear) viscous gradient with a linear friction term. (see also Van Gent, 1992, and Van Gent, 1993):

$$\frac{1}{\rho g}\frac{\partial p}{\partial x} = i = a u_f + b u_f |u_f| + c\frac{\partial u_f}{\partial t} \quad \text{with:} \quad a = \alpha\frac{(1-n)^2}{n^3}\frac{\upsilon}{g d_{(n)50}^2}$$

$$b = \beta\frac{(1-n)}{n^3}\frac{1}{g d_{(n)50}}$$

(5.3)

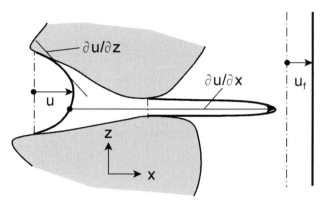

Figure 5-2 Velocities, gradients and averaging

For stationary flow, $\partial u_f/\partial t = 0$, this is the classical Forchheimer-equation. The influence of the third term is usually small and can be neglected in most cases. For more information, also on the values for c, see Van Gent, 1993. In non-stationary, oscillatory flow, the coefficient b in the second term has to be corrected with a factor which depends on the wave cycle, see Van Gent, 1993. The expressions for the coefficients a (s/m) and b (s^2/m^2) have been derived theoretically, while the dimensionless coefficients α and β (depending, among other things, on the grading and the shape of the grains) have to be determined experimentally. Without any further information, $\alpha \approx 1000$ and $\beta \approx 1.1$ can be used as a first estimate (with possible values twice as low or twice as high). For more accurate values, it is better to measure a and b in a laboratory or field test.

Figure 5-3 shows measured gradients i and filter velocities u_f for various materials ranging from fine sand ($d_{50} = 130$ μm) to rubble 60-300 kg ($d_{n50} \approx 0.4$ m). This double logarithmic graph clearly shows that the flow in fine material is laminar: the relation between i and u is linear, indicating that the relation is adequately described by the first term of the right-hand side of equation (5.3). For rock, the relation is quadratic (slope 1:2 on double logarithmic scale), indicating that the second term dominates. In between, as for gravel, both terms play a role.

Note: The transition between laminar and turbulent flow is more gradual than in open-channel flow; beside turbulence, convective terms also play a role in the averaged formula. Despite this difference, the flow types will be called laminar and turbulent.

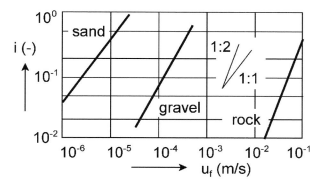

Figure 5-3 Relation between filter velocity and gradient for various materials

The relation between velocity and pressure gradient is sometimes written as follows:

$$u_f = k(i)^{\frac{1}{p}} \qquad (5.4)$$

in which k is the permeability (m/s) of the porous material. For laminar flow, $p = 1$, the Forchheimer-equation reduces to Darcy's law and k is the inverse of a in equation (5.3). For turbulent flow, $p = 2$. The following table gives an order of magnitude of the permeability, k, as defined in equation (5.4), of some materials:

Table 5.1 Values of k for various materials

Material	d_{50} ($< 63.10^{-3}$ m) or d_{n50} (m)	Permeability, k (m/s)	Character of flow
Clay	$< 2.10^{-6}$	10^{-10} - 10^{-8}	laminar
Silt	2.10^{-6} - 63.10^{-6}	10^{-8} - 10^{-6}	laminar
Sand	63.10^{-6} - 2.10^{-3}	10^{-6} - 10^{-3}	laminar
Gravel	2.10^{-3} - 63.10^{-3}	10^{-3} - 10^{-1}	transition
Small rock	63.10^{-3} - 0.4	10^{-1} - 5.10^{-1}	turbulent
Large rock	0.4 - 1	5.10^{-1} - 1	turbulent

5.2.2 Laminar flow

In case of laminar flow, equation (5.3) reduces to Darcy's law. It is sometimes more convenient to use heads instead of pressures:

$$h = z + \frac{p}{\rho_w g} \qquad (5.5)$$

where h is the piezometric head or potential and z is the elevation above a reference level. The Darcy relations for two dimensions now read:

$$u_f = -k_x \frac{\partial h}{\partial x} \qquad w_f = -k_z \frac{\partial h}{\partial z} \tag{5.6}$$

Darcy's law is the equation of motion for laminar groundwater flow. To compute a velocity and pressure distribution in a soil mass we also need to take into account the continuity equation:

$$\frac{\partial u_f}{\partial x} + \frac{\partial w_f}{\partial z} = 0 \tag{5.7}$$

Combining Darcy's law and the continuity equation, the Laplace equation is found:

$$k_x \frac{\partial^2 h}{\partial x^2} + k_z \frac{\partial^2 h}{\partial z^2} = 0 \tag{5.8}$$

Equation (5.8) can be solved, but only for very simple geometries, analytical, approximate solutions exist. For more complex situations, the equations can be solved numerically (using either finite elements or finite differences), graphically (with a rectangular flow net pattern) or with an electric analogon (based on the analogy between Darcy's law and Ohm's law for conductivity). Numerical solutions are easily available nowadays, but for a first idea, a sketch of a graphical solution can be useful. In the following sections, the numerical program MSEEP (GeoDelft, 1993) has been used in some examples.

The complete solution of equation (5.8) is often presented as a rectangular pattern of streamlines and equipotential lines, along which the piezometric head is constant, see Figure 5-4.

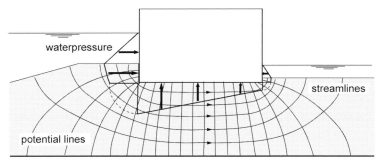

Figure 5-4 Groundwaterflow under a caisson

Between two streamlines, the total amount of flow is equal. Equation (5.6) shows that there will be no flow when the gradient of the piezometric head is 0. Hence, all streamlines are perpendicular to equipotential lines (the bottom is always an equipotential line since the piezometric level in the water is constant). With some effort and skill it is possible to draw a flow net iteratively by hand leading to a rough insight into the flow pattern and the pressures in the subsoil which are important for the stability of structures and protections, see Figure 5-4.

Flow force

The structure of grains and pores cause friction for the flowing water. The other way around, the water exerts a force on the grains (action = reaction). This is the flow force (sometimes called the flow pressure, but its dimension is force per unit volume), which is important in the stability of the grains at the outer boundaries of the soil. For the *x*-direction (other directions similarly) it is given by:

$$F_f = \rho_w\, g\, i = \rho_w\, g\, \frac{\partial h}{\partial x} \tag{5.9}$$

5.3 Stability of closed boundaries

5.3.1 Impervious bottom protections

In the case of a head difference across a bed protection, the pressures can often be easily determined as shown in Figure 5-5. A flow net, drawn either by hand or with a computer program, will be necessary when the flow geometry is more complex.

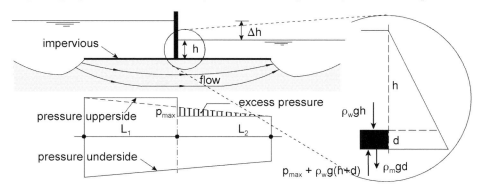

Figure 5-5 Pressures in case of an impervious bottom protection

For a stable bottom protection, the vertical equilibrium has to be considered. The maximum excess pressure depends on the location of the closure gate with respect to the upstream and downstream ends of the bed protection:

$$p_{max} \approx \frac{L_2}{L_1 + L_2}\, \rho_w\, g\, \Delta h \tag{5.10}$$

The upward force per m^2 at the most unfavourable spot of the protection is $p_{max} + \rho_w g(h + d)$ and the downward force: $\rho_m gd + \rho_w gh$ leading to:

$$\left(\rho_m - \rho_w\right)g\,d \ge p_{max} \tag{5.11}$$

Note: When the downstream waterlevel is lower than the upperside of the protection, the weight of the water on top becomes zero. With a downstream waterlevel equal to the underside of the protection, ρ_m has to be used in equation (5.11) instead of $(\rho_m - \rho_w)$. For a downstream waterlevel between the upperside and the underside of the protection, only the use of all downward and upward forces that lead to equation (5.11), see also Figure 5-5, gives a correct result. The use of all forces is therefore preferred above the use of Archimedes-type equations, in order to avoid mistakes.

Example 5-1

A 0.8 m thick protection with $\rho_m = 2000$ kg/m^3 is situated as in Figure 5-5 (gate located halfway the protection). The downstream waterlevel is equal to the underside of the protection. The upstream level is 3 m higher. The downward force per m^2 is then: 0.8*2000*g ≈ 16 kN. The upward force is: 1/2*3*1000*g ≈ 15 kN, hence, the situation is stable. The use of equation 5.11, using ρ_m, gives the same result.

Now, the downstream waterlevel is raised 1 m. The downward force then becomes: 0.8*2000*g + 0.2*1000*g ≈ 18 kN. The upward force becomes: 1/2*2*1000*g + 1*1000*g ≈ 20 kN so, now there is no equilibrium at the most unfavourable spot. Equation 5.11 and 5.10 give: (2000-1000)*g*0.8 ≈ 8 kN which is less than p_{max} (= 10 kN), so, the same result is achieved. The head difference was reduced with 33%, but the situation is no longer stable. Although a somewhat hypothetic example, it clearly demonstrates the influence of buoyancy.

To design the bottom protection for the location with p_{max} is a conservative approach and if the material has some strength (e.g. concrete), a protection that can resist a smaller p may be acceptable because other parts, where the excess pressure is less, can contribute to the stability. When the material is easily distorted (e.g. asphalt), it will be lifted locally, relieving the excess pressure with minimal movement. This lifting acts like a valve and the system "breathes". As long as no sediment from under the protection is lost and the protection does not crack, *occasional* uplifting during extreme loads can be tolerated, provided inspections are carried out frequently. For an asphalt protection with *permanent* loading, equation (5.11) should be obeyed (asphalt is a fluid!).

5.3.2 Impervious slope protections

The determination of the pressures against an impervious slope protection is more complicated than for an impervious bottom protection. Figure 5-6 shows the flow net and pressures for a slope of 1:2 with an impervious layer of negligible thickness, reaching 1 m below the outside waterlevel and a head difference inside of 1.5 m. Figure 5-6a shows the potential lines and streamlines as determined using MSEEP. Figure 5-6b gives the "pressures" (p divided by ρg) directly under the protection, compared with the hydrostatic values inside and outside the slope.

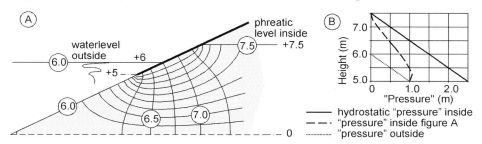

Figure 5-6 Flow net and pressures under an impervious layer on a slope

The values under the protection are lower due to the open slope below the protection. The excess pressures follow from the difference between the pressure inside and outside. The maximum excess pressure occurs at the outside waterlevel and is, according to MSEEP, about 0.7 m water.

Approximate, analytical solutions are available, see Figure 5-7. Under a protection, for $h_1/(h_1 + h_2) < 0.8$, the maximum excess pressure with regard to the waterlevel on the slope, is given in m water by (see TAW, 2000):

$$H = \frac{h_1}{\pi} \arccos\left[2\left(\frac{h_1 + d\cos\alpha}{h_1 + h_2} \right)^{\frac{\pi}{\arctan(\cot\alpha)+\pi/2}} - 1 \right]$$

(5.12)

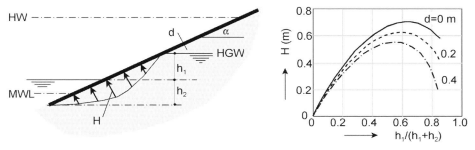

Figure 5-7 Pressures under impervious slope protection

For the same slope as in Figure 5-6, the maximum excess pressure, H, is given in Figure 5-7 as a function of $h_1/(h_1+h_2)$ and of the layer thickness d. The graph shows for $d = 0$ and $h_1/(h_1+h_2)$: $H \approx 0.7$ m, which is equal to Figure 5-6. With a protection of a certain thickness, the pressure at the bottom end will not be equal to the piezometric level downstream, see also section 5.3.1, reducing the excess pressure.

With equation (5.12) the maximum excess pressure can be computed when the phreatic level inside the soil mass is given. As a first approximation, this level can be approached with:

$$HGW = a\,(HW - MWL) \qquad (5.13)$$

in which HGW is the maximum ground water level, HW the maximum level outside and MWL the average level outside. In a tidal area, $a \approx 0.5$ and in a river area $a \approx 0.3$. For a detailed design, the use of a numerical model is recommended.

Note: the maximum excess pressure in equation (5.12) occurs when $h_1/(h_1+h_2) \approx 0.5$ (not to be confused with a in equation (5.13) which only determines the position of HGW). In the case when this maximum is reached for a level $< MWL$, MWL should be used to determine h_1.

Figure 5-8 shows the forces that play a role in the equilibrium of an impervious layer on a slope. The protection is loaded with the excess water pressure, H, which can be determined, either with equation (5.12) and (5.13) or with a numerical model. The weight, W, must counterbalance this load, either directly or via the shear force between the layer and the slope.

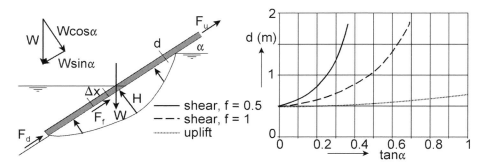

Figure 5-8 Stability of impervious layer on slope

Shear

The friction between protection and subsoil must be enough to balance the component of the weight parallel to the slope. The friction depends on the effective weight perpendicular to the slope. Stability for a length of slope Δx just below the outside water level is ensured when:

$$f\left[\left(\rho_m - \rho_w\right)g\,d\,\Delta x\cos\alpha - H\rho_w g\,\Delta x\right] \geq \left(\rho_m - \rho_w\right)g\,d\,\Delta x\sin\alpha \qquad (5.14)$$

in which *f* is the friction between layer and slope. This leads to a layer thickness:

$$\frac{H}{\Delta d} = \frac{f \cos \alpha - \sin \alpha}{f} \tag{5.15}$$

Uplift

The protection is stable against uplift when there is equilibrium perpendicular to the slope:

$$(\rho_m - \rho_w) g \, d \, \Delta x \cos \alpha \geq H \rho_w g \Delta x \quad \rightarrow \quad \frac{H}{\Delta d} = \cos \alpha \tag{5.16}$$

Note: Above the waterlevel ρ_m should be used instead of $(\rho_m - \rho_w)$ in these formulas.

Figure 5-8 gives the necessary layer thickness for slopes between 0 and 45 degrees using a friction coefficient, $f = 0.5$ and 1 respectively. For the other parameters, see Example 5-2. From this we can see the following:
- For a horizontal layer ($\alpha = 0$), there is no difference between shear and uplift, or in other words: shear does not play a role.
- The shear criterion gives thicker layers than the uplift criterion. When $\tan \alpha$ becomes *f*, the necessary *d* becomes infinite, which, with $f \cos \alpha = \sin \alpha$, directly follows from equation (5.15).
- A larger friction, *f*, is favourable, compare $f = 0.5$ with $f = 1$. For a very high value of *f*, the results for shear and uplift will coincide. The protection is "nailed", so to speak, to the slope. But *f* is limited by the properties of the slope. It can never be greater than $\tan \phi$, the internal friction of the slope material, which becomes the weakest spot in the stability. For many cases, $2/3 \tan \phi$ can be used as a first estimate for the value of *f*.

When the shear criterion is violated, part of the layer will hang on the upper part of the protection with force F_u in Figure 5-8. The tension in the layer can be computed by: $F_u = (\rho_m - \rho_w) \cdot g \cdot d \cdot \sin \alpha \cdot L$, where *L* is the length over which the shear criterion is violated. Another possibility is that the protection is going to lean on the lower part, which is again supported by the toe of the slope, with F_d. Actually, the whole protection should be taken into account and all forces should be integrated. Dependent on the strength of the layer, shear failure should not occur too often to avoid fatigue. E.g. under springtide conditions, with a frequency of occurence in the order of magnitude of weeks, the shear criterion should be met.
Uplift is worse: sand can move under the protection when it is lifted and the layer can deform. The uplift criterion should be met in design circumstances, e.g. a fast fall of water level after a storm.

Example 5-2

For a dike in a tidal area, a slope of 1:4 has to be protected with a layer of asphalt concrete, ρ_m = 2450 kg/m³, ρ_w = 1020 kg/m³, hence, Δ ≈ 1.4. The friction coefficient between asphalt and sand, f = 0.5. The protection reaches from MWL – 1 m up to 2 m above storm HW (see definitions in Figure 5-7). For a storm, HW is MWL + 4 m, while during springtide, HW = MWL + 1.5 m. What is the necessary layer thickness to avoid shear failure and uplift?
With equation 5-13 we find HGW during storm = 0.5*4 = MWL + 2 m, hence h_1 + h_2 in Figure 5-7 = 3 m. With equation 5-12 and a first estimate of d = 0.5 m we find a maximum excess pressure, H ≈ 0.65 m for $h_1/(h_1 + h_2)$ ≈ 0.5. During an extreme storm, only uplift is taken into account, hence, from equation 5-15 we find: d ≈ 0.65/(1.4*0.97) ≈ 0.48. With this value, the calculation can be repeated, finally leading to d ≈ 0.46.
This is checked for shear during springtide. HGW then is MWL + 0.75 m and h_1 + h_2 = 1.75 m. With d = 0.46 m, we now find a maximum H ≈ 0.3 m for h_1/h_1 + h_2 ≈ 0.5. From equation 5-14 we find d ≈ 0.44 m. So, the situation during storm is normative.

Note: The equations (5.15) and (5.16) give a rather conservative approach, as no cohesion in the asfalt layer is assumed and the thickness is determined by the most unfavourable spot of the revetment. For the layer as a whole, the thickness can be taken parabolic along the protection length with the maximum thickness derived as outlined above and the minimum thickness determined by construction aspects (0.1 – 0.2 m).

5.4 Stability of open boundaries

5.4.1 Heave and piping

When there is a head difference across a hydraulic structure, there will always be some seepage, as soil is never completely impervious. When the flow force in a porous flow becomes greater than the weight of the grains, these grains will float and some sediment transport can occur. This failure mechanism is called *heave*, while the process is often called *fluidisation* of the soil. Problems get out of hand when erosion goes on and a continuous channel, a "pipe", develops, see Figure 5-9. When the load remains, this process will lead to progressive failure because the seepage length shortens continuously. This failure mechanism is called *piping*.

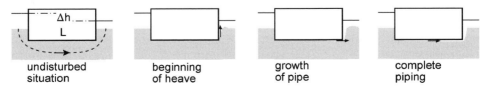

| undisturbed | beginning | growth | complete |
| situation | of heave | of pipe | piping |

Figure 5-9 Heave and piping under structure

Theoretically, heave starts at the downstream end when the upward flow force, see equation (5.9), is equal to the weight of the soil:

$$\rho_w g \, i \approx (1-n)\left(\rho_g - \rho_w\right) g \tag{5.17}$$

which, with a specific mass of the grains, $\rho_g \approx 2650$ kg/m^3 and $n \approx 40\%$ gives $i_C \approx 1$. In practice however, lower values for the critical gradient are found, due to irregularities in the porous medium which lead to a concentration of the flow. The same result will be found when the specific weight of saturated soil is used: $(1-n)(\rho_g - \rho_w) = 0.6*(2650-1000) \approx 1000 \approx \rho_s - \rho_w \approx$ 2000-1000 kg/m^3, see also Intermezzo 5-1. Equation (5.17) can also be expressed as the equilibrium of any soil layer, with thickness d:

$$h_u - h_t \le \frac{1}{\gamma} d \, \frac{\rho_s - \rho_w}{\rho_w} \tag{5.18}$$

h_u is the piezometric level under the soil layer, and h_t above, $(h_u - h_t)/d = i$. γ is a safety coefficient, e.g. 2, due to irregularities of the soil. Equation (5.18) can be used for heave of cohesionless soils, like sand, or for a bursting clay layer.

The classical approach to these phenomena is by Bligh and Lane dating back to the first decades of this century, see Figure 5-10. No clear distinction is drawn between heave and piping in these classical formulae. The critical head difference is given by:

$$\frac{\Delta h_c}{L} \le \frac{1}{C_{creep}} \tag{5.19}$$

In terms of a stability parameter, Δh is the load and L is the strength.

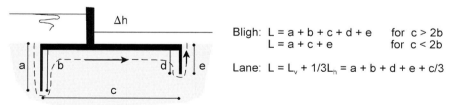

Bligh: L = a + b + c + d + e   for c > 2b
       L = a + c + e   for c < 2b

Lane: L = L$_v$ + 1/3L$_h$ = a + b + d + e + c/3

Figure 5-10 Definitions of seepage length in the classical approach

Intermezzo 5-1

Water and grains: different approaches

There are several ways to deal with forces and stresses in soil. These different approaches can all be correct, but should not be mixed up. To illustrate some of these ways, two simple examples are given. First, a vessel of 1 m² and 2 m high filled with water and with 1 m³ of sand is considered. With a porosity of 40% and a specific mass of 2650 kg/m³ of the grains, the specific mass of wet sand becomes: 0.6*2650 + 0.4*1000 ≈ 2000 kg/m³. The total weight of water and sand in the vessel is 30 kN, resulting in a (total) stress on the bottom of 30 kN/m².

A second example is a system of two communicating vessels, which are open at the top. The first, 1 m high, is filled with water and sand. The second, 2 m high, is filled with water. The head difference causes a flow which is replenished with Q.

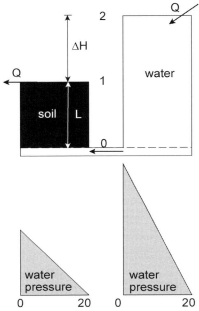

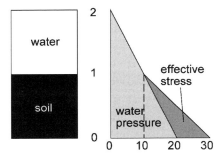

The strength of soil depends on the effective stresses. According to Terzaghi these stresses are: $\sigma' = \sigma - p$ in which σ is the total stress, σ' is the effective stress and p the waterpressure. Since the waterpressure is omnipresent, the effective grain stress on the bottom of the vessel is 30 – 20 = 10 kN/m². The same result is obtained with Archimedes' law: The effective weight of the grains on the bottom result in an effective grain stress: 0.6*(2650 – 1000)*g ≈ 10 kN/m². A third way to arrive at this value is to think of 1 m³ of wet sand packed in plastic foil under water. The effective grain stress then becomes: (2000 – 1000)*g = 10 kN/m².

The total stress at the bottom is 20 kN/m2 (either from the wet sand: 1*2000*g or from the water: 2*1000*g). The effective grain stress at the bottom of the vessel with sand is now: $\sigma' = \sigma - p$ = 20 – 20 = 0 kN/m². So, there is no effective grain stress and the soil has no strength. It is liquefied and acts like a fluid. The same result is found when the flow force is considered. According to equation (5.9):
$F_f = \rho_w * g * \Delta H / L = 1000*10*1/1 = 10$ kN/m³. This compensates again the effective weight of the grains under water: 10 kN/m³. The flow force also doubles the gradient of the water pressure in the vessel with sand.

Table 5.2 Creep coefficients for Bligh and Lane formulae

Material (interpolation for other diameters)	Lane C_{creep}	Bligh C_{creep}
Silt	8.5	18
Fine sand (150 – 200 μm)	7	15
Coarse sand (300 – 1000μm)	5	12
Fine gravel (2 - 6 mm)	4	9
Coarse gravel (> 16 mm)	3	4

Piping new method

In connection with the raising of the dikes in the Netherlands, much research has been carried out concerning piping. The generally accepted analysis is by Sellmeijer, 1988. Sellmeijer's model consists of a numerical solution of the relevant differential equations describing the phenomenon of piping. Not only the threshold of piping, but also the transportation of grains through an already formed channel was taken into account. Based on Sellmeijer's method new safety rules concerning piping in dikes in the Netherlands have been established. In most situations, a reduction of the required dike width (= seepage length) appeared to be possible. For a dike that needs to be raised, this can mean avoiding the demolition of the houses near the dike. By means of curve-fitting, the results of this model (see Calle/Weijers, 1994) the following formula was obtained:

$$\frac{\Delta h_c}{L} \leq 0.87\, \alpha\, c \left(\frac{\rho_g - \rho_w}{\rho_w} \right) (0.68 - 0.10 \ln c)$$

$$\text{with:} \quad \alpha = \left(\frac{D}{L} \right) \left(\frac{D}{L} \right)^{\frac{0.28}{\left(\frac{D}{L}\right)^{2.8} - 1}} \quad \text{and:} \quad c = 0.25 d_{70} \left(\frac{g}{v\, kL} \right)^{\frac{1}{3}}$$

(5.20)

In equation (5.20) the quotient D/L plays an important role, see Figure 5-11. When the thickness of the permeable layer, D, becomes small, L can be reduced.

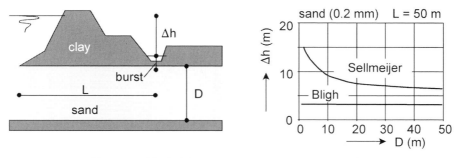

Figure 5-11 Definitions and results new piping method

Example 5-3

A dike is situated on a 50 m long clay layer with a 20 m thick layer of sand (d_{50} = 200µm) underneath. The clay layer under the ditch is 2 m thick, see Figure 5-11. To avoid piping, what is the maximum safe head difference?

Applying Bligh's rule, we find: Δh = 50/15 = 3.3 m. With Lane this would be: Δh = (50/3)/7 = 2.4 m, since only one third of the horizontal seepage length is taken into account. For piping under a dike with only a horizontal seepage length, Bligh is normally applied. For Sellmeijer, there is more work to do. With D/L = 0.4, α becomes 1.32. For c we need the permeability, k, which is derived from equation 5.3. The flow in sand is laminar, porosity is taken 0.4 and with ν = 1.33*10^6 (temperature groundwater 10^0 C), k becomes 5.2*10^5 m/s. With d_{70} estimated to be 1.25*d_{50} we find c = 0.088. The permissible head difference then becomes: 0.87*1.32*0.088*1.65*(0.68-0.1*-2.66)*50 = 7.7 m. With a safety factor of 1.2 this becomes 6.4 m which is still almost twice the value found with Bligh (for which a safety factor is assumed to be included already). The permissible head difference can still be enlarged with the heave or burst strength of the clay layer under the ditch, which gives an extra 0.5*2 = 1m.

Figure 5-11 gives a comparison between equation (5.20) and the rule according to Bligh (see Figure 5-10) for sand (same parameters as in Example 5-3). For small values of *D*, the critical head difference is much higher following Sellmeijer's rule than following Bligh, while for large values, there is less difference.

5.4.2 Micro-stability of slopes

With micro-stability the stability of the outer grains on a slope is meant, in contrast to the stability of the slope as a whole, the macro-stability, see section 5.5. Figure 5-12 shows the streamlines and equipotential lines in a dike with a head difference. The phreatic surface coincides partly with the downstream slope, which is practically always the case. That means that there is always a part of the downstream slope where groundwater flows through the slope. This is important in determining the stability.

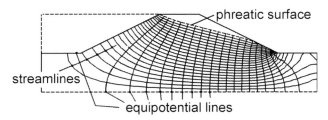

Figure 5-12 Porous flow in a dike (results MSEEP)

The equilibrium of grains on a slope in flow has been discussed before, but the load now comes from the inside of the slope. Here we cannot consider the stability of a single grain, since we have averaged the flow over all pores by means of u_f, see section 5.2.1. By doing so, we have given up all information on velocities in the pores and instead we consider a unit volume of soil.

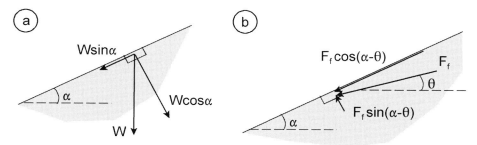

Figure 5-13 Forces on slope with porous flow

The equilibrium of forces on a unit volume of soil, without porous flow and cohesion, simply gives, again, see Figure 5-13a:

$$\rho_s g \cos\alpha \tan\phi \geq \rho_s g \sin\alpha \quad \rightarrow \quad \phi \geq \alpha \tag{5.21}$$

This equation is valid above and under water, since ρ_s, the mass density of the soil, vanishes from the equation. It can be the density of dry or wet soil, with or without effects of the buoyancy. Indeed, above and under water, a cautiously built up slope, can have a maximum angle, α, equal to the angle of repose, ϕ.

Now with porous flow, see Figure 5-13b, the soil being under water, equation (5.9) is valid and the situation can be described as follows:

$$\left[\left(\rho_s - \rho_w\right)g\cos\alpha - \rho_w gi\sin(\alpha - \theta)\right]\tan\phi \geq$$
$$\left(\rho_s - \rho_w\right)g\sin\alpha + \rho_w gi\cos(\alpha - \theta) \tag{5.22}$$

With $\rho_s \approx 2000$ kg/m^3 and $\rho_w = 1000$ kg/m^3 this simplifies to:

$$\tan\phi \geq \left[\frac{\sin\alpha + i\cos(\alpha - \theta)}{\cos\alpha - i\sin(\alpha - \theta)}\right] \tag{5.23}$$

Without porous flow ($i = 0$), this again gives $\phi \geq \alpha$ for stability. Some special cases of the equilibrium with porous flow are treated in the next paragraphs.

Horizontal seepage

When seepage flows horizontally through a slope, $\theta = 0$ and $i = \tan\alpha$, see Figure 5-13a. This situation can result from seepage through a dike, from a sudden fall of the waterlevel outside a revetment or from heavy rainfall on a revetment.

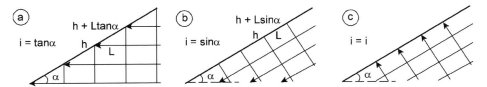

Figure 5-14 Flow gradients and micro-stability

The demand for stability from equation (5.23) then becomes:

$$\tan\phi \geq \frac{\sin\alpha + \tan\alpha\cos\alpha}{\cos\alpha - \tan\alpha\sin\alpha} = \frac{2\sin\alpha}{\cos\alpha - \sin^2\alpha / \cos\alpha} =$$
$$= \frac{2\sin\alpha\cos\alpha}{\cos^2\alpha - \sin^2\alpha} = \tan 2\alpha \quad \rightarrow \quad \phi \geq 2\alpha$$

(5.24)

in other words: given a soil type, the permissible slope angle is half the angle of repose. For example, with $\phi = 30°$, this leads to $\alpha \approx 15°$ or a slope of 1:3.5.

Seepage parallel to the slope

The gradient of the flow now becomes: $i = \sin\alpha$ and $\theta = \alpha$. With equation (5.23) this simply gives:

$$\tan\phi \geq \frac{\sin\alpha + \sin\alpha}{\cos\alpha} \quad \rightarrow \quad \tan\phi \geq 2\tan\alpha$$

(5.25)

which is a little more favourable than equation (5.24). This situation can occur if water flows over a dike or/and if a relatively permeable layer is situated on less permeable soil.

Seepage perpendicular to the slope

When the seepage takes place under water, the slope surface is an equipotential line and the flow will be perpendicular to the surface. In that case, $\theta = \alpha - 90°$ giving:

$$\tan\phi \geq \frac{\sin\alpha}{\cos\alpha - i}$$

(5.26)

For $\alpha = 0$, this again gives $i = 1$ for fluidisation of the soil.
Note: One of the assumptions in equation (5.23) was no cohesion. A slope consisting of clay or loam can, of course, be much steeper.

5.5 Macro stability of slopes

The stability of a slope as a whole is usually referred to as the macro-stability, in contrast to the micro-stability which deals with the outermost grains on a slope, see section 5.4.2. Micro-stability is a limit case of macro-stability, as we will see later on.

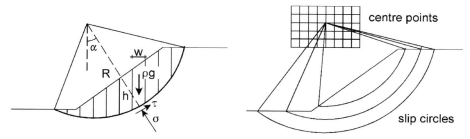

Figure 5-15 Slip circle approach of macro-stability

Macro-stability is usually approached with a slip-circle analysis, for which the Bishop-method is widely used, see Verruijt, 1999. The general idea is again load versus strength, the load being the weight of the soil mass within the circle and the strength the shear along the circle, see Figure 5-15. Since both load and strength can now be expressed in the same units (force, considering the equilibrium around the centrepoint of the circle), the stability can be expressed by a coefficient, F, which is the strength divided by the load. When $F > 1$ the slip circle is stable and when no circles with $F < 1$ can be found, the slope is stable (within the limits of accuracy of the calculation method).

As known from soil mechanics, shear stress depends on the normal stress, so the resisting shear will vary along the circle. A practical solution for the analysis is thus to divide the soil above the circle into slices. The shear strength in the soil with regard to the maximum possible shear is then given by:

$$\tau = \frac{1}{F}(c + \sigma_n' \tan \phi) = \frac{1}{F}\left(c + \left(\sigma_n - p\right)\tan\phi\right) \tag{5.27}$$

where c is the cohesion of the soil, p the water pressure and σ_n and σ_n' the total and effective stresses normal to the slip circle, respectively, see also Intermezzo 5-1. The equilibrium around the centre point of the circle, for all slices, gives (per m width), see Figure 5-15:

$$\sum \rho_s g h w R \sin \alpha = \sum \tau \frac{w}{\cos \alpha} R \tag{5.28}$$

Combining equations (5.27) and (5.28) gives the total strength divided by the load:

$$F = \frac{\sum\left[\left(c + \sigma_n' \tan\phi\right)/\cos\alpha\right]}{\sum \rho_s g h \sin \alpha} \tag{5.29}$$

Note: Wet soil is heavier than dry soil and the effective grain-stresses are influenced by the pore water pressures ($\sigma_n' = \sigma_n - p$), so pore water influences both the loading and the strength. A critical situation occurs after high sea or river levels, when the waterlevel suddenly drops and the slope is filled with water, increasing the load while the water pressures decrease the strength. The same can happen due to heavy rainfall.

In the Bishop method, the effective normal stresses are computed from the vertical equilibrium in every slice (see also Figure 5-15 and Verruijt, 1999):

$$\rho_s g h w = \sigma_n \cos\alpha \frac{w}{\cos\alpha} + \tau \sin\alpha \frac{w}{\cos\alpha} \quad \rightarrow \quad \rho_s g h = \sigma_n' + p + \tau\tan\alpha \qquad (5.30)$$

With equation (5.27) we find:

$$\sigma_n'\left(1 + \frac{\tan\alpha\tan\phi}{F}\right) = \rho_s g h - p - \frac{c}{F}\tan\alpha \qquad (5.31)$$

Substituting σ_n' from equation (5.31) into equation (5.29) finally leads to:

$$F = \frac{\sum \dfrac{c + \left(\rho_s g h - p\right)\tan\phi}{\cos\alpha\left(1 + \tan\alpha\tan\phi / F\right)}}{\sum \rho_s g h \sin\alpha} \qquad (5.32)$$

Going from equation (5.29) to (5.32), eliminates the unknown σ'. Now, F can be computed from all slices iteratively, since F occurs on both sides of the equation. This iterative computation is valid for one slip circle. To determine the minimal value of F for a given slope, several centre points with shallow and deep circles have to be evaluated, see Figure 5-15.

Usually many slip circles with many slices have to be evaluated, so this is typically a task for a computer. Many computercodes are available for this purpose, in the Netherlands e.g. STABIL of Delft University of Technology or MSTAB of GeoDelft. The following examples are based on MSTAB.

First we check the relation with micro-stability. Micro-stability can be approximated with a very shallow slip circle. A 10 m high slope 1:3 has a phreatic level that coincides with the slope for 5 m. Coinciding with the slope means having a gradient equal to $\tan\alpha$, which is equivalent to the case of horizontal outflow. From micro-stability we know that, in that case, the equilibrium is reached when $\alpha = \phi/2$, giving $\phi \approx 37°$.

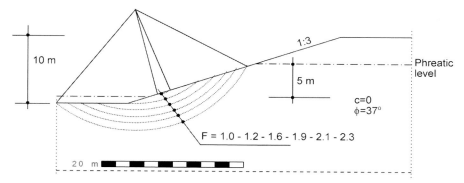

Figure 5-16 F-values for various slip circles with one centre point

Figure 5-16 gives the results for various shallow and deep slip circles for this case. The very shallow circle gives $F \approx 1$. The other circles give higher values, since more favourably situated adjoining slices make the situation more stable. So, indeed, the seepage surface is the most unstable part of the slope and the micro-stability rules should be obeyed there. Other parts of the slope are less vulnerable. Note that the pore pressures from the phreatic level in the computation give the same result as the flow force approach in section 5.4.2, see also Intermezzo 5-1. So, both approaches are appropriate and it would be wrong to take both mechanisms into account.

When there is a considerable inhomogenity in the soil, sliding will take place along the weakest layers, where c and/or $\tan\phi$ are low and hence the shear strength is low. In Figure 5-17, the same geometry as in Figure 5-16 is applied, with the same phreatic level, but with a 1 m thick layer with a much lower strength ($\phi = 20°$). The minimum value for F is now 0.85 for a circle crossing the bad soil layer. Around the given centre point, lines of equal F have been drawn for other centre points. Actually, the failure will now be determined by the position of the weak layers and the failure surface is not necessarily a circle anymore, but can be any shape. However, to find weak spots, an approach with circles can still be very useful. So, when an embankment is constructed with such a bad layer in the subsoil (possibly caused by some recent sedimentation of mud), it should always be removed, for even a gentle slope of 1:3 is not stable in this case.

The same can occur when σ' is low due to high pore pressures. Again for the same geometry as in Figure 5-16, this is demonstrated in Figure 5-18. Now, a clay layer of high strength ($\phi = 37°$, $c = 20$ kN/m^2) is present under the slope, but under this layer, the piezometric level has the same value as upstream inside the slope (assumed to be the result of some connection between the upstream side and the subsoil). This gives high pore pressures in the deeper subsoil, causing zero shear stresses ($\sigma' = \sigma - p$, with $\sigma'_{min} = 0$ and $\tau = c + \sigma' \tan\phi$). Most shear comes from the clay layer, due to the cohesion, and some shear is present above the clay layer where there is no excess pressure. The resulting F is now very low: 0.32.

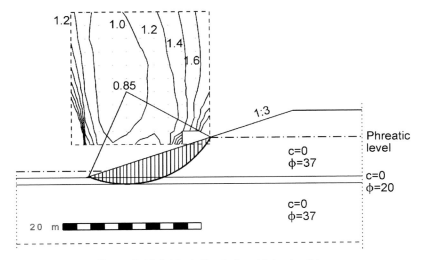

Figure 5-17 Critical slip circle with bad soil layer

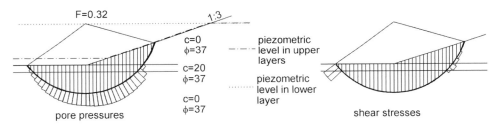

Figure 5-18 Slip circle with high pore pressures

Summarizing: when no computercode for stability is available, an approach with micro-stability gives a safe slope angle for homogeneous soil. This angle should always be applied on the part of the slope with seepage, where micro-stability is an important failure mechanism. Inhomogenity of the soil can give, usually unpleasant, surprises when no soil data are available.

For a more detailed treatment of soil mechanical aspects, see Verruijt, 1999 or CUR 162 (1992).

5.6 Load reduction

Porous flow can cause problems in two different ways: by pressures or by flow. These are closely related as follows from the basic equations and the presentation of streamlines and potential. When the porous flow velocities are too high and piping forms a threat, the seepage length should be increased. This, however, can lead to higher pressures in the underground. When the pressures are too high, a drain can be constructed, which can again lead to erosion. So, load reduction in porous flow is ambiguous.

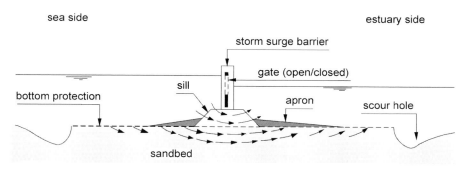

Figure 5-19 Flow under gate

As an example we consider a situation, based on the Eastern Scheldt storm surge barrier. With a head difference of more than 5 m and a gate structure with a width of the same order of magnitude, the flow through the sill reaches very high values, see Figure 5-19, with the danger of erosion and piping. For some schematized cases we will demonstrate the various options for load reduction. For a 40 m thick sandbed with $k = 0.0001$ m/s and a head difference of 5 m across a 5 m wide gate, the equipotential lines and streamlines are computed with the finite element program MSEEP, see Figure 5-20.

Note: In this example, only the porous flow aspect will be treated. Scouring or stone stability under or behind an open gate is not considered.

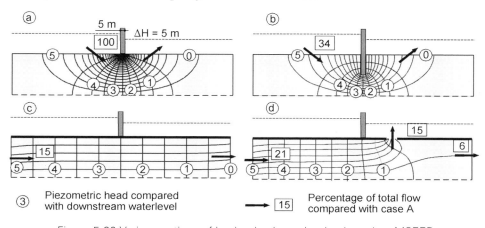

Figure 5-20 Various options of load reduction at barrier, based on MSEEP

Figure 5-20a shows the situation without any measures. Where the gate touches the bottom, a high concentration of streamlines and equipotential lines is found, indicating large velocities and gradients. The total number of stream lines in the four cases in Figure 5-20 differ such that the distance between the streamlines represents more or less the same amount of flow for all cases; since the total resistance in case a is much smaller than in the other cases, many more streamlines have been drawn for case a, see also the total discharges, indicated in the figure as a percentage of the discharge in case a. The maximum gradient of the piezometric head in case a is about 100% (5m/5m), which will definitely lead to erosion due to piping, see section 5.4.1. A reduction of this load can be achieved by adding an impervious horizontal or vertical protection, creating a larger seepage length. Figure 5-20b shows the effect of a vertical continuation of the gate into the bottom to a depth of 20 m. The total seepage is about 3 times smaller and the maximum gradient about 5 times smaller than in case a. This means a significant reduction of the porous flow load for this case. Special care will be needed during the construction of such a screen, since any leak can bring back the situation of case a.

Figure 5-20c gives the results for a 100 m long impervious bottom protection at each side of the gate. The total remaining seepage is now only 15% of that in case a. The subsoil now simply becomes a horizontal tube with vertical potential lines at equal distances, which can easily be calculated without the aid of a computer. Immediately downstream of the gate, the piezometric level in the subsoil is 2.5 m compared with the downstream waterlevel. This also means an excess pressure of 2.5 m water under the bottom protection. To avoid lifting, the weight of the protection should be large enough, see section 5.3.

A hole, 10 m wide, 25 m downstream of the gate acts as a drain or valve and reduces the maximum excess pressure with more than 1 m, see Figure 5-20d. At the same time, it introduces a concentrated outflow through the hole. A filter on top of the hole may be necessary to avoid piping, but this seems much more manageable than in the situation of case a. One should be certain, however, that the filter functions properly. Dirt or the growth of plants or shellfish, can reduce the permeability, leading again to higher pressures.

Note: An impervious protection layer of sufficient length upstream from the gate, without any protection downstream, will prevent piping while not causing any excess pressure trouble downstream. For other reasons, this is not a practical solution: downstream of a gate, a protection will be needed against scouring in case of flow through the gate while in a tidal area, upstream and downstream reverses all the time.

5.7 Summary

Porous flow, the flow through pores, can be described with the same laws as open channel flow, but can only be described adequately when averaged over many pores. The basic equation is the (extended) Forchheimer equation:

$$\frac{1}{\rho g}\frac{\partial p}{\partial x} = i = a u_f + b u_f |u_f| + c\frac{\partial u_f}{\partial t}$$

in which a laminar part and a turbulent part can be discerned, as in open channel flow. The flow through fine material, like clay or sand, is laminar for which the Forchheimer relation reduces to Darcy's law. In combination with the continuity equation, this can be translated, either graphically or numerically, into equipotential lines and streamlines. With this information about loads (pressures and velocities), the stability of various porous flow situations can be judged. For turbulent porous flow, as in rock, the knowledge is often still insufficient and empirical relations have to be used in design computations.

Impervious protections have to withstand occurring pressures. Weight and friction are the parameters that determine the stability of protections without internal strength. When weight and friction of a protection on a slope are insufficient, tension forces will occur in the protection, which the material has to withstand. The internal strength of e.g. concrete slabs against pressures perpendicular to the slab, is beyond the scope of this book.

Open boundaries have to withstand the flow forces in the soil, for which the equilibrium is expressed as follows:

$$\rho_w \, g \, i \;\approx\; (1-n)\left(\rho_s - \rho_w\right) g$$

Heave is the phenomenon which occurs when this equilibrium is violated and piping is the subsequent ongoing erosion downstream of a relatively impervious structure. To avoid heave and piping, the seepage length has to be sufficiently great, to keep the gradients in the subsoil low. On a slope, seepage can cause micro-instability if the slope is too steep. To avoid this, the slope angle should be about half the value of the angle of repose of the slope material, ϕ.

Macro-instability occurs when the sliding forces in a slope exceed the friction forces, determined by soil properties, such as cohesion, c, and angle of repose, ϕ. High water pressures inside have an unfavourable influence on the stability. The same is true for a layer in the subsoil with low values of c and ϕ.

6 POROUS FLOW
Filters

Filter test with flow parallel to interface (Delft Hydraulics Laboratory)

6.1 General

Filters can have different functions: preventing erosion of the covered subsoil or preventing pressure build-up in the covered subsoil (drainage) or a combination of both. Figure 6-1 gives some examples of filters.

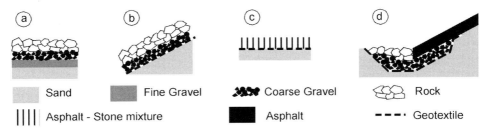

Figure 6-1 Examples of filters

Filters against erosion

Filters as a protection against erosion are particularly needed in situations with large gradients on the interface between soil and water, either the bank or the bottom. Examples are a dam with a head difference, such as a caisson dam or a storm surge barrier, where gradients of more than 10 % parallel to the interface are possible ($\Delta H \approx$ (O)m, $L \approx$ (O)10 m). That gradient is usually two or three orders of magnitude smaller in open water flow (slope in rivers $\approx$ (O) 0.001 - 0.0001). In open channel flow, however, the flow velocity and the turbulence of the flow can penetrate through the large stones of a bottom protection. The main function of protective filters, is to prevent the washing away of the underlying material. That means that the underlying grains (usually referred to as the *base layer*) should not pass the pores of the upper layer (the *filter layer*). This can be prevented either by one or more layers of grains of varying diameter, a *granular filter*, see Figure 6-1a, or a *geotextile*, see Figure 6-1b. Figure 6-1c shows an example of a prefabricated combination, where stones with a diameter of a few cm are "glued" together with asphalt on a geotextile, reinforced with steel cables if necessary. With these relatively modern protection mats, it is possible to realise a reliable and large protection relatively fast. The possible combinations of granular material, geotextiles, glues and reinforcements seem almost unlimited nowadays, see also appendix A - Materials. It should be kept in mind, however, that many of these highly specialized products only become economically attractive when very large quantities are used. So, for many projects, protections with granular material and standard geotextiles, constructed on the spot, should be considered first in a design. Moreover, for protections that have to last very long (say more than 100 years), the lifetime of most geotextiles cannot be guaranteed yet. For this reason, the sill in the Rotterdam Waterway Storm Surge Barrier (1994), was completely constructed with granular material.

Drainage filters

A filter for drainage, see Figure 6-1d, can help to prevent high excess pressures under an impervious revetment, which is the second possible main function of a filter.

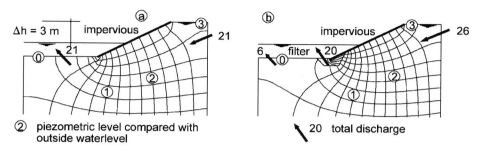

Figure 6-2 Influence of filter on flow under impervious dike revetment

Figure 6-2 shows the influence of a filter at the toe of a dike (sand with $k = 0.0001$ m/s) with an impervious revetment. A 1 m thick and 2 m wide gravel filter ($k = 0.1$ m/s) is placed at the toe. The filter reduces the maximum excess pressure under the revetment with 0.5 m (see lines of equal piezometric level: without filter, near the watersurface, the potential ≈ 1.5 m while with filter the potential ≈ 1 m). Another effect is the decrease of the gradient in the sand. Although the filter "attracts" seepage, it is less concentrated than without filter (see the distance between streamlines at the toe in Figure 6-2a and at the transition between sand and gravel in Figure 6-2b). Of course, the filter itself should be able to resist the concentrated upflow and to prevent piping, which is also a type of erosion, for which extra layers of material can be necessary between the sand and the gravel.

For filters against erosion the emphasis is on *stability*, while *permeability*, in order to prevent pressure build-up, is the main property of drainage filters. But actually, both aspects play a role, as illustrated by the above example of the drainage filter, which had to prevent erosion due to piping. In fact, both stability and permeability are important in any filter design. In a filter that is only meant to prevent erosion, pressure build-up due to insufficient permeability also has to be avoided in order to prevent lifting or sliding of a filter layer. Being closed for the base material and permeable for water are opposing demands; a solution has to be found within the boundaries of possibility. The usual approach in design is to establish the necessary diameter for the stability of the top layer in waves and currents, to check the filter relations between the top layer and the original (base) material and to add a geotextile or as many granular layers in between as necessary to meet the filter demands. In the coming sections, rules to dimension these filters will be given.

An overview of the most relevant aspects of modern filter design is published in CUR, 1993.

6.2 Granular filters

6.2.1 Introduction

The design of a granular filter can be based on different criteria, two of which will be discussed here. Figure 6-3 presents the relation between the critical gradient in porous flow, above which grains from the base layer are no longer stable under the filter layer, and the quotient of some representative diameters, to be defined later on, of the the two layers. The gradient is, again a relation between load and strength: $I = \Delta H/L$, where the head difference is the load and the length can be interpreted as the strength. As with other load and strength parameters presented before, I can be seen as a mobility parameter and I_c as a stability parameter.

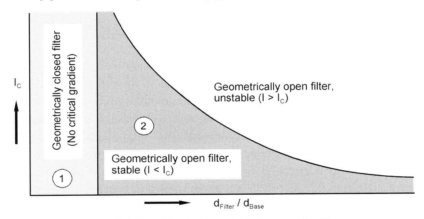

Figure 6-3 Possible design criteria for granular filters

In this figure two areas without erosion can be discerned:

1 Geometrically closed: These are the classical filter rules formulated by Terzaghi in the thirties. The stability of the filter depends on the geometrical properties of the materials, the sieve curves. The size of the grains is chosen such that they cannot move in the filter. This often leads to a conservative design, since the filter can stand any load: there is no critical gradient for the transition between the two layers. The filter layer diameters are a few times to one order of magnitude larger than the base layer diameters and the permeability of the filter layers has to be checked carefully.

2 Geometrically open: The grains of the filter layer are much larger than those of the base layer, thus allowing grains in the latter layer to move in and through the filter layer. But, like stability in open channel flow, there will only be erosion when the load is higher than some critical value. A critical filter gradient can now be determined and a filter is stable, when this critical gradient is larger than the occurring gradients in the structure. So, now the hydraulic loads are taken into

account, leading to a more economic design. The disadvantage, compared with a geometrically closed filter, is that detailed information on the loading gradients is now necessary. Permeability is usually assured as these filters are more open than the geometrically closed filters.

Sometimes, some loss of material is accepted within the limits of admissable settlements, which leads to even larger quotients of filter and base diameters. This design approach is rather complicated and will not be treated.

6.2.2   Geometrically closed filters

The idea behind a geometrically closed filter is the following, see Figure 6-4. The space between packed grains is much smaller than the grains themselves. For spheres with equal diameter D, this space can be blocked by a sphere with a diameter which is about 6 times smaller than D, see Figure 6-4a.

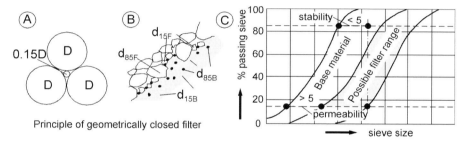

Figure 6-4 Blockage of particles and sieve curves for geometrically closed filters

In a layer of grains with varying diameter, the space between the grains is governed by the smaller grains, with say d_{15}, the sieve diameter which is passed by only 15% of the mass of grains. So, the escape route in the filter layer is determined by d_{15F} in Figure 6-4b. The largest grains of the base layer, d_{85B} in the figure, get stuck in the pores of the filter layer and block the passage of all other grains of the base layer, *provided the base layer is internally stable*. This means that the range of grain diameters in the base layer should not be too large, so that the larger grains can block the smaller ones.

In order to prevent pressure build-up, the permeability of a filter layer should be larger than the permeability of the base layer. Since the permeability is also governed by the smallest grains, this leads to a relation between the d_{15} of both layers. In words, we just described three relations for a geometrically closed granular filter: *stability* between filter layer and base layer, *permeability* and *internal stability*. In formula, based on experiments:

$$\text{Stability: } \frac{d_{15F}}{d_{85B}} < 5 \quad \text{Int.Stability: } \frac{d_{60}}{d_{10}} < 10 \quad \text{Permeability: } \frac{d_{15F}}{d_{15B}} > 5 \qquad (6.1)$$

The demands for stability and permeability are contradictory: according to the stability rule d_F should be *smaller* than $5d_B$ and for permeability *larger* than $5d_B$. However, the use of two different diameters for the base material in the denominator (the 15% largest and the 15% smallest, respectively) gives the margin to be able to design a filter, see Figure 6-4c. The three parts of equation (6.1) form a unit and are actually all stability rules: the first rule prevents movements of the larger grains from the base layer, which together with the third rule guarantees the stabilty of all grains in the base layer; the second rule prevents pressure build-up and guarantees the stability of the filter layer as a whole. The permeability rule is especially important when there is a large gradient perpendicular to the interface. See also Intermezzo 6-1.

Intermezzo 6-1

The permeability rule is meant to prevent pressure build-up at the interface between two layers. Why should d_{15F} be 5 times larger than d_{15B} when the only aim is to have a greater permeability in the filter layer than in the base layer in order to prevent pressure build-up? Why is two times greater not enough? According to this rule, pressure build-up could even occur *within* a layer where after all "d_{15F}/d_{15B}" = 1.

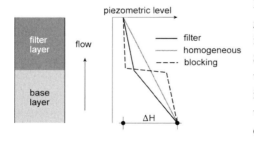

The answer has to be found in the stability and the migration of grains. Without any migration, the piezometric levels would look like the solid line ("filter"): a larger gradient in the base layer than in the filter layer, due to a difference in permeability. In a single layer, where no migration occurs when it is internally stable, it would look like the dotted line ("homogeneous"). The stability rule in equation (6.1) prevents the larger base layer grains to move. Migration of smaller grains at the interface is possible until the internal stability closes off the base layer. Some migration of grains at the interface will always occur and the permeability rule is meant to prevent blocking of the interface with these grains, which would cause the dashed line ("blocking") and possibly lifting of the filter layer.

With these rules, a stable and permeable filter can be designed, either starting from the original (in situ) soil as base layer or from the top layer which is necessary for stability. This can lead to more than two layers in the filter, sometimes even to 4 or 5. When using the relations in equation (6.1), the finest layer of each pair is named the base layer and the coarsest the filter layer.

Example 6-1

What filter material is needed on top of sand with d_{50} = 0.4 mm when the geometrically closed filter rules have to be obeyed?
First we have to assume some gradation of the sand. A width (d_{85}/d_{15}) of 2 is quite normal (always check!), giving e.g. $d_{15} \approx$ 0.3 mm and $d_{85} \approx$ 0.6 mm. The stability rule gives: d_{15F} < $5.d_{85B}$ = 3mm and the permeability rule: d_{15F} > $5.d_{15B}$ = 1.5 mm. So, d_{15} of the filter material should lie between 1.5 and 3 mm. The internal stability rule, finally, says something about the possible width of the gradation. A d_{60}/d_{10} ratio of 10 is more or less equivalent with a d_{85}/d_{15} ratio of 12 - 15. So, fine gravel with a d_{15} of 2 mm and a d_{85} of 10 mm will be suitable ($d_{50} \approx$ 5 mm). A wider gradation (a d_{85} of 20-25 mm is possible) can reduce the number of filter layers.

6.2.3   Geometrically open filters

As mentioned in section 6.2.1, the idea behind a geometrically open filter is that the grains of the base layer can erode through the filter layer, but that the occurring gradient is below the critical value. Via the Forchheimer equation, see chapter 5, the gradient is related to the velocity in the filter, which is again related to the forces acting on a grain. When the loading forces on a grain of the base layer are smaller than the resisting forces, there will be no erosion. Although the flow through a granular filter can be in any direction, it is advantageous for design and research purposes, to discern the two main directions: perpendicular and parallel, see Figure 6-5. The critical gradient in the case of parallel flow is defined in the *filter layer* and with perpendicular flow in the *base layer*, which is also related to a different erosion mechanism.

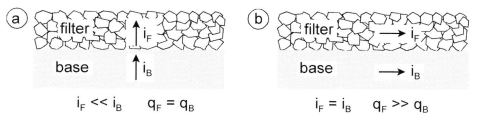

Figure 6-5 Perpendicular and parallel flow in granular filter

When the water flows parallel to the interface, the gradient in both layers is about the same, causing the flow velocity in the filter layer to be much higher than in the base layer, due to the greater permeability. At the interface a velocity gradient will exist, inducing a (shear) stress on the upper grains in the base layer. The situation is very much like incipient motion in an open channel and the critical shear stress, hence the critical velocity, hence the critical gradient is defined in the filter layer. With perpendicular flow, there is a serial system where-in the flow through base layer and filter layer is equal, causing a much larger gradient in the base layer, because of the larger permeability of the filter layer. Erosion of the base layer will take place due to fluidization, hence the gradient in the base layer determines the stability.

Perpendicular flow

With flow perpendicular to an interface in a filter we have to take gravity into account. When the base layer is on top of the filter layer and the flow is downward, the finer grains will easily fall through the filter layer and it is therefore recommended to follow the rules for geometrically closed filters. In the case of upward flow through a base layer lying under a filter layer, there is always a lower limit for erosion: the fluidization criterion, for which the gradient is about 1 or 100% which follows from the vertical equilibrium, see chapter 5. The upper limit, again, is a geometrically closed filter.

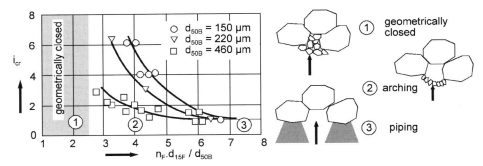

Figure 6-6 Perpendicular flow through geometrically open filters (de Graauw,1983)

Figure 6-6 shows the results of some tests (from de Graauw et al.,1983). The diameter ratio on the horizontal axis is somewhat different from the one used in the stability relation of equation (6.1): instead of d_{85B}, d_{50B} is now used and the porosity in the filter layer is also taken into account. The relation between the two can be estimated: assuming $n_F \approx 0.35\text{-}0.4$ and $d_{50B} \approx 0.7\text{-}0.8 \cdot d_{85B}$ gives: $n_F \cdot d_{15F} / d_{50B} \approx 0.35\text{-}0.4 \cdot d_{15F} / (0.7\text{-}0.8 \cdot d_{85B}) \approx 0.45\text{-}0.55 \ast 5 = 2.2\text{-}2.7$ as an estimate for the geometrically closed limit, area 1 in the figure. The lower limit is formed by fluidization (piping) in which case the filter grains are so large that the base acts as if there is no filter (area 3 in the figure). In between, in area 2, several mechanisms play a role. Grains that flow from the base layer into the filter layer, arrive in much larger pores with much lower flow velocities, so the forces on such a grain decrease and it will not be transported further upward. Another mechanism is arching, where bridges of fine grains block the filter, although the geometrical criteria are not met. Note that the results are relatively more favourable for finer grains.

Parallel flow

The stability of base material in a flow parallel to the interface of a filter can be seen as the stability of a grain on the bottom of a very small channel. It is attractive to link the Forchheimer equation with the threshold of movement as given by Shields, see chapter 3, in order to establish a relation for the critical gradient in porous flow. This approach was used by de Graauw,1983 and resulted in an empirical relation. The most important parameters appeared to be the diameter of the base material (as a measure of the stability) and the diameter and porosity of the filter material (both as a measure of the flow through the filter). The experimental results finally yielded:

$$I_C = \left[\frac{0.06}{n_F^3 \, d_{15F}^{4/3}} + \frac{n_F^{5/3} \, d_{15F}^{1/3}}{1000 \, d_{50B}^{5/3}} \right] u_{*c}^2 \qquad (6.2)$$

in which u_{*c} is the critical shear velocity according to Shields. The two terms in equation (6.2) come from the idea of a laminar and a turbulent part as found in the Forchheimer equation.

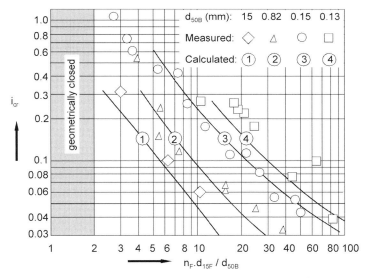

Figure 6-7 Critical parallel gradient for geometrically open filter layers

Figure 6-7 shows a comparison between some computed lines and the experiments. Again, for a given ratio of $n_F \, d_{15F} / d_{50B}$, the critical gradient is seen to be greater for fine base material than for coarse base material. This can be explained by the lower filter velocities in fine pores: a relatively larger gradient is needed in the filter layer to reach the critical velocity. With base material larger than about 5 mm, no further reduction of the critical gradient occurs (compare the Shields parameter for large grains in chapter 3).

Note 1: When the filter is on a slope, u_{*c}^2 in equation (6.2) has to be corrected in the same way as is done in open channel flow, with $\sin(\phi - \alpha) / \sin\phi$, see chapter 3.

Note 2: When there is a combination of perpendicular and parallel flow, the stability rules for parallel flow can be used, provided the perpendicular gradient is < 0.5.

Example 6-2

What is the critical parallel gradient of stone class 60-300 kg lying on stones of 30-60 mm?

Appendix A gives for 60-300 kg: $d_{15} \approx 0.35$ m (sieve diameter instead of nominal diameter) and for 30-60 mm: $d_{50} \approx 45$ mm. u_{*c} comes from the Shields relation using $\psi = 0.06$ (see chapter 3) giving: $u_{*c}^2 = \Delta g d \psi = 1.65*9.81*0.045*0.06 = 0.044$ (m/s)^2. With $n_F = 0.4$ and equation 6.2 this gives: $I_c = [0.06/(0.4^3*0.35^{4/3}) + 0.4^{5/3}*0.35^{1/3}/ (1000*0.045^{5/3})]*0.044 = 16\%$

Cyclic loading

Although stability in waves will not be treated before chapter 8, the influence of cyclic loading will be dealt with here, in order to complete the overall picture of filters.

In tests with parallel flow, for cycle periods > 2 s, the critical filter velocities and gradients were found to be the same as for stationary flow, see De Graauw et al, 1983. This confirms what was claimed in chapter 5: the inertia term in the Forchheimer equation appears to be of minor importance. So, for cyclic parallel flow in filters, equation (6.2) can be used to determine the critical gradient. The loading gradient in parallel filters can be estimated roughly with wave equations, see chapter 7.

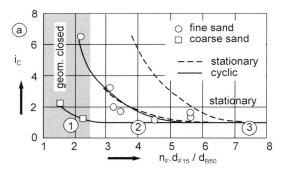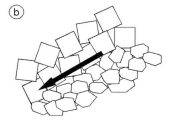

Figure 6-8 Filters in unstationary flow

Cyclic perpendicular loading, see De Graauw et al., 1983, produced considerably different results compared with stationary tests, see Figure 6-8a. The critical gradient appeared to be lower, for both fine and coarse sand. An explanation is offered by the arches, see Figure 6-6, which can not build up or are destroyed during reversed flow. For coarse sand there appeared to be a critical gradient of about 200% inside the geometrically closed region. For a permeable dike toe, one does not have to worry about a geometrically closed filter drain. In case of heavy dynamic loading, e.g.

"rocking" of an offshore structure on a filter foundation, specific hydraulic and soil mechanic research should be carried out.

For the transition between an armour layer and a first filter layer in breakwaters, relations like $W_F / W_B < 10$ - 15 are often mentioned in manuals. With $W \propto d^3$ this comes down to $d_F / d_B < 2.2$ - 2.5, which is a much heavier demand than the geometrically closed criteria (with a width ratio $d_{85}/d_{15} \approx 2$ and the stability rule $d_{15F}/d_{85B} < 5$, $d_{50F}/d_{50B} < 10$). The relations between diameters in geometrically closed filters are such that the grains can not move inside the filter, so this should also be good enough in unstationary conditions. This is indeed the case: only in very heavy cyclic loading perpendicular to the interface, the geometrical limits are sometimes exceeded, see Figure 6-8a. The strictest demands on breakwaters are not dictated by filter rules, but by mechanical considerations. Often, the individual elements in the armour layer, see Figure 6-8b, are too large to be able to speak of a real layer. With the geometrically closed rules, the surface would be too smooth to form a stable foundation for the large elements. Another advantage of larger elements under the armour layer, is that the wave energy can be absorbed better, leading to a more stable armour layer, see chapter 8. Even then, $W_F / W_B < 10$ - 15 seems very strict and there are examples of ratios of more than 25 ($d_F / d_B \approx 3$) which function satisfactory.

Note: One could worry about the permeability rule when larger elements are applied under the armour layer and $d_{15F}/d_{15B} > 5$ will clearly be violated. With such large elements, usually with a narrow gradation, there is little chance of blocking the pores as described in Intermezzo 6-1. Beside that, there will not be a large gradient pushing the stones of the underlayer through the armourlayer.

Bed protections

Bed protections in rivers or behind outflows are a special case. The (parallel) gradient in open channel flow is very small (order of magnitude $I = 10^{-3}$ or less) and for a stable filter, according to equation (6.2) and Figure 6-7, the ratio between filter diameter and base diameter could be very large. In this case, however, turbulence can penetrate into the top layer and increase the average gradient at the interface, just as in open channel flow.

Figure 6-9 shows tests carried out at Delft University of Technology (Van Os, 1998). A caisson is placed in a flume with adjustable distance between the caisson and a filterbed of stones ($d_{50} = 20$ mm) on a sandbed ($d_{50} = 0.1$ mm). Two filterbed thicknesses were examined: 60 mm and 100 mm. If the caisson is placed directly on the filter (gap height = 0), the critical gradient for motion at the interface between filter and sand is about 5.5%, which is in line with equation (6.2). There is hardly any influence of the filter thickness. With increasing gap height, the velocity above the filterbed increases rapidly for a given gradient. The critical gradient decreases with increasing gap height. The thickness of the filter clearly plays a role in the value of the critical gradient. A thicker filter gives a larger critical gradient, which is to be expected. Example: a 7 cm gap gives a critical gradient of 2.8 % for a filter of 60 mm and 3.7 % for a 100 mm thick filter.

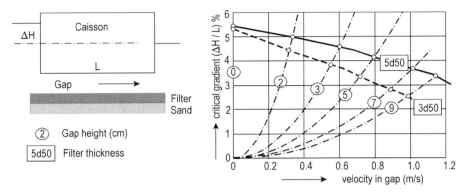

Figure 6-9 Tests with flow over bottom protection with filter

Turbulent velocities have been measured inside the pores of the filterbed and analysed. It appears that velocity and turbulence diminish rapidly inside the filter, but remain constant at levels lower than $1.5d_{50}$ inside the filter. Hence, a thicker filter is not a better protection against erosion due to the decrease of the turbulent velocities inside the filter! The explanation has to be found in the longer path for the sand through the filterbed. Theoretical considerations and analysis of the velocity signal has led to the thesis that fluctuations with a high frequency are responsible for incipient motion of the sand grains. Fluctuations with a low frequency transport the grains through the filterbed. A larger gap allows larger eddies in the current above the filter, leading to a lower critical gradient. In a thicker filter, a sand grain needs more powerful eddies to be transported to the surface of the filter, hence a larger critical gradient.

For provisional application, Wörman, 1989, can be used:

$$\frac{d_{15T}}{d_{85B}} = 5 \frac{\left(1-n_T\right)}{n_T} \frac{\Delta_B}{\Delta_T} \frac{D}{d_{50T}} \tag{6.3}$$

in which D is the thickness of the top layer. For materials with the same density and a normal porosity this leads to $d_{15T} / d_{85B} \approx 8$ - $10 * D/d_{50T}$, which with a minimum value of $D / d_{50T} = 2 - 3$, gives about 3 - 6 times more favourable ratios than the rules for geometrically closed filters. When D becomes very large, it is recommended to use the relation of equation (6.2) as a lower limit.

Note: Always check whether there is a parallel gradient active at the interface, also in open channel flow. E.g. under or near a groyne or pier, the accelerations and pressure differences can give considerable gradients!

6.3 Geotextiles

6.3.1 Introduction

'Geotextile' is a generic term for all kinds of foil- and cloth-like synthetic materials which are becoming increasingly important in civil engineering. They are used to armour soil in foundations and slopes, as membranes to prevent seepage or to protect the environment against pollution from a dump area, and as filters in hydraulic engineering. The latter use is the subject of this section. Geotextiles come in many shapes, depending on the production technique. The two main types are woven and non-woven, see annex A. A disadvantage of geotextiles is that they may weather e.g. as a result of ultra-violet light, and that they are susceptible to wear and tear by chemical, biological or mechanical processes. These aspects are beyond the scope of this book. For an extensive treatment of geotextiles the reader is referred to Veldhuijzen van Zanten, 1986.

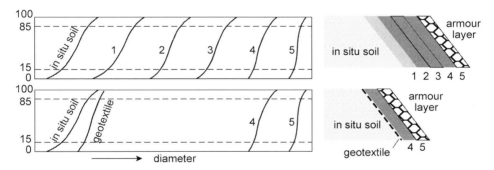

Figure 6-10 Possible effectiveness of geotextile in filter: sieve curves and situation

Geotextiles are considered to be a breakthrough in filter design, both in cost effectiveness for large projects and in possibilities to design a sandtight filter with a limited thickness, see Figure 6-10. They are especially useful in cases where there is no space for a granular filter consisting of many layers or where it is difficult to construct such a filter. The geotextile in this figure is completely responsible for the filter function. The main functions of the layer between the top or armour layer and the geotextile are to prevent damage to the geotextile by individual large stones and to prevent flapping of the cloth, which can cause loss of the base material.

As for granular filters, the two main items are again stability (sandtightness) and permeability. Another similarity is the existence of geometrically closed and open filters. Only geometrically closed filters will be considered in this chapter. Another item that can play a role is the stability of a top layer on top of a geotextile on a slope. As for impervious revetments, friction between the geotextile and the slope has to be checked.

Intermezzo 6-2 Improvisation

Improvisation should not play a very important role in a project of some quality. It would indicate a lack of reflection on design and construction. Nevertheless, a situation may arise where improvisation is inevitable, e.g. in a remote area where material and equipment is not always reliable, creativity of people on the spot is paramount.

After the Second World War, the dike breaches in Walcheren, the Netherlands, were repaired with highly improvised filters. The dikes had been bombed to inundate the island and the breaches were closed with caissons and ships, both being leftovers from the war in Normandy. The operation had to be completed under high pressure and with minimal means in the days just after the war. Mattresses were sunk to protect the bottom against erosion and to function as a foundation for the caissons.

When one of the tidal gaps was almost closed and the head difference across the dam was high, the caissons tilted due to piping. The bottom was uneven and, as the mattresses did not cover the bed completely, eroded due to the strong currents. Torpedonets (other war-leftovers, curtains consisting of steel rings to prevent submarines from entering a harbour) were dumped into the holes and that solved the problem.

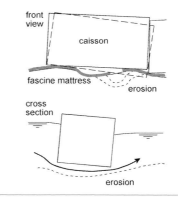

6.3.2 Filter stability

As for granular filters, the sandtightness depends on the relation between the proportions of the base material and the holes in the filtertextile. To characterize the openings of the textile, a kind of reversed sieve test is performed with uniform grains of different, known, diameter. The textile is used as a sieve in a standardized procedure (which can be done in many ways, see Veldhuijzen van Zanten,1986; here the Dutch procedure, as developed by Delft Hydraulics, is presented). When X % of the grains with a certain diameter remain on the textile, that diameter is taken as O_X. (100 - X) % passes the "sieve", this is the reverse of a normal sieve procedure! Figure 6-11a gives the result of such a test. Important numbers in sandtightness are O_{90} and O_{98}, sometimes called O_{max}, a measure for the largest holes in the textile.

Note that, of course, the escape route is now determined by the largest holes instead of the finest grains!

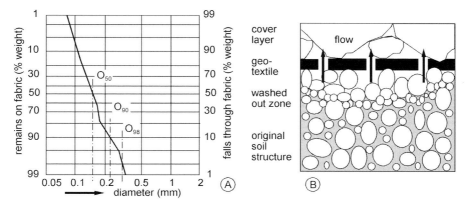

Figure 6-11 Definitions apertures geotextile and migration of fine particles

The most strict geometrical filter rule is that the smallest particles may not pass the largest opening in the textile. This should be applied when the filter is cyclically loaded and no loss of material is acceptable. In many situations, some loss of fine material is not detrimental to the functioning of the filter, since a small layer under the textile can act as part of the total filter system, see Figure 6-11b. The finer parts are washed through the textile and the coarser particles act as a filter for the remaining soil, provided the subsoil is internally stable, see also granular filters.

The stability rule for stationary flow through geometrically closed geotextiles is simply:

$$O_{90} < 2 \cdot d_{90B} \tag{6.4}$$

For cyclically loaded geotextiles 2 to 4 times lower values for the apertures should be used, depending on the permissibility of some sediment loss through the textile.

6.3.3 Permeability

To prevent pressure build-up, a geotextile should be more permeable than the subsoil. A simple rule is that the permeability, k, of the geotextile should be more than 10 times larger than that of the subsoil. The permeability of a geotextile can be measured in the same way as for soils, using Darcy-type relations, but to characterize the permeability of geotextiles the permittivity parameter is often used, defined as follows:

$$P = \frac{u_f}{\Delta h} = \frac{k}{e} \tag{6.5}$$

in which Δh is the head difference, e is the thickness of the geotextile and k the "normal" permeability coefficient. P can be seen as the permeability per m thickness of geotextile and is a property of the material, regardless of the thickness of the geotextile.

The following table gives an idea of the permittivity of some geotextiles:

Type	O_{90} (mm)	P (1/s)
Mesh net	0.1 - 1	1 - 5
Tape fabric	0.05 - 0.6	0.1 - 1
Mat	0.2 - 1	0.05 – 0.5
Non-woven	0.02 - 0.2	0.01 – 2

Usually, a geotextile is more permeable than the subsoil, but two phenomena can decrease the permeability: blocking and clogging.

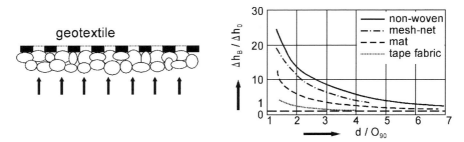

Figure 6-12 Blocking of geotextile

Blocking occurs when large particles seal the openings in the textile. In that case, the permeability can decrease dramatically as is shown in Figure 6-12, indicating results of permeability tests for which a uniform base material was used (all grains were of about the same size). When the diameter of the base material has the same order of magnitude as the O_{90}, the head difference over the textile will increase by a factor 10-20 (compared with the situation without grains). Note that not every type of textile is equally sensitive to this phenomenon.

Clogging is the trapping of (very) fine particles in the openings of the textile, also leading to a decrease of permeability. This can happen when water is contaminated with chemicals, e.g. iron. In contrast to blocking, clogging is a time-dependent process. The idea is that it stabilizes at a certain level after a certain time, but not much experimental evidence is available to support this thesis. One should realize that clogging does not occur exclusively in geotextiles; it also takes place in granular filters.

When the permeability of the textile is 10 times greater than that of the subsoil, there is usually no significant pressure build-up, not even when clogging takes place.

However, when used in very contaminated groundwater, one should be very careful. When the danger of blocking or clogging can be ruled out, a permeability ratio (with the subsoil) of 2 or 3, is sufficient.

A better requirement would be to link the permeability or permittivity directly to the absolute pressure build-up under the geotextile in order to prevent sliding along a slope, see next section.

6.3.4 Overall stability

Basically the same forces are active as on an impervious layer, see chapter 5. In Figure 6-13 a geotextile lies on a slope and is covered with a filter layer and a top layer.

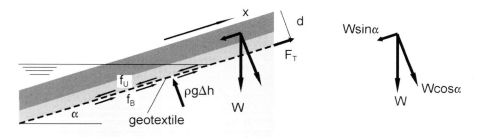

Figure 6-13 Overall stability of filter with geotextile

To prevent the stone layers on the geotextile from sliding down the slope, friction is needed between the filter layer and the geotextile. The amount of friction follows from:

$$f_U \, W \cos\alpha \geq W \sin\alpha \qquad (6.6)$$

where f_U is the friction factor between stones and geotextile, giving $f_U > \tan\alpha$, a requirement which is usually met. W disappears from the equation after dividing, so this requirement is valid for a protection above or under water.

The equilibrium demand for the whole protection, including geotextile, on the slope reads:

$$f_B\big[\big(\rho_{mEFF}\big)g\,d\,\Delta x \cos\alpha - \rho_w g\,\Delta h\,\Delta x\big] + F_T \geq \big(\rho_{mEFF}\big)g\,d\,\Delta x \sin\alpha \qquad (6.7)$$

in which ρ_{mEFF} is the effective density of the layer as a whole. Above water, with $\rho_S \approx 2650$ kg/m^3 and porosity $n \approx 0.4$, this is equal to $(1-n)\,\rho_s \approx 1600$ kg/m^3. Under water this has to be reduced by ρ_w, giving $\rho_{mEFF} = (1-n)(\rho_s - \rho_w) \approx 1000$ kg/m^3.

Δh is the pressure difference across the geotextile, expressed in m water, which can be neglected in most cases, but has to be checked when the permittivity (see equation

(6.5)) is very low, e.g. by clogging. F_T is the tension force in the geotextile (support from the toe is neglected since a geotextile can not develop a pressure force). F_T can be determined by estimating f_B, the friction factor between geotextile and subsoil, which is usually in the range (0.6 - 0.9)$\tan\phi$. Low water determines the maximum tension force, since the uplift effect from the water (Archimedes) is then minimal and the maximum unsubmerged weight "hangs".

6.4 Summary

Filters can have two functions: preventing erosion under a top layer or drainage to prevent pressure build-up. In both cases filter stability and permeability play a role. Stability concerns the grains that should remain stable in the filter, while the permeability is important to avoid pressure build-up, which can cause instability of a layer as a whole. The two main types of filters are granular filters and geotextiles.

Granular filters fall into two categories: geometrically closed and geometrically open. Geometrically closed filters are stable, regardless of the porous flow through the filter. In fact there is no critical gradient because the relation between the diameters in the filter is such that the finer grains cannot move through the pores of the coarser grains. Geometrically closed filters have to be designed within the margins of stability and permeability. The design rule consist of three inseparable parts:

$$\text{Stability: } \frac{d_{15F}}{d_{85B}} < 5 \quad \text{Int.Stability: } \frac{d_{60}}{d_{10}} < 10 \quad \text{Permeability: } \frac{d_{15F}}{d_{15B}} > 5$$

Stable geometrically open filters have a critical gradient, above which the base material is no longer stable. As for open-channel flow, the occuring gradient needs to remain below the critical value. This means that both these gradients have to be known. Section 6.2.3 treats the critical gradients for flow perpendicular and parallel to the interface of the layers. The occuring gradient has to be derived from calculations or measurements inside the filter in a test facility. In bottom protections with a low gradient but high turbulence, the thickness of the layers becomes important, which is expressed in equation (6.3).

Geotextiles can replace several filter layers, resulting in a thinner filter. A simple rule for stationary flow is: $O_{90} < 2 \cdot d_{90B}$. For cyclic flow, the apertures should be 2 to 4 times smaller, depending on the permissibility of sediment loss.

The permeability of a geotextile is usually expressed as the permittivity, which is defined as the permeability per m thickness of geotextile. The permeability of a geotextile can be reduced by blocking (sudden) or clogging (a gradual closure of the holes in the textile). With a permeability 10 times larger than that of the subsoil, the permeability is usually enough to prevent pressure build-up. On slopes it may be necessary to check the stability of the entire combination of filter layers and geotextiles. The friction between geotextile and the top layers or subsoil has to be sufficient and/or the geotextile has to be strong enough to withstand any possible tension forces.

7 WAVES
Loads

Spilling breaker on dike slope (Zealand)

7.1 Introduction

A "wave" is the generic term for any (periodic) fluctuation in water height, velocity or pressure. This chapter will be restricted to wind-generated waves. The term *sea* is often used for "fresh" waves, where the driving wind force is still active, in contrast to *swell*. Swell is the name for waves that were caused by wind, but possibly long ago (days) and far away on the ocean (thousands of km), travelling on with the, slowly dissipating, energy gained from the wind.

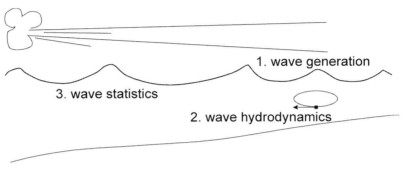

Figure 7-1 Wave issues

The technical treatment of (wind) waves can be divided roughly into three categories, see Figure 7-1:

1. **Generation**
   The generation of waves by wind is described with relations of the type: H, $T_{characteristic} = f(u_{wind}, h, fetch)$. When for a coastal project no measured wave data and no advanced software to predict waves are available, appendix 0 gives some useful relations for a first estimate. See also SPM,1984

2. **Hydrodynamics**
   Velocities and forces in waves are, of course, important when dealing with erosion and protection. These parameters are described with relations of the type: u, p, $\tau = f(H,T,h)$. The wave generating forces no longer play a role, the starting point is now wave height and period. Phenomena like refraction and diffraction, also being part of wave hydrodynamics, are beyond the scope of this book.

3. **Statistics**
   The water surface of wind waves is irregular because the driving force, the wind, is turbulent. It is therefore necessary to characterize a wave field by means of statistical parameters. Relations of the type: $p(H) = f(H_{characteristic}, distribution\ function)$ give the probability of a certain wave height in a wave field. Wave spectra have become the most important description of the joint distribution of wave heights and periods in wave fields.

These three issues in wave description each have their own clear-cut function and sometimes lead to confusion. Hydrodynamics is not limited to regular waves, although most relations have been formulated for periodical fluctuations. Wave statistics of irregular waves do not replace hydrodynamics, but are only a mathematical tool to describe the fluctuations of the water surface. For any individual wave in an irregular wave field, forces can be derived from hydrodynamical relations, while the statistics give the probability of occurence of those forces.

Appendix 7.7.2 gives some information on wave statistics. Here, it is sufficient to say that a wave field is usually characterized by the significant wave height, H_s, and the peak period, T_p. H_s is defined as the average height of the highest third part of the waves in a wave field, while T_p is the peak period of the wave spectrum, the period with the maximum energy density. H_s, in combination with the Rayleigh distribution, characterizes the state of the sea at a certain moment. This state of the sea can change every hour or so, giving different values for H_s, which also has a distribution in time, the so-called long-term distribution. For maintenance and design considerations, this long term distribution is important in the choice of representative loading conditions.

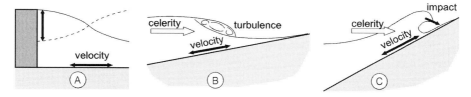

Figure 7-2 Examples of wave loads

Figure 7-2 shows how waves can load protections. Starting point is a given wave near the location to be considered. Interaction between the waves and the structure plays an important role. Figure 7-2a gives a vertical wall, which is reflecting a progressive wave, resulting in a *standing wave*, with pressure fluctuations against the wall and a load on the bottom due to velocities in the wave. Figure 7-2b shows a *breaking wave* on a mild slope, a so-called *bore*, which is very much like a hydraulic jump. Due to the breaking, the motion in the wave becomes very turbulent. The wave in Figure 7-2c, breaks on a steeper slope, causing steep pressure gradients and possibly impacts due to collisions between protection and waves.

7.2 Non-breaking waves

7.2.1 General

Wind waves in deep water are, from a hydronamical point of view, short waves. That means waves where vertical accelerations can not be neglected. In long waves, that exist in tides, storm surges, tsunamis and river floods, the negligible vertical accelerations make the pressure distribution hydrostatic. Waves in shallow water, especially swell, can be seen as a transition from short to long waves.

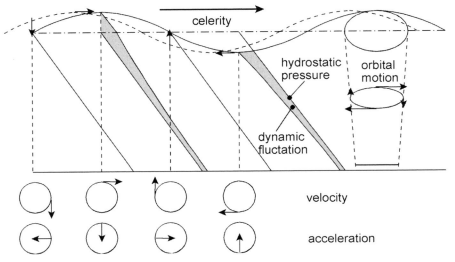

Figure 7-3 Wave motion in periodic, unbroken wave

Figure 7-3 shows the main characteristics of pressures and velocities in a simple, periodic, unbroken sine-wave. The pressures deviate from the hydrostatic values, due to the vertical accelerations. The particle velocities are related to the orbital motion. In general these are ellipses, in deep water they are circles and at the bottom they are straight lines. One has to bear in mind, that a wave has two velocities: the celerity or propagation speed, which is the speed of the wave shape, and the particle velocity. A clear demonstration of the difference between these velocities can be seen during sporting events, when people just rise and sit down again, while "The Wave" travels through the crowd.

Water in motion can be fully described by the Navier-Stokes equations, but, in different situations, different terms can be neglected. In open-channel flow, friction has an important influence on the flow and the thickness of the boundary layer is more or less equal to the water depth. The loads by long waves on protections can be seen as a succession of more or less stationary flow situations, which have been treated in chapter 2. Nowadays, pressures (water levels) and velocities of long waves, are easily computed numerically for long periods at many locations. A numerical solution for a complete irregular wave field, is still not possible for practical reasons, mainly due to the large number of calculation points in time and space, although techniques do improve rapidly. Fortunately, however, the fluid motion in short waves, excluding a thin boundary layer (see next section), can be described as irrotational. Given the shape of the wave, this makes analytical solutions for pressures and velocities in a wave possible.

A different wave shape gives different relations for the water motion in the waves. Figure 7-4 gives an overview of the validity of various wave theories, based on various wave shapes, adapted from LeMéhauté, 1976. The horizontal axis represents the water depth, made dimensionless with the acceleration of gravity and the square of the wave period, which is a measure of the wavelength. The vertical axis represents the wave height, again made dimensionless with the wavelength. The vertical axis is therefore a measure of the wave steepness ($\propto H/L$) and the horizontal axis a measure of the relative waterdepth ($\propto h/L$). Upwards, the validity (or better: the very existence of waves) is limited by breaking due to steepness and to the left by breaking due to restricted water depth, see section 7.3.1.

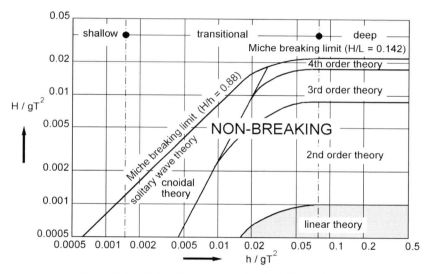

Figure 7-4 Validity of wave theories (LeMéhauté, 1976)

The shape of waves with increasing steepness in deep water (upper right corner in Figure 7-4) has to be described with more sine components, leading to more complex solutions of the equation of motion. Waves with considerable wave height in shallow water, can be described with the cnoidal wave theory and, on the verge of breaking, with the solitary wave theory. Here only the linear or first-order wave theory will be used, in which the shape of the wave is a simple sine. In appendix 7.7.1, an overview of the formulae in this theory is given. According to Figure 7-4, this theory may only be applied with relatively small waves in deep water (another name for this theory is the small-amplitude wave theory). The approximation of waves by a simple sine function is a crude simplification. For an adequate understanding, however, the linear theory is very useful and attractive, since it gives a simple, but complete description of the pressure and velocity field. Calculated values outside the range of validity can serve as a first indication.

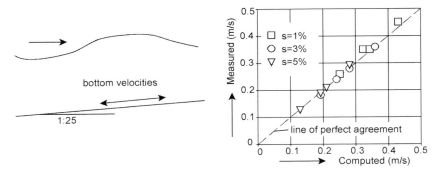

Figure 7-5 Application linear wave theory

Figure 7-5 gives an example of the application of the linear wave theory on a mild slope, outside the limits of validity (from Schiereck et al, 1994). With (regular) wave heights of 0.2 - 0.3 m, water depths of 0.4 - 0.6 m and a wave steepness (defined as $s = H/L_0$) ranging from 0.01 - 0.05, the values of H/gT^2 are in the range 0.002 - 0.008 and those of h/gT^2 in the range 0.004 - 0.02, which is far outside the permissible range in Figure 7-4. Despite that, the similarity between measured and computed values is remarkably good. So, in many cases, the linear theory can be used for a first estimate of maximum bottom velocities, which are quite important for bottom protections. **Note:** This is not true for parameters such as the surface profile and the surface velocity, see also LeMéhauté, 1969.

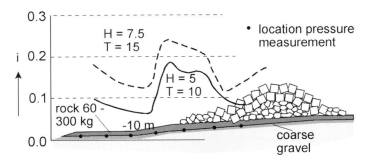

Figure 7-6 Gradient in filter under breakwater

An even more amazing example of the value of the linear wave theory is shown in Figure 7-6. The pressures in the transition between coarse and fine filter layers under the Hook of Holland breakwater (Port of Rotterdam) have been measured in a laboratory model, from which the gradient was determined. This is a porous flow situation for which the Forchheimer relation (see chapter 5) should be applied. Assuming that the coarse grains do not influence the pressure gradient near the

bottom and using the linear wave theory to calculate the pressure gradient at the
bottom gives (expressed in piezometric gradient, see Table 7.1 in appendix 7.7.1):

$$\frac{\partial\left[\dfrac{H}{2\cosh(kh)}\right]}{\partial x}\sin\left(\omega t - kx\right) = \frac{-kH}{2\cosh(kh)}\cos(\omega t - kx) \rightarrow i_{max} = \frac{kH}{2\cosh(kh)} \quad (7.1)$$

For a waterdepth $h = 10$ m, a wave $H = 5$ m and $T = 10$ s this gives: $i_{max} \approx 0.14$
(measured 0.18) and for $H = 7.5$ m and $T = 15$ s: $i_{max} \approx 0.15$ (measured 0.22). For a
detailed design of a large breakwater this is certainly not accurate enough, but for a
preliminary design, it gives an order of magnitude. Taking into account the steeper
wave front near the breakwater, compared with a simple sine wave, it is even
possible to give a better estimate, e.g. by using the maximum steepness in a
(measured) wave profile divided by the maximum steepness in a sine wave as a
correction factor.

7.2.2 Shear stress

Chapter 2, showed that in wall flow, without a fully developed boundary layer, the
shear stress is higher than in uniform flow, due to accelerations. In waves, a similar
phenomenon occurs, but the accelerations are local: varying in time but not in place.
In that case the growth of the boundary layer is approximated by (see Booij, 1992):

$$\frac{d\delta}{dt} \approx \kappa u_* \approx 0.4 u_* \quad (7.2)$$

κ is the von Karman constant and u_* is the shear velocity which is much smaller than
the "driving" velocity of the orbital motion at the edge of the boundary layer, $u_* \approx 0.1$
u_b. In a wave, the boundary layer can only grow during half the wave period and has
to start from scratch again when the flow reverses. With $T = 5$ s and $u_b = 1$ m/s, this
gives $\delta \approx 0.1$ m. In this boundary layer, the flow is highly rotational and the
assumptions on which the wave equations are based are not valid. The boundary
layer is, however, small compared with normal water depths, see Figure 7-7, so the
wave equations can be used for the greatest part of the water depth. This thin
boundary layer, however, is responsible for a high shear stress under short waves.
In a tidal wave with $u \approx 1$ m/s and $T \approx 45000$ s, the same approach shows that it takes
5 - 10 minutes for the boundary layer to grow as large as the waterdepth. So, the
assumption that tidal flow can be seen as a succession of quasi-stationary flow
situations is realistic when we look at bottom shear, which is important for bed
protections.

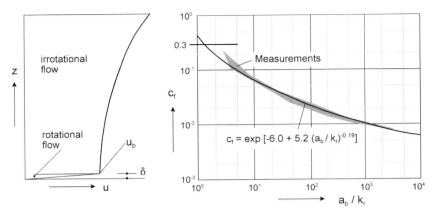

Figure 7-7 Friction under waves

In short waves the boundary layer never grows to the order of magnitude of the water depth, hence high friction factors can be expected. The friction factor for a wave is defined as for flow:

$$\hat{\tau}_w = \tfrac{1}{2}\rho\,c_f\,\hat{u}_b^2 \quad \text{with: } \hat{u}_b = \omega\,a_b = \frac{\omega\,a}{\sinh kh} \quad \text{and: } u = \hat{u}_b \sin\omega t \tag{7.3}$$

where the subscript w stands for wave and b for bottom. The average value of the shear stress over a wave period is half the value of the maximum (the average of $\sin^2\omega t$ is 1/2). Jonsson, 1966, found an expression for turbulent flow over a rough bed based on experiments, it was rewritten in a more practical form by Swart, see CUR/CIRIA,1991:

$$c_f = \exp\left[-6.0 + 5.2\left(a_b\,/\,k_r\right)^{-0.19}\right] \quad \text{with: } c_{f\max} = 0.3 \tag{7.4}$$

Comparing the values in Figure 7-7 with those for flow, one notices that shear stress under short waves can be much higher than in flow. c_f in uniform flow is g/C^2 (see chapter 2) which results in values of (O) 0.01, while in very short waves, where the amplitude at the bottom has the same order of magnitude as the roughness, values of 0.1-0.3 seem possible. 0.3 is usually considered the maximum value.

Example 7-1

What is the shear stress under a wave, 3 m high with a period of 7 s, in a waterdepth of 8 m and with a bottom rouhness of 0.1 m?
The wave length in 8 m deep water becomes 55.2 m (found iteratively from Table 7.1 or from Figure 7-24). k then becomes $2\pi/L = 0.114$ and $\omega = 2\pi/T = 0.898$. $a_b = \tfrac{1}{2}H\,/\sinh(kh)$
$= 1.44$ m and $u_{bmax} = \omega a_b = 1.29$ m/s. c_f then becomes $\exp(-6+5.2(1.44/0.1)^{-0.19} = 0.057$.
The shear stress is then: $\tfrac{1}{2}\cdot 1000\cdot 0.057\cdot 1.29^2 = 47.4$ N/m².

Waves and currents

In most cases, both waves and flow will be present. When waves are dominant, the approach mentioned above is recommended, but when flow is dominant, according to Bijker, 1967, the influence of waves can be taken into account by adding the current and orbital velocity vectorially at a level where:

$$u_{c-t} = \frac{\sqrt{g}}{\kappa C} u_c \quad \text{and} \quad u_{b-t} = \frac{1}{\kappa} \sqrt{\frac{c_f}{2}} u_b \sin(\omega t) \tag{7.5}$$

in which $\kappa \approx 0.4$ and u_b is the maximum orbital velocity at the bottom. The results are:

$$u_r = \sqrt{\frac{g}{\kappa^2 C^2} u_c^2 + \frac{c_f}{2\kappa^2} u_b^2 \sin^2(\omega t) + 2 \frac{\sqrt{g}}{\kappa C} u_c \frac{1}{\kappa} \sqrt{\frac{c_f}{2}} u_b \sin(\omega t) \sin(\phi)} \tag{7.6}$$

where ϕ is the angle between wave and current direction. The resulting shear is:

$$\tau_r = \rho \kappa^2 u_r^2 \tag{7.7}$$

According to equation (7.6) the resulting shear stress is maximum when the wave direction is parallel to the flow ($\phi = 90°$, see Figure 7-8). **Note:** For sediment transport, equation (7.6) is averaged over a wave cycle. The third term then vanishes and the resulting shear becomes independent of ϕ. For bottom protections, the maximum value determines incipient motion.

Figure 7-8 shows the results of equation (7.7) for the wave used in Example 7-1 ($H = 3$ m, $T = 7$ s, $u_b = 1.29$ m/s and $c_f = 0.057$) with a current of 1 m/s and a Chezy value of 50 $\sqrt{}$m/s. The shear stress only due to the current is 4 N/m^2, only due to waves it is 47.4 N/m^2 (as in Example 7-1). Waves perpendicular to the current ($\phi = 0$) give a shear stress equal to the sum of both (51.4 N/m^2). Waves in the direction of the current ($\phi = 90°$), give a maximum shear stress of 78.6 N/m^2.

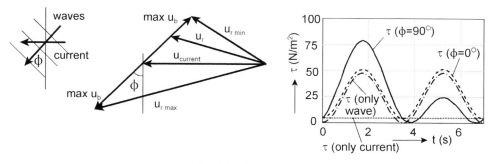

Figure 7-8 Combined wave-current action

7.3 Breaking waves

7.3.1 General

A wave breaks as a result of instability which develops when the wave can no longer exist. When the particle velocity u exceeds the celerity c, the particles "leave the wave profile". This instability (with air entrainment) makes the flow highly complex and the problem hardly lends itself to computation. Breaking occurs because a wave is very steep (on deep water) or because the water is very shallow or a combination of these reasons. Both limits are described with the breaking criterion by Miche:

$$H_b = 0.142 \, L \tanh\left(\frac{2\pi}{L} h\right) \tag{7.8}$$

For deep water, this simply becomes: $H_b/L = 0.142$ ($h/L > 0.5 \rightarrow \tanh(2\pi \, h/L) \approx 1$). This is the maximum possible steepness for an individual wave. For H_s in irregular waves, about half this value is found, since H_{max} in irregular waves is about $2H_s$, see the Rayleigh distribution in appendix 7.7.2. In practice, the steepness, H_s/L_{0p} (deep water wavelength related to the peak period), in a sea (fresh wind waves) is seldom more than 0.05. For swell, much lower values occur: 0.01 or less. For shallow water, equation (7.8) leads to $H_b/h = 0.88$ ($h/L < 0.1 \rightarrow \tanh(2\pi \, h/L) \approx 2\pi \, h/L$). The solitary-wave theory leads to a slightly different limit: $H_b/h \approx 0.78$. Applied to H_s, this results in a maximum value of $H_s/h \approx 0.4$ - 0.5, corresponding with the breaking of the larger waves (1.5 - 2 H_s). If there is a shallow foreshore and no wave measurements or computations are available, this ratio is convenient to give a rough estimate of the design waves. See also section 0.

Breaker types

For waves breaking on a slope, the dimensionless Iribarren number or surf similarity parameter is of crucial importance in all kinds of problems in shore protection. The parameter is defined as:

$$\xi = \frac{\tan\alpha}{\sqrt{H/L_0}} \tag{7.9}$$

where α is the angle of the slope. $\sqrt{(H/L_0)}$ can be seen as the wave steepness (a rough approximation, because H is the local wave height and L_0 is the deep-water wavelength). ξ represents the ratio of slope steepness and wave steepness. L_0 represents the influence of the wave period, since $L_0 = gT^2/2\pi$. In irregular waves, H_s and L_{0p} are normally used in equation (7.9).

For different values of ξ, waves break in a completely different way. Figure 7-9 shows the various types. The transition between breaking and non-breaking lies around ξ ≈ 2.5 - 3. For higher values the wave surges up and down the slope with minor air entrainment. The behaviour of the waves from ξ ≈ 3 to 5, is therefore often called surging breaker although it can be questioned whether this is a breaking or a standing wave. A collapsing breaker is between breaking and non-breaking. The most "photogenic" breaker is of the plunging type (ξ ≈ 0.5 to 3). In plunging breakers the crest becomes strongly asymmetric; it curls over, enclosing an air pocket and impinges on the slope like a water jet. With decreasing slope angles, the crest of a plunging breaker becomes less asymmetric and the water jet projected forward from the crest becomes less and less pronounced. This leads to the spilling breaker type (ξ < 0.3). The transition between the various breaker types is gradual and these values for ξ are just an indication. For more detail, see Battjes, 1974. After breaking, the wave travels on with a celerity equal to about √(gh) and the wave behaves like a bore or a moving hydraulic jump, see intermezzo 7.1. For more detail, see Fredsøe, 1992.

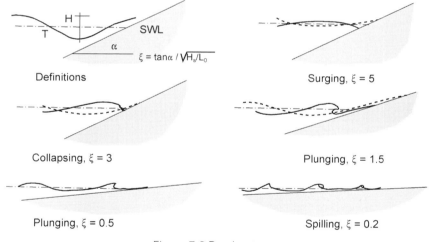

Figure 7-9 Breaker types

The parameter ξ indicates that, for a slope, the notions "steep" and "gentle" are relative. A slope of 1:100 is usually thought to be very "gentle", but for a tidal wave it can be just as steep as a vertical wall is for a wave with a period of a few seconds. This is illustrated by the fact that a tidal wave on a beach does not break and is completely reflected (the phenomena described in this paragraph are completely different from those in tidal waves; the example only shows the relativity of all notions in hydrodynamics). ξ also indicates that an incoming wave that is already very steep will easily break, even on a steep slope.

Breaking also means energy transformation via turbulence and friction into heat. Turbulence in wind waves has only been studied recently and much is yet unknown. For flow we made a distinction between wall turbulence, e.g. induced by friction along the bed, and free turbulence, e.g. in mixing layers. The same can be done for waves.

Intermezzo 7-1

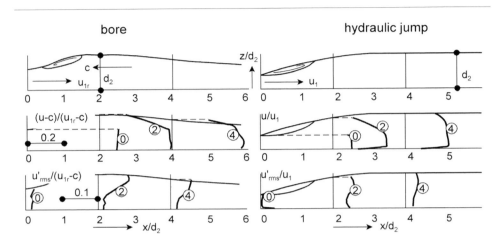

This figure shows a comparison between velocities and turbulence in a bore and a hydraulic jump (from Stive, 1984). The hydraulic jump data are from the same source as the example in chapter 2, but now $Fr = 2$.

Velocities in the jump are related to the supercritical flow upstream of the jump (u_1) and the distances are related to the downstream water depth (d_2). To make the comparison, the velocities in the bore are related to the difference between the encounter velocity in the wave trough and the celerity of the bore.

With this velocity difference, a Fr-number can be defined, 1.6 for the bore in this case. The figure illustrates that the same phenomena occur in breaking waves as in hydraulic jumps. However, no quantitative relations between wave characteristics and turbulence are available for stability rules for protections in breaking waves. Equation (7.10) is an approximation of the total energy loss.

In non-broken waves, the only cause of turbulence is bed friction. In breaking and broken waves much turbulent kinetic energy is produced by the friction between the surface roller of the bore and the body of water moving with the wave. The total energy loss in a bore can be approximately computed using the analogy with a hydraulic jump with (see Battjes/Janssen, 1978):

$$D = \rho g \frac{H^3}{4Th} \qquad (7.10)$$

in J/(s·m^2) where *H*, *T* and *h* are the wave height, the period and the water depth respectively.

7.3.2 Waves on a foreshore

This book deals with the protection of banks and shores, so we could start with the wave characteristics at the toe of the shore: a dike, a sea-wall or the like. But as promised, the focus is on the understanding of processes rather than on formulas. Waves travelling from deep water to a shore undergo all kinds of alterations, so it is useful to know the history of waves arriving at the toe of a coast.

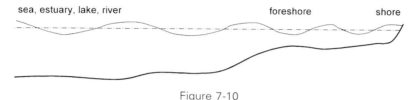

Figure 7-10

Of course, a *wind* blowing to the shore keeps transferring energy into the waves. But for wave generation, fetch is also important, see appendix 0; usually the wind can be neglected when starting at the foreshore. As the introduction of this chapter states, *refraction* will not be treated. The discussion will be limited to one-dimensional processes going from deep to shallow water. The *shoaling* process causes the wave height to increase and can be described with the linear wave theory. The bottom *shear*, see section 7.2.2, causes the wave height to decrease. *Breaking* completes the energy transfer, since waves cannot go on in water of zero depth.

Figure 7-11 shows some results of computation with SWAN, a numerical model for wave propagation in shallow water (see Booij, N. (1999)). A wave field with H$_s$= 4 m (defined as 4√m$_0$, see appendix 7.7.2) and a peak period T$_p$= 8 s travels from a depth of 20 m towards the shore. Two different slopes, 1:500 and 1:50, are used. Figure 7-11a shows the decreasing wave heights for the two slopes. The difference is partly due to friction and partly due to breaking.

(**Note**: the water depths are related to the original Still Water Level. Due to wave set-up the water level rises near the shore leading to a wave height ≠ 0.)

On a slope of 1:500 the waves travel a distance ten times longer and experience more friction. The method of Battjes/Janssen, 1978, was implemented in SWAN for the breaking of waves in shallow water. Based on the analogy between a bore and a hydraulic jump the energy dissipation in this method is given by:

$$D = \frac{1}{4}Q_b \rho g \frac{H_m^2}{T_p} \qquad\qquad (7.11)$$

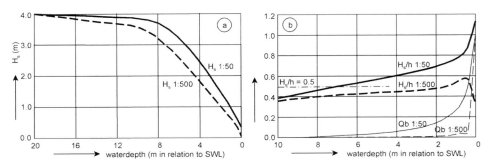

Figure 7-11 Wave decrease on different slopes

in which H_m is given by equation (7.8). Q_b is the relative number of all (irregular) waves that are broken and can be computed from:

$$\frac{1-Q_b}{\ln Q_b} = -\left(\frac{H_{rms}}{H_m}\right)^2$$

(7.12)

indicating that, when the wave present at some location (represented by H_{rms}) is higher compared with the maximum possible wave height (H_m), there will be more breaking. This also shows that the Battjes/Janssen-method contains a mixture of wave physics and wave statistics.

Figure 7-11b shows Q_b and the wave height relative to the water depth (including the wave set-up) on both slopes. On the 1:500 slope, it appears that wave breaking causes an energy loss which is less important than friction. Only in the last stretch of the slope breaking becomes important (1 m on the horizontal axis in Figure 7-11b represents 500 m for the 1:500 slope and 50 m for the 1:50 slope!) On the 1:50 slope, wave breaking is already important from a depth of 5 m.

Figure 7-11b also shows that the relation H_s/h is not constant along the slope, indicating that breaking takes time, hence distance. Breaker depths depend on the slope angle, see also also section 0. But for a foreshore of some dimension, e.g. a sand bar along the coast, a simple relation can be used for a preliminary design if no wave data are available:

$$H_s \approx 0.5h$$

(7.13)

Wave height distribution and spectra

Breaking of waves on a shallow foreshore also causes deviations from the Rayleigh distribution for wave heights. The larger waves break first as can be observed at any beach. Battjes/Groenendijk, 2000, established a relatively easy procedure to create an adapted wave-height distribution. Appendix 7.7.2 gives the outlines of this approach.

The shape of the wave spectrum will also change considerably on a shallow foreshore. The consequences of all this for the stability of protections are not clear yet.

7.4 Waves on slopes

7.4.1 General

As stated in section 7.3.1, the surf similarity parameter, ξ, plays an important role in the behaviour of waves on a slope. Reflection, breaker depth, wave run-up but also stability of material on a slope can often be expressed as a function of ξ.

Reflection

Vertical walls fully reflect incoming waves. Reflection is defined as the height of the reflected wave in relation to the incoming wave height: K_R (reflection-coefficient) = H_R/H_I. It can be rationalised that K_R is proportional to ξ^2, see Battjes, 1974. Experimentally, $K_R \approx 0.1\xi^2$ was found for values of ξ below the breaking limit, see Figure 7-12. For $\xi > 2.5$, K_R slowly tends to 1, the value for total reflection.

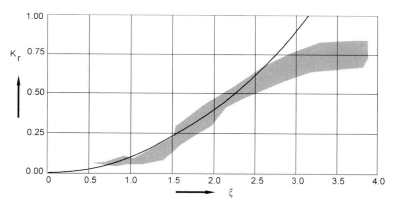

Figure 7-12 Reflection as function of ξ (Battjes, 1974)

Absorption

Energy that is not reflected or transmitted, is absorbed on the slope and this absorption is always at the expense of the protection. A low ξ means little reflection, see Figure 7-12, so, given a certain wave, a gentle slope (tan α low) gives more absorption which is an unfavourable loading situation for a protection. However, the energy absorption per m^2 is lower and the final result is therefore more favourable. From Figure 7-13 it seems reasonable that the plunging breaker, which is common for dikes and seawalls, forms a heavier load for a protection than the spilling type.

Another aspect of the heavy loading by plunging breakers is the jet-like behaviour of the plunging water. A jet acts locally on the bed, doing much more harm than diffusing turbulence from a surface roller. Stability relations show that collapsing breakers cause the most damage, see the next chapter. This can possibly be explained by the fact that both surface roller turbulence and a (beginning) jet act directly on the slope (no body of water between roller or impinging jet and the bed to act as a buffer).

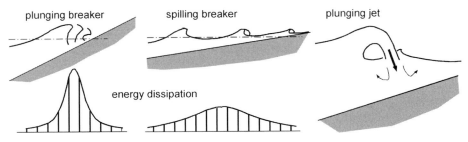

Figure 7-13 Energy dissipation for two breaker types

Breaker depth

For a horizontal bottom, the solitary-wave theory gives a relation between the wave height and the waterdepth at which waves break: $\gamma_b = H/h \approx 0.78$, while equation (7.8) gives $\gamma_b = 0.88$. For slopes, γ_b depends on ξ. Figure 7-14 shows some experimental results (using ξ_0 with the deep-water wave height H_0, since H is hard to define at the point of breaking).

There clearly is a relation with ξ; for steeper slopes, the waves break at a smaller depth. Apparently, the waves need time to break, during which they can travel to a smaller depth. Compare also the difference between the two slopes in Figure 7-11b.

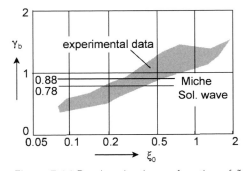

Figure 7-14 Breaker depth as a function of ξ

7.4.2 Run-up and run-down

Run-up regular waves

Run-up is defined as the maximum water level on a slope during a wave period and the run-down is the minimum level (both relative to the still water level), see Figure 7-15.

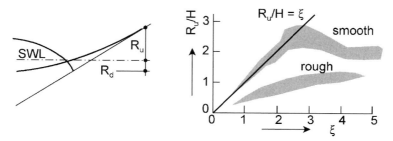

Figure 7-15 Run-up and run-down, definitions and run-up data

For breaking waves (ξ < 2.5-3) on smooth slopes, the Hunt's formula gives:

$$\frac{R_u}{H} = \xi \tag{7.14}$$

Rip-rap slopes give values which are about 50% lower, see also Figure 7-15. Run-up appears to be maximum around $\xi = 2.5$-3, which means just at the transition between breaking and non-breaking. For $\xi \rightarrow \infty$ the theoretical value is $R_u/H = 1$ (standing wave). So, the momentum of the wave motion on a slope leads to much higher levels.

Run-up irregular waves

Most wave run-up formulae for irregular waves use a value in the distribution that is only exceeded by a small number of the waves, because the considered slope is usually meant to protect something (not too much water should pass the slope). In The Netherlands, $R_{u2\%}$ is used for the design of sea dikes. Ongoing research in physical models and prototype has resulted in a number of successive formulae. In recent years especially attention has been paid to shallow foreshores and wide (and double-peaked) spectra. All research has been summarised in TAW, 2002, and figure 7-16:

$$\frac{R_{u2\%}}{H_{m_0}} = \min\left\{ A\gamma_b\gamma_r\gamma_\beta\xi_0, \ \ \gamma_r\gamma_\beta\left(B - \frac{C}{\sqrt{\xi_0}}\right)\right\} \tag{7.15}$$

$$\overline{A} = 1.65 \qquad \overline{B} = 4.0 \qquad \overline{C} = 1.5$$

$$\sigma_A \quad = 0.07\overline{A} \qquad \sigma_B \quad = 0.07\overline{B} \qquad \sigma_C \quad = 0.07\overline{C}$$
$$= 0.1155 \qquad \qquad = 0.28 \qquad \qquad = 0.105$$

$$A_{design} = 1.75 \qquad B_{design} = 4.3 \qquad C_{design} = 1.6$$

For values of $\xi_p < 1.8$ the left-hand side of the "min" statement is valid, for the other values the right hand side becomes the minimum. The average values and the standard deviations should be used in probabilistic computations, the design values in case of a deterministic calculation. The formula is valid in the range $0.5 < \gamma_b\xi_0 < 10$.

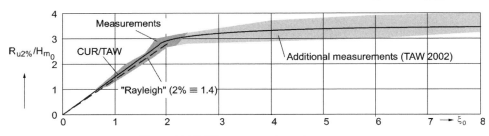

Figure 7-16 Wave run-up irregular waves

For large ξ, the run-up shows a rather constant value, as is the case for regular waves, see figure 7-15. **Note**: When the Rayleigh distribution for the 2% highest waves is applied to the expression for regular waves, equation (7.14), an expression similar to the left side of equation (7.15) is found with a factor 1.4 instead of 1.65.

Correction factors

Roughness

The run-up formulae are for smooth slopes like asphalt of smooth concrete blocks ($\gamma_r = 1$). For grass slopes, γ_r is between 0.95 and 1, but for design purposes TAW (2002) recommends to use a value of 1. For a rubble slope is one layer of riprap a reduction of $\gamma_r = 0.7$ is recommended, while for a double layer of $\gamma_r = 0.55$ may be used.

Angle of attack

When the wave attack is not perpendicular to the slope (i.e. $\beta \neq 0$), a reduction factor is applied. For long-crested waves, like swell or ship waves, $\gamma_\beta = \sqrt{\cos\beta}$ (from the energy flux, which is $\approx H^2\cos\beta$) with a minimum of 0.7. For short-crested waves (wind waves, which are not unidirectional like swell): $\gamma_\beta = 1 - 0.0022\,\beta$ (β in degrees) with a minimum of 0.8.

Berm

A "berm" reduces the wave run-up. Figure 7-17 gives the parameters which play a role in the berm reduction factor, γ_B. At both sides of the berm, the slope is intersected at a vertical distance H_s from the horizontal centre plane of the berm, giving a length L_B. h_B is the distance between SWL and the berm level (can be negative or positive). γ_B finally becomes:

$$\gamma_B = 1 - \frac{B_B}{L_B}\left[0.5 + 0.5\cos\left(\pi\frac{h_B}{x}\right)\right]$$
$$x = z_{2\%} \quad \text{for } z_{2\%} > -h_B > 0 \quad \text{(berm above SWL)}$$
$$x = 2H_s \quad \text{for } 2H_s > h_B \geq 0 \quad \text{(berm below SWL)}$$

(7.16)

with limits: $0.6 \quad \gamma_B < 1$. Equation (7.16) shows that a berm on SWL is most efficient. For more information, see TAW (2002).

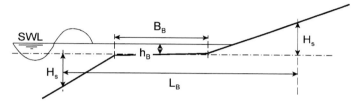

Figure 7-17 Definitions berm reduction

Foreshore

On very shallow foreshores wave break; the wave spectrum transforms and is flattened. Also the waves are no longer Rayleigh distributed. According to TAW (2002) one should use for the calculation of ξ not the T_{m0}, but the $T_{m-1,0}$, which is m_{-1}/m_0 (the first negative moment of the spectrum divided by the zeroest moment of the spectrum). In case no detailed spectrum information is available, one may use $T_p = 1.1T_{m-1,0}$. For the wave height on the shallow foreshore one should use the H_{m0} on the foreshore.

Example 7-2

A grass dike with slopes 1:3 and a 2 m berm at design level is attacked by perpendicular (swell) waves with $H_s = 1$ m and a steepness of 0.01. What is the wave run-up? Starting point is equation 7.14. $\gamma_r = 1.0$, γ_B is found from equation 7.15. The berm height is equal to the waterlevel, so, $h_B = 0$ and $L_B = 2 H_s \cot\alpha + 2 = 8$ m, hence, $\gamma_B = 0.75$. γ_β is 1, we do not assume a very shallow foreland. The surf similarity parameter is $\tan\alpha/0.1 = 3.33$. The wave run-up finally, is then: $R_{u2\%} = H_s * Min\{1.65*0.75*1.0*1.0*3.33, 1.0*0.75(4.0-1.5/\sqrt{3.33}\} = 1*Min\{4.12, 2.38\} = 2.38$ m above design level. However, for a design different values for the constants have to be used, so $R_{u2\%} = H_s * Min\{1.75*1.0*0.75*1*3.33, 1.0*0.75(4.3-1.6/\sqrt{3.33}\} = 1*Min\{4.40, 2.57\} = 2.57$ m

Run-down regular waves

In Battjes,1974 the run-down is:

$$R_d = R_u(1-0.4\xi) = H(1-0.4\xi)\xi \qquad (7.17)$$

This implicates that for "real" breaking waves, the water level does not drop below the still water level because the water in a wave that flows down on a slope, meets the water running up from the next wave (for gentle slopes like beaches this phenomenon dominates and is called set-up). Figure 7-18 again shows experimental results for smooth slopes.

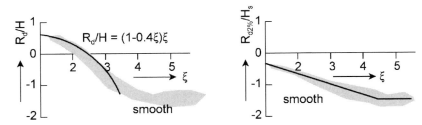

Figure 7-18 Run-down in regular and irregular waves

Run-down irregular waves

The run-down in irregular waves is (see CUR/CIRIA, 1991):

$$R_{d\,2\%} = -0.33H_s\xi_p \quad (R_{d\,2\%\,max} = -1.5H_s) \qquad (7.18)$$

Note: in irregular waves, the run-down is always below SWL.
No run-down reduction, due to roughness etc. is taken into account. CUR/CIRIA, 1991 gives more details.

7.4.3 Wave impacts

The pressure under a wave increases and decreases with the wave cycle as long as the water keeps in touch with the point where the pressure is considered. This is often called the quasi-static wave load, see point 1 in Figure 7-19.

When water from the wave collides with the surface, a very short, very high, impact pressure will occur, the dynamic wave load, wave impact or wave shock, see point 2 in Figure 7-19. (If you have difficulties imagining the difference between quasi-static wave pressures and wave impacts you can easily measure them in a swimming pool. Going down to the bottom, you feel the quasi-static pressures on your ears. Next, fall flatly from the high diving-board: you will never forget the difference again.)

A first approximation, see Figure 7-19 and TAW, 2000, is:

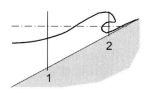 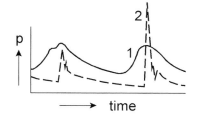 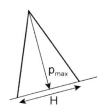

Figure 7-19 Wave impact on slope

$$p_{\text{max 50\%}} \approx 8\,\rho_w\,g\,H_s\tan\alpha \qquad p_{\text{max 0.1\%}} \approx 16\,\rho_w\,g\,H_s\tan\alpha \qquad (7.19)$$

in which $p_{\text{max 0.1\%}}$ is the maximum pressure exceeded by 1 in 1000 waves. This expression gives values several times higher than follows from the quasi-static pressures in the wave itself. The shape of the impact pressure distribution is assumed to be a triangle with H as base length.

7.5 Load reduction

There are many ways to reduce waves. The best wave reductor is probably a shallow foreshore where waves loose energy due to bottom friction and breaking, see section 0. In this section, the starting point is the wave in front of the considered structure. The two principal mechanisms are *reflection* and *absorption.* Reflection is total when an incoming wave is returned and a standing waves occurs, see appendix 7.7.1. In that case, the wave *transmission* is zero. Absorption can be achieved in many ways. The first kind of absorption occurs when the water in the waves has to flow through and around many construction parts: stones, piles, bars, trunks and roots etc. The same occurs at the edges of dams or plates which are meant as wave reductors. The second kind of absorption is the internal work in structures: floating mattresses, bending reed etc. In general, the relation between incoming, transmitted, reflected and absorbed wave energy is:

$$\frac{1}{8}\rho g H_I^2 = \frac{1}{8}\rho g H_T^2 + \frac{1}{8}\rho g H_R^2 + \text{absorption} \qquad (7.20)$$

The effectiveness of a wave reducing device is often expressed by means of the transmission coefficient:

$$K_T = \sqrt{F_T / F_I} = H_T / H_I \approx \left(H_T\right)_{1/3}\big/\left(H_I\right)_{1/3} \qquad (7.21)$$

in which F_T and F_I is the energy flux of respectively the transmitted and the incoming wave. For irregular waves, the expression with significant wave heights is a first approximation. A short review of wave reductors will now follow.

Pile screens

A wall that consists of vertical piles can act as a wave reductor if there is little space between the piles. The linear wave theory can be used, assuming that the transmitted energy is proportional to the relative spacing between the piles, see Figure 7-20a.

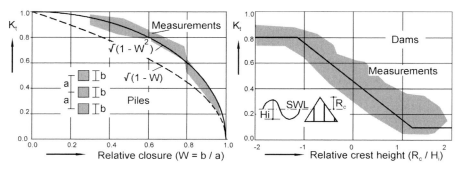

Figure 7-20 Wave transmission through pile screen and over dam

In that case, with $E = 1/8 \rho g H^2$:

$$H_T = \sqrt{(1 - W) H_I^2} \quad \rightarrow \quad K_T = \sqrt{(1 - W)} \qquad (7.22)$$

Figure 7-20a shows experimental results (Grune/Kohlhase, 1974). Equation (7.22) underestimates the transmission, probably because the diffraction of the reflected energy is greater than the energy losses in the wakes around the piles (both are neglected in (7.19)). An equally simple and satisfying empirical expression is:

$$K_T = \sqrt{1 - W^2} \qquad (7.23)$$

Dams

A very common type of wave reductor is a dam or reef with a crest around SWL. Now, the linear wave theory leads to a solution that makes no sense. In the *small-amplitude* wave theory, energy transmission is zero when the crest of a dam is equal to the water level, see Figure 7-20b. In practice, $K_T \approx 0.5$ is found for that case, indicating that about 25% of the wave energy is transmitted. Figure 7-20b shows the results of a great number of measurements, relating the wave transmission to the relative crest height, R_C / H_{Si} (R_C measured from the still water level). When $R_C / H_{Si} \approx$ 1, considerable transmission is still possible. These points in the figure represent small-amplitude waves which easily penetrate through the large elements on the top of a porous dam. For realistic waves in design conditions, $K_T \approx 0.1$ for R_c / H_i as indicated in the figure. For more information, see Van der Meer/d'Angremond, 1991.

Floating breakwaters

A floating ship, high and deep, can effectively block wave energy. Here, only low structures are considered. There are many types of floating breakwaters: rigid, flexible, porous etc. They have in common that they are only effective when the length in the wave direction is greater than the wavelength. The mooring forces on the anchors are a problem, as they may become extremely high, see Van der Linden, 1985. Floating breakwaters are probably only of practical use to reduce waves temporarily, e.g. during construction. Figure 7-21 gives a theoretical curve and measurements for a floating slab. λ is the length of the protection, L the wave length.

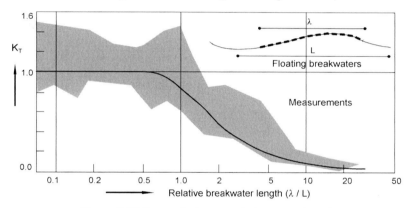

Figure 7-21 Transmission vs. relative wave length

7.6 **Summary**

For the pressures and velocities in non-breaking, short waves, the linear wave theory, see appendix 7.7.1, is a valuable and comprehensive tool to get an idea of orders of magnitude and of the influence of parameters such as wave height, period, water depth etc. For the description of an irregular wave field, H_s and T_p are mostly used when working with single parameters instead of a whole spectrum.

To determine the shear stress under non-breaking waves, section 7.2.2 should be consulted (also for a combination of current and waves). Wave heights are limited due to breaking. A general limit for individual waves is given by Miche:

$$H_b = 0.142 \, L \, \tanh\left(\frac{2\pi}{L}h\right)$$

For irregular waves, the limits are about $H_s/L_0 \approx 0.05$ in deep water and $H_s/h \approx 0.5$ in shallow water.

In breaking waves, the character of the flow in the waves changes completely and becomes highly turbulent. Unfortunately, until now, little quantative information regarding turbulence in the surf zone is available, so, most relations are based on empirical data. In many of these empirical relations, the surf similarity parameter, ξ, plays an important role:

$$\xi = \frac{\tan\alpha}{\sqrt{H/L_0}}$$

Parameters such as breaking type, breaker depth, reflection, wave run-up and run-down are related to ξ.

Wave load reduction is possible with various devices e.g. dams, screens, floating breakwaters etc. For an effective wave reduction, these structures must have large dimensions compared with water depth or wavelength. Their costs should always be in balance with their effectiveness, otherwise dimensioning the protection for the full load can be more cost effective.

7.7 Appendices

7.7.1 Linear wave theory

Progressive waves

Starting point again are the Navier-Stokes equations, neglecting boundary effects. Outside the boundary layer viscosity can be neglected and the flow can be considered irrotational. In that case there are no Reynolds-stresses and a velocity potential, ϕ, can be defined (limited here to two dimensions):

$$u = \frac{\partial\phi}{\partial x} \qquad w = \frac{\partial\phi}{\partial z} \tag{7.24}$$

The basic equations (not shown here) can be solved when a surface profile is assumed (see also Figure 7-22a) and the boundary conditions at the bottom and the surface are taken into account:

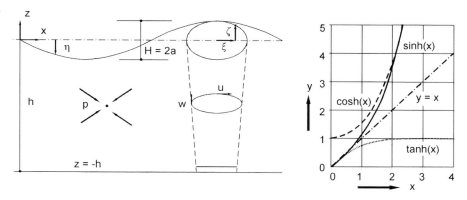

Figure 7-22 Definitions and behaviour hyperbolic functions

$$\eta = \frac{H}{2}\sin\left(\frac{2\pi t}{T} - \frac{2\pi x}{L}\right) = a\sin(\omega t - k x) = a\sin\theta \qquad (7.25)$$

where the amplitude, a, is half the wave height, H. The velocity potential becomes:

$$\phi(x,z,t) = \frac{\omega a}{k}\frac{\cosh(h+z)}{\sinh(kh)}\cos(\omega t - k x) \qquad (7.26)$$

Note: $z = 0$ represents the still water level and z is positive upwards.

All other quantities in the wave, as a function of x, z and t, can be derived from this expression for the velocity potential ϕ. Table 7.1 presents the various parameters. In the table, the extremes for deep and shallow water follow from the values for transitional water depth, since for deep water: $kh \rightarrow \infty$, so: tanh$kh = 1$, sinh$kh = \infty$ and cosh$kh = \infty$ and for shallow water: $kh \approx 0$, so: tanh $kh = kh$, sinh $kh = kh$ and cosh $kh = 1$, see Figure 7-22b.

Standing waves

Vertical walls act like a mirror: the incoming waves are reflected, the result is superposition of two progressive waves with celerity $+c$ and $-c$. This leads to a standstill of the wave, hence standing wave. The wave height doubles.

Figure 7-23 shows the pattern of a standing wave (or the often used French word "clapotis"). Note that the orbits of the particle movement have degenerated into straight lines. At the antinodes there is only a vertical movement and at the nodes movement is horizontal. The wave profile for a standing wave is given by:

$$\eta = a_s \cos kx \ \sin \omega t = 2 \ a_i \cos kx \ \sin \omega t \qquad (7.27)$$

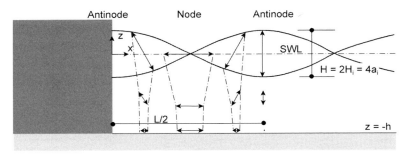

Figure 7-23 Standing wave

in which a_i is the amplitude of the incoming wave. Other characteristic expressions for standing waves are:

$$u = \omega\, a_s \frac{\cosh k(h+z)}{\sinh kh} \sin kx \cos \omega t \qquad w = \omega\, a_s \frac{\sinh k(h+z)}{\cosh kh} \cos kx \sin \omega t$$

$$p = -\rho\, g\, z + \rho\, g\, a_s \frac{\cosh k(h+z)}{\cosh kh} \cos kx \sin \omega t \qquad\qquad (7.28)$$

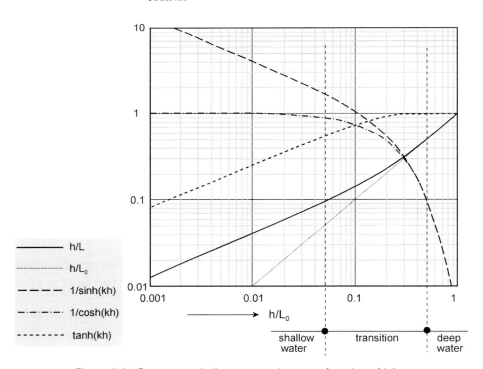

Figure 7-24 Parameters in linear wave theory as function of h/L_0

When the wall slopes backwards, the vertical motion of the water at the interface increases: the maximum water level exceeds H_i (which is the maximum level above

SWL in Figure 7-23). The standing wave pattern remains till the waves break (depending on the slope and the wave steepness).

Relative depth Characteristics	Shallow Water $\dfrac{h}{L} < \dfrac{1}{20}$	Transitional water depth $\dfrac{1}{20} < \dfrac{h}{L} < \dfrac{1}{2}$	Deep Water $\dfrac{h}{L} > \dfrac{1}{2}$
Wave Celerity	$c = \dfrac{L}{T} = \sqrt{g h}$	$c = \dfrac{L}{T} = \dfrac{g T}{2 \pi} \tanh kh$	$c = c_0 = \dfrac{L}{T} = \dfrac{g T}{2 \pi}$
Wave Length	$L = T\sqrt{g h}$	$L = \dfrac{g T^2}{2 \pi} \tanh kh$	$L = L_0 = \dfrac{g T^2}{2 \pi}$
Group Velocity	$c_g = c = \sqrt{g h}$	$c_g = n c = \dfrac{1}{2}\left[1 + \dfrac{2 k h}{\sinh 2 k h} \right] * c$	$c_g = \dfrac{1}{2} c_0 = \dfrac{g T}{4 \pi}$
Energy Flux (per m width)	$F = E c_g =$ $= \dfrac{1}{2}\rho g a^2 \sqrt{g h}$	$F = E c_g = \dfrac{1}{2}\rho g a^2 n c$	$F = \dfrac{T}{8 \pi}\rho g^2 a^2$
Particle velocity			
Horizontal	$u = a\sqrt{\dfrac{g}{h}}\sin\theta$	$u = \omega a \dfrac{\cosh k(h+z)}{\sinh kh}\sin\theta$	$u = \omega a e^{kz}\sin\theta$
Vertical	$w = \omega a\left(1 + \dfrac{z}{h}\right)\cos\theta$	$w = \omega a \dfrac{\sinh k(h+z)}{\sinh kh}\cos\theta$	$w = \omega a e^{kz}\sin\theta$
Particle displacement			
Horizontal	$\xi = -\dfrac{a}{\omega}\sqrt{\dfrac{g}{h}}\cos\theta$	$\xi = -a\dfrac{\cosh k(h+z)}{\sinh kh}\cos\theta$	$\xi = -a e^{kz}\cos\theta$
Vertical		$\zeta = a\dfrac{\sinh k(h+z)}{\sinh kh}\sin\theta$	$\zeta = a e^{kz}\sin\theta$
Subsurface pressure	$p = -\rho g z +$ $+ \rho g a \sin\theta$	$p = -\rho g z +$ $+\rho g a \dfrac{\cosh k(h+z)}{\cosh kh}\sin\theta$	$p = -\rho g z + \rho g a e^{kz}\sin\theta$

Table 7.1 Summary of Wave Characteristics

$$a = \frac{H}{2} \qquad \omega = \frac{2\pi}{T} \qquad k = \frac{2\pi}{L} \qquad \theta = \omega t - k x$$

7.7.2 Wave statistics

Wave spectrum

Wind is a turbulent flow with irregular velocity variations. So, when wind blows over a water surface, the resulting waves will be irregular too, see Figure 7-25a. An irregular wave field is best described with a so-called variance- or energy-density spectrum, see Figure 7-25b.

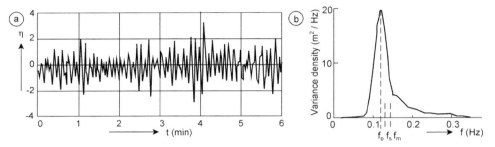

Figure 7-25 Wave registration and wave spectrum North Sea

A variance-density spectrum can be used for the statistical description of any fluctuating signal, e.g. the deviation of a sailing ship from its course, the turbulent velocity fluctuation or the surface elevation at one location, as shown in Figure 7-25a. Without bothering about the mathematical backgrounds, it is assumed that at any moment, the surface elevation, η, can be seen as the sum of an infinite number of simple cosine waves with a random phase α:

$$\underline{\eta}_t = \sum a_i \cos\left(2\pi f_i t + \underline{\alpha}_i\right) \tag{7.29}$$

Every component i has a distinct value for the amplitude a and the frequency f. $\underline{\alpha}_i$ is the only random variable. A single component i is related to the spectrum via:

$$\frac{1}{2} a_i^2 \approx E(f_i)\, \Delta f_i \tag{7.30}$$

where $E(f_i)$ is the variance density and $E(f_i)\Delta f_i$ the variance. The unit for frequency is Hz and for variance m², hence the unit for variance density is m²/Hz which is somewhat peculiar at first sight. The physical meaning of a spectrum is clear when one realizes that the variance is identical to the energy in waves ($= \frac{1}{2}\rho g a^2$) reduced by a factor ρg. The physical interpretation of a wave spectrum is thus *the distribution of energy over the various wave frequencies*, hence the name energy-density spectrum.

Figure 7-26 shows some examples of registrations of a fluctuating variable with the accompanying variance-density spectra. The first is a simple sine, which has only one period. The variance density is infinitely high and narrow, see equation (7.30). For pure noise, the spectrum is very wide. A registration like the one in Figure 7-25, gives a spectrum with a clear peak, where the variance density has a maximum.

In section 7.1 "fresh" waves were called *sea*, while *swell* was said to consist of "old" waves. The last example in Figure 7-26 is for a situation where both occur, leading to a so-called double-peak spectrum. Both peak frequencies are different and the swell part of the spectrum is usually rather narrow.

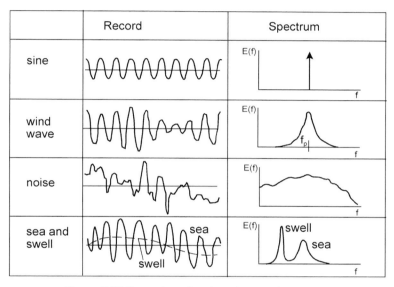

Figure 7-26 Examples of registrations and spectrum

Wave heights

The total area of the spectrum is equal to the total variance:

$$m_0 = \int_0^\infty E(f)df = \text{variance}(\eta_t) = \sigma_\eta^2 \tag{7.31}$$

Multiplied with ρg it represents the total energy of the waves. Of course there is a relation between the variance and some characteristic wave height, hence between m_0 and the wave height. But first an unambiguous definition of wave height is necessary, for what is actually a wave in Figure 7-27a? By agreement, the period of an individual wave is defined as the time between two downward zero-crossings, T_z, and the wave height H is the highest crest minus the lowest trough between these zero-crossings.

Preferably, the wave heights are characterized by H_{rms}, as it is representative for the wave energy and directly related to σ, see also turbulence in chapter 2. It can be theoretically derived that $H_{rms} = 2\sqrt{2}m_0$. There is, however, a history with regard to characterizing a wave field.

Important elements in this history are visual wave observations and the Second World War. Until some decades ago, visual observations were the only way to obtain wave information and even today much information concerning ocean waves is gathered from observations by crew members on navy and merchant ships. A visually estimated wave height, H_{visual}, compared with recordings, appears to be equal to the average height of the 1/3 part of the highest waves, $H_{1/3}$. This wave height is also called the "significant" wave height, H_s, but signifying for what? In the second World War, many beach landings took place in Europe and the Pacific. The visually observed wave height was significant for the commander of an amphibian craft who had to decide whether to go on with the operation or not. Related to (narrow) spectra, $H_{1/3} \approx 4\sqrt{m_0}$, in modern wave measurements and computations, the significant wave height is therefore often indicated as H_{m0}. On shallow foreshores the relation between H_{rms} and m_0 is no longer constant, so, the only good measure for wave heights would be $\sqrt{m_0}$. But the "significant" wave height has become customary and it is hard to replace.

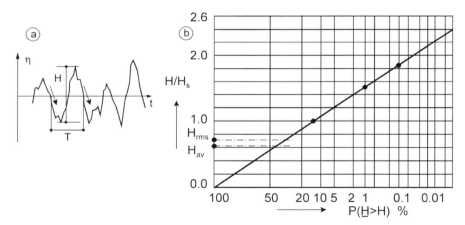

Figure 7-27 Wave definitions and wave height distribution

From the spectrum a wave height distribution can also be derived. For a (theoretically narrow) spectrum, the distribution of wave heights can be described with a Rayleigh-distribution, see Figure 7-27b:

$$P\{\underline{H} > H\} = \exp\left[-\left(\frac{H}{H_{rms}}\right)^2\right] = \exp\left[-2\left(\frac{H}{H_s}\right)^2\right] \qquad (7.32)$$

both expressions are equivalent, since $H_s = H_{rms}\sqrt{2}$.

This distribution is valid in water that is deeper than $\sim 3H_s$. In shallow water the highest waves break and the distribution deviates. From equation (7.32) follows directly that H_s is exceeded by 13,5 % of the waves and H_{rms} by 37 % of the waves. Extreme values are $H_{1\%} \approx 1,5H_s$ and $H_{0.1\%} \approx 1.85H_s$.

Attention is drawn to the fact that *five* indications for the characteristic wave height have already been used, all of them common in wave literature:

$$H_s \equiv H_{visual} \equiv H_{1/3} \equiv H_{13.5\%} \equiv H_{m0} \approx 4\sqrt{m_0} \qquad (7.33)$$

Note: A significant wave height, H_s, with a Rayleigh-distribution is used to characterize waves in a record of about half an hour, the so-called *short-term wave height distribution*, representative for the state of the sea at a certain moment. For design purposes, a *long-term wave height distribution* is needed to estimate the probability of a certain sea state. Many different values for H_s over a long period are used to form a long term distribution, see chapter 10.

Wave periods

Wave periods in an irregular wave field are defined as shown in Figure 7-27a. Many attempts have been made to relate wind wave heights and periods. The relation between individual height and period is, however, weak. To define a characteristic wave period, there are, again, some possibilities. The average value can be used. It has the advantage of simplicity: x waves during y seconds gives $T_m = y/x$. A better measure is T_p, the so-called peak period, where the energy density is maximal, see Figure 7-25b. T_s or $T_{1/3}$ (defined similarly to H_s: the average period of the 1/3 highest waves) is also still widely used. $T_s \approx 0.9T_p$ can be used, rather independent of the spectrum shape. The relation between the mean period and the peak period depends on the spectrum shape and values for $T_m/T_p = 0.7 - 0.9$ have been found. As a first approximation: $T_m \approx 0.8T_p$ can be used.

For "fresh" wind waves (sea), a wave steepness of $s \approx 0.05$ is often found. From this, a relation between H_s and T_p can be derived: $H_s/1.56T_p^2 = 0.05 \rightarrow T_p \approx 3.6\sqrt{H_s}$, which can serve as a first estimate when only wave height data are available. Of course, this must be checked by measurements. In swell, a much smaller wave steepness is found, e.g. $s = 0.01$ giving $T_p \approx 8\sqrt{H_s}$.

Shallow water

In shallow water, waves experience many alterations, which will also effect the wave statistics. For the wave height distribution, Battjes/Groenendijk, 2000, made a so-called composed Weibull-distribution. Below a transitional value of the wave height (H_{tr}), the Rayleigh distribution remains valid. Above this value the exponent in the distribution function has a different value (≈ 3.6):

$$\Pr\{\underline{H} \le H\} = \begin{cases} F_1(H) = 1 - \exp\left[-\left(\dfrac{H}{H_1}\right)^2\right] & H \le H_{tr} \\[3mm] F_2(H) = 1 - \exp\left[-\left(\dfrac{H}{H_2}\right)^{3.6}\right] & H > H_{tr} \end{cases} \tag{7.34}$$

H_{tr} at a certain water depth is found from the spectral area, m_0 (known from computations like SWAN, see section 0), the foreshore slope angle and the waterdepth. H_{rms} is also a function of m_0 and the water depth. Figure 7-28a shows various wave heights as a function of H_{tr}, all made dimensionless with H_{rms}.

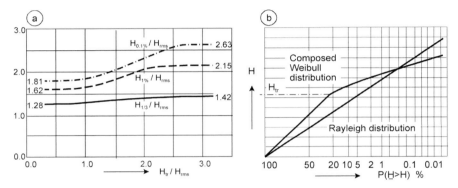

Figure 7-28 Wave height distribution in shallow water (Battjes/Groenendijk,2000)

Figure 7-28a shows that the ratios between various wave heights and the H_{rms} change when the waves enter shallow water. For very shallow water, $H_{1\%} \approx 1.6 H_{rms}$ instead of $2.15 H_{rms}$ as would follow from equation (7.32). These corrections for the Rayleigh distribution can be used in equations for wave run-up (see section 7.4.2) or stability (see chapter 8). For more detail see Battjes/Groenendijk, 2000.

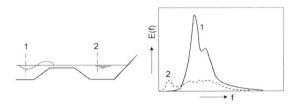

Figure 7-29 Wave spectra across a shallow bar

The spectrum shape also changes when waves travel from deep to shallow water. Figure 7-29 shows measured spectra of waves travelling across a shallow bar. The spectrum area decreases considerably, the graph flattens while a new small peak for low frequencies emerges. On top of the bar, the wave height distribution deviates from the Rayleigh distribution, but behind the bar it recovers again!

7.7.3 Wave generation

Knowledge of wave generation by wind is still mainly empirical. A simple and important method is the Sverdrup-Munk-Brettschneider method, see e.g. SPM, 1984. The most important parameter in wave generation is the wind velocity, u_w. When there are no limitations in water depth or fetch, the relation for waves are simply:

$$\frac{gH_s}{u_w^2} = 0.283 \quad \text{and} \quad \frac{gT_s}{2\pi u_w} = 1.2 \tag{7.35}$$

in which H_s is the significant wave height and T_s the significant wave period (the average period of the 1/3 highest waves). u is the wind velocity (usually 10 m above the water surface and averaged over 1 hour).

Limiting parameters are the fetch (the length over which the wind is effective in generating waves), the water depth and the storm duration. With the water depth and the fetch, F, as limiting factors, equation (7.35) becomes:

$$\frac{gH_s}{u_w^2} = 0.283 \tanh\left[0.578\left(\frac{gh}{u_w^2}\right)^{0.75} \right] \tanh\left[\frac{0.0125\left(\dfrac{gF}{u_w^2}\right)^{0.42}}{\tanh\left[0.578\left(\dfrac{gh}{u_w^2}\right)^{0.75} \right]} \right]$$

$$\frac{gT_s}{2\pi u_w} = 1.20 \tanh\left[0.833\left(\frac{gh}{u_w^2}\right)^{0.375} \right] \tanh\left[\frac{0.077\left(\dfrac{gF}{u_w^2}\right)^{0.25}}{\tanh\left[0.833\left(\dfrac{gh}{u_w^2}\right)^{0.375} \right]} \right] \tag{7.36}$$

This may look rather grim, but including a hyperbolic tangent is only a convenient way to include the influence of a parameter that vanishes automatically above some value. For instance, the limiting influence of the depth on the wave height vanishes when $gh/u^2 > 5$-10; above this depth, $\tanh \approx 1$. Figure 7-30 gives some examples of calculated wave heights for fetches of 10 and 100 km and water depths of 5 and 50 m. "Waves unlimited" represents equation (7.35), while the other curves have been drawn using equation (7.36). For very low wind velocities ($u < 1.5$ m/s), the applied depth and fetch do not limit the wave height. For $1.5 < u < 3$ m/s a fetch of 10 km is limiting but a water depth of 5 m is not. For winds of upto 15-20 m/s a fetch of 10 km is more limiting than a depth of 5 m. For stronger winds, a depth of 5 m is more limiting than a fetch of 10 km.

Figure 7-31 shows wave periods for the same examples of fetch and water depth. For the whole range of wind velocities, a longer fetch gives higher periods. For a sea (F

= 100 km, h = 50 m), typical values are $T_s \approx$ 8-10 s and for an estuary (F = 10 km, h = 5 m), $T_s \approx$ 4-5 s.

Limitation by storm duration is treated as an equivalent fetch in equation (7.36), see SPM, 1984. Figure 7-32 gives an example for two fetches and two wind velocities.

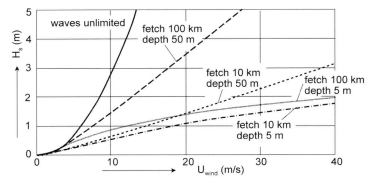

Figure 7-30 Wave height as a function of wind, depth and fetch

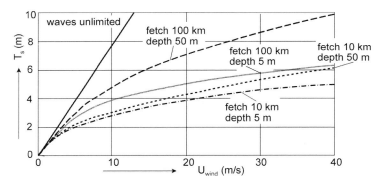

Figure 7-31 Wave period as a function of wind, depth and fetch

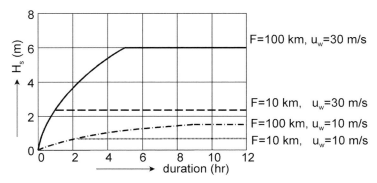

Figure 7-32 Wave height as function of duration, fetch and wind velocity

8 WAVES
Erosion and stability

Dike with basalt columns and concrete blocks (Zealand),
courtesy Rijkswaterstaat

8.1 Erosion

Before considering the necessity of protecting a bank, a shore or a bottom against waves, one should have an idea of the possible erosion. In this book, erosion due to waves is limited to local erosion. For banks and shores this will be treated on the basis of beach and dune erosion research; for bottom erosion, breakwater investigations will be used. Erosion due to long-shore or cross-shore sediment transport is part of coastal morphology and will not be treated.

8.1.1 Erosion of slopes

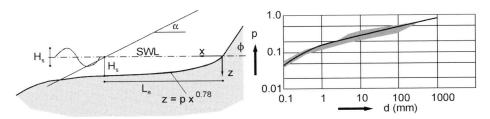

Figure 8-1 Erosion of slope by waves

Waves acting on an unprotected slope cause a step profile. This is the natural equilibrium profile for all slopes composed of loose material (see e.g. Vellinga, 1986). Theoretically, monochromatic waves lead to a parabolic beach with $z \propto x^{2/3}$. For practical situations, Vellinga proposes:

$$z = 0.39 w^{0.44} x^{0.78} = p x^{0.78} \tag{8.1}$$

in which w is the fall velocity of the particles; for the definition of z and x, see Figure 8-1. p is expressed in $m^{0.22}$!

This is an equilibrium profile; development in time is not considered. The erosion depth below SWL is about H_s. From equation (8.1) the intrusion length of the waves in the profile is then found by:

$$L_e = p^{-1.28} H_s^{1.28} \tag{8.2}$$

Above the still water level the slope is assumed to be equal to the angle of repose. Figure 8-1 gives a relation for the value of p in the profile formula. The given relations can serve only as a first indication and should be applied for wave attack well beyond the limit of stability, see section 8.3.2.

Example 8-1

A sand slope, with grain size 0.5 mm, is temporarily unprotected. What is the erosion length if this slope is attacked by waves with H_s = 1.6 m?
From Figure 8-1 we find for grains of 0.5 mm, $p \approx 0.1$. From equation 8.2 we then find: L_e = $0.1^{-1.28} * 1.6^{1.28} \approx 35$ m.

8.1.2 Bottom erosion

Bottom scour due to waves can be important in front of walls where a standing wave pattern is possible. From chapter 7 we know that the highest velocities occur in the nodes of such a wave and indeed, for fine sediments, the maximum scour is found there. For coarse sediments, however, the maximum scour is found between the node and antinode, probably due to a different eddy and ripple pattern. The standing wave pattern for irregular waves is less distinct than for regular waves. This is also found in the scouring pattern, which decreases strongly with the distance from the wall, see Figure 8-2a and b.

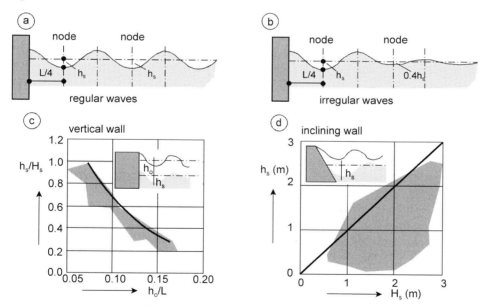

Figure 8-2 Bottom erosion in front of wall due to waves

The depth of the scour hole can be estimated roughly with Figure 8-2c (from Xie Shi-Leng, 1981). For shallow water (small h_0/L), $h_s \approx H_s$ can serve as a first guess. For deeper water, the scour is considerably less.

When the wall is not vertical but inclining, the maximum scour will occur at the foot of the wall, due to the backflow from the sloping wall, see Figure 8-2d. The erosion is roughly proportional to the reflection coefficient, so a vertical wall gives maximum values. As a first guess for all cases, $h_s \approx H_s$ can be used as an upper limit.

8.2 Stability general

Several stable protections against waves are possible. In this book they will be divided into three main categories, with the following keywords:

1   Loose grains, rip-rap, rock, open, permeable.
2   Coherent, semi-permeable, placed block revetment.
3   Impervious, asphalt, concrete.

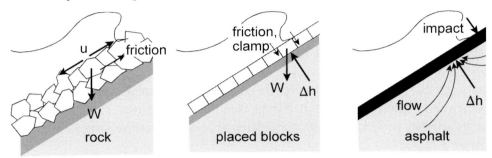

Figure 8-3 Three main types of protection against waves

The loads on these three archetypes of protection are the same, but completely different mechanisms determine the dimensions of the protection layers. The differences lie in the transfer functions from the external to the internal load and from the internal load to the response of the structure (strength). This can be illustrated by considering the so-called leakage length, Λ.

Figure 8-4 Definition leakage length

Consider the flow in Figure 8-4a. In the filter layer, the flow is assumed to be parallel to the interface, while in the top layer it is perpendicular. When there is no filter layer (Figure 8-4b), some assumption has to be made for the thickness where parallel flow can be expected. These assumptions may not be completely true for all revetment types in Figure 8-3, but the concept is used for illustration purposes only.

Starting with these flow directions, the flow resistance of each layer can be determined. The leakage length, Λ, is now defined as the length of protection in which the flow resistance through top layer and filter layer are the same, see Figure 8-4a. This definition can be expressed by (see also De Groot et al., 1988):

$$\frac{d_T}{k_T \Lambda} = \frac{\Lambda}{k_F d_F} \quad \rightarrow \quad \Lambda = \sqrt{\frac{k_F d_F d_T}{k_T}} \tag{8.3}$$

in which k_F and k_T are the permeability of filter and top layer, respectively, and d_F and d_T the thickness. So, a large leakage length means a relatively impermeable top layer compared with the filter layer.

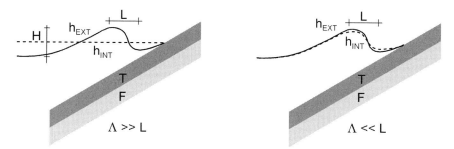

Figure 8-5 Influence leakage length

Λ is a measure for the exchange between external and internal loads, see Figure 8-5. L is some characteristic length of the external load (not necessarily the wave length, but e.g. the length of the wave front). When $\Lambda \gg L$, there is little exchange between the outside and the inside and the piezometric level inside the filter cannot follow the level outside, leading to a large head-difference. When $\Lambda \ll L$, the exchange is easy and there is hardly any head-difference: the pressure gradients in the load are only determined by L.

Table 8-1 Typical values for Λ and L for different protection types

Parameter	"Rock"	"Blocks"	"Asphalt"
d_T (m)	0.5	0.25	0.25
d_F (m)	0.25	0.2	2
k_T (m/s)	0.5	0.001	"0"
K_F (m/s)	0.1	0.05	0.0001
Λ (m)	0.15	1.5	"∞"
L (m)	1-2	1-2	1-2

Table 8-1 gives some typical values for the three protection types of Figure 8-3. For a protection of loose rock $\Lambda \ll L$, for a placed block revetment $\Lambda \approx L$ and for an asphalt protection $\Lambda \gg L$. This may explain why these protections behave completely differently, while the strength is also different.

In the case of loose rock, $\Lambda \ll L$ and there is no head difference across the top layer. L determines the pressure gradients, leading to uprush and downflow velocities, which in turn cause (drag)forces on the individual stones. **Note**: Obviously, the assumption of flow perpendicular to the top layer (Figure 8-4) is no longer valid.

The strength of this revetment is a result of friction between the stones, hence of their weight, see Figure 8-3a. Porous flow has some influence on the stability, but this is of minor importance compared with the external flow forces. Thus, the stability of a slope with a rock top-layer is governed by the flow caused by the waves around the stones, analogous to the stability in flow as described in chapter 3. Section 8.3 deals with stability of loose grains. A filter is usually necessary for stability, since the top layer is very open and the underlying grains can be washed away without the filter.

For an impervious protection, $\Lambda >> L$ and the head-difference across the top layer is approximately $H/2$, see Figure 8-5. However, the transfer function from the wave load to the response of the asphalt layer is such that this load is usually negligible. The head difference causes a force which will try to lift the protection layer. The load is local while the layer is coherent and heavy and, as a consequence, difficult to move. Moreover, the cavity between protection and subsoil has to be filled with water within part of a wave period, for which the porous flow is usually not fast enough. Wave impact can cause damage, if it occurs very frequently, see section 8.4.2. The stability of the layer is governed by the pressure differences due to variations in the waterlevel as described in chapter 5. A filter for sandtightness is unnecessary. A filter for drainage can be applied when the phreatic level inside the slope, due to tides or surges, will cause high pressures under the revetment.

In the case of placed blocks, $\Lambda \approx L$. In the cases where $\Lambda >> L$ or $\Lambda << L$, it is possible to focus on a dominant mechanism, but when $\Lambda \approx L$, the situation is more complex. The surface is usually too smooth for the water to exert drag forces. The load comes from inside, caused by the pressure difference under and above the blocks. The strength is the result of the cooperation between the blocks, either by friction or by clamping. Now the porous flow is paramount for the stability and the filter plays a crucial and complex role, see section 8.4.1.

8.3 Stability of loose grains

As a logical continuation of chapter 3 (Stability in flow), we will start to examine the stability of loose material under waves and pretend that the loading situation and stability under waves are not different from those in flow. After all, the water in a non breaking wave, is just flowing to and fro, albeit with a higher shear stress. In breaking waves, we can not expect this approach to be succesful in a quantitative way, but it is always instructive to try and compare the results.

8.3.1 Stability in non-breaking waves

Chapter 7 gave Jonsson's expression for the shear stress under a wave and with this shear stress we could use the stability relation for flow, e.g. the one by Shields, which relates the dimensionless shear stress to the grain diameter. The results of this approach deviate from the original Shields' results, probably due to different boundary development in an oscillating flow. Figure 8-6a shows stability measurements in oscillating flow presented by various authors, summarized by Sleath, 1978.

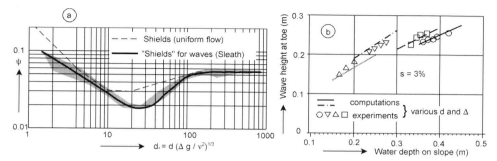

Figure 8-6 Modified Shields-diagram for waves and stability in non-breaking waves

Sleath uses the same dimensionless shear stress as Shields does (ψ), and the same dimensionless grain diameter as Van Rijn for flow, see chapter 3. So, the results can be compared directly, see Figure 8-6a. For large d_* (turbulent boundary layer), Sleath found a value for ψ to be 0.055, which is almost the same value as Shields. In stationary flow, this value implies a relatively large amount of stones in motion. How logical is it that the same value is found in oscillating flow? Two factors can play a role. The criterion for the threshold of movement in oscillating flow is not quite clear, but a potentially unstable stone in oscillating flow can not move a long distance, hence the subjective threshold will probably be assumed rather high compared with the threshold in steady flow. Another aspect may be the phenomenon we have seen in an accelerating steady flow, which is that ψ increases as τ increases. The graph by Sleath is based on several investigations. One of them is by Rance/Warren,1968. Their results can be summarized by the equation (see Schiereck et al.,1996):

$$\frac{a_b}{T^2 \Delta g} = 0.025 \left(\frac{a_b}{d_{50}} \right) \quad \rightarrow \quad d_{n50} = 2.15 \frac{\hat{u}_b^{2.5}}{\sqrt{T}(\Delta g)^{1.5}} \tag{8.4}$$

in which a_b is the orbital stroke at the bottom and $\hat{u}_b$ the maximum orbital velocity, where d_{n50} is assumed to equal $0.84 d_{50}$. Equation (8.4) has the advantage that no iteration is necessary between the bottom roughness and the shear stress as is the case in the Jonsson-Sleath approach.

Figure 8-6b shows a comparison between stability calculations with orbital velocities from linear wave theory, Jonsson's friction factor and $\psi = 0.055$ for stones on a slope 1:25 and laboratory tests with regular non-breaking waves, see Schiereck et al, 1994.

The computed wave heights at the toe of the slope, for which the material starts to move, correspond quite good with the measured values. Equation (8.4) gives similar results.

8.3.2 Stability in breaking waves

Most research on the stability of stones on a slope has been carried out in the field of breakwaters, which is very much related to slope protections and revetments, but it is not the same. An important difference in the stability is the porosity of the entire structure. Breakwaters usually have a porous core, while in dike revetments the core is made of clay or sand. This has a significant influence on the stability of the protecting armour layer as we will see later on in this section. We will start, however, with basic principles and shortcomings and extensions will be discussed later on.

Encouraged by the succesful results of the modified Shields-curve for waves, the next step is using a slope correction factor and taking the breaking of the waves into account. We will use the same slope factor as for flow in the slope direction (see section 3.3. For the 1:25 slope in the previous section, this factor ≈ 1). For the velocity in a breaking wave, no reliable expression is available. As a first guess, we will assume that the velocity in a breaking or broken wave on a slope is proportional to the celerity in shallow water with the wave height as a representative measure for the waterdepth: $u \propto \sqrt{gH}$. Following the same reasoning as for stability in flow, see chapter 3, we find:

$$\underbrace{\rho_w g H d^2}_{\text{"drag" force}} \quad \propto \quad \underbrace{(\rho_s - \rho_w)g d^3}_{\text{resisting force}} \quad \underbrace{(\tan\phi\cos\alpha \pm \sin\alpha)}_{\text{slope correction}} \tag{8.5}$$

Note: + and - in the slope correction are for uprush and backwash, respectively. By raising all terms to the third power and working with the mass of the stone ($M \propto \rho_s d^3$) we find, as was already proposed by Iribarren, 1938:

$$M \propto \frac{\rho_s H^3}{\Delta^3 (\tan\phi\cos\alpha \pm \sin\alpha)^3} \tag{8.6}$$

Many tests were performed (mostly by Hudson,1953) to find the constants of proportionality in equation (8.6). For practical reasons, Hudson finally proposed another formula:

$$M = \frac{\rho_s H_{sc}^{\ 3}}{K_D \Delta^3 \cot\alpha} \quad \left(\text{or: } \frac{H_{sc}}{\Delta d} = \sqrt[3]{K_D \cot\alpha}\right) \tag{8.7}$$

Note: The use of $H_{sc}/\Delta d$ as a stability parameter is convenient and it resembles the stability parameter in flow situations, $u_c^2/\Delta d$. The subscript c in the stability parameter (to discern stability from mobility) is not always used consequently.

The slope correction in Irribarren's formula is now reduced to cot α. This means that the validity of Hudson's formula is limited, because cot α is insufficient to describe friction and equilibrium on a slope: when $\alpha = 0$, $M = 0$ and when $\alpha > \phi$, M still has a finite value, which is nonsense. The range of α for which Hudson is valid is about $1.5 < \cot \alpha < 4$. The Hudson-formula was tested for waves that did not break at the toe of the slope and did not overtop it. For other cases, extra corrections for K_D are sometimes applied. Hudson's formula is simple and is used worldwide. K_D is again a 'dustbin-factor' in which the accepted degree of damage is implicitly included. K_D has different values for different kinds of elements (3-4 for natural rock to 8-10 for artificial elements like tetrapods and tribars). The Shore Protection Manual (SPM, 1984) gives values for K_D for various circumstances. The simplicity of the Hudson-formula has its price. Some limitations of the presented formulae have already been mentioned. The most important limitations are:

Wave period

In the Irribarren-formula this parameter is absent. There are two ways in which the period influences the stability. On the one hand, the period is related to the wave-length, hence to the wave steepness and hence to the breaking pattern on the slope, which definitely plays a role. On the other hand, inertia forces on a grain may play a role, which depend on du / dt, hence on the wave period.

Permeability

The permeability of the structure must play an important role. The assumptions on which Irribarren's formula is based, only include a kind of drag force on the slope. The forces under and behind a grain are certainly partly responsible for the equilibrium. It is easy to imagine that a homogeneous mass of stones reacts differently from a cover layer of stones on an impermeable core. In the first case, a lot of wave energy is dissipated in the core, while in the latter the pressure build-up under the cover layer can be considerable.

Number of waves

All model tests in the fifties and sixties were carried out with regular waves. In those tests it appeared that the equilibrium damage-profile was reached in, say, one half hour. The wave height in the tests was then usually declared to be equivalent to H_s. It appeared from tests with wave spectra that the number of waves has some influence, which is logical, as more waves mean a greater chance of a large one occurring.

Damage level

As for stones in flow, the threshold of motion for stones in waves is not always clear. The K_D-values in the Hudson-formula are supposed to be valid for 5% damage, but the definition of damage is not very clear.

All these objections have led to new research activities, which have increased following some failures of breakwaters. In a period of about ten years the recommended coefficients for the Hudson formula in the Shore Protection Manual for breakwater design have increased by about 200%. In The Netherlands, extensive model tests were carried out to overcome the limitations of the Hudson-formula. The results of curve-fitting the outcome of these large and small scale tests with irregular waves, finally led to the following equations, see Van der Meer,1988:

$$\frac{H_{sc}}{\Delta d_{n50}} = 6.2\, P^{0.18} \left(\frac{S}{\sqrt{N}}\right)^{0.2} \xi^{-0.5} \qquad \text{(plunging breakers)}$$

$$\frac{H_{sc}}{\Delta d_{n50}} = 1.0\, P^{-0.13} \left(\frac{S}{\sqrt{N}}\right)^{0.2} \xi^{P} \sqrt{\cot \alpha} \qquad \text{(surging breakers)}$$

(8.8)

in which: P is a measure for the permeability of the structure, S a measure for the damage and N the number of waves. The transition between the two expressions is found by equating them, giving:

$$\xi_{\text{transition}} = \left[6.2\, P^{0.31} \sqrt{\tan \alpha}\right]^{\left(\frac{1}{P+0.5}\right)} \qquad (8.9)$$

When $\xi > \xi_{\text{transition}}$ the equation for surging breakers has to be used, for $\xi < \xi_{\text{transition}}$, the equation for plunging breakers. In practice, for $\cot \alpha \geq 4$ surging waves do not exist and only the expression for plunging waves is recommended for use.

Compared with Hudson's equation, these equations are a step forward, because more parameters are included. The physical base of the Van der Meer equations is, however, still weak and in stone stability under waves there is still much to be understood. As a stability parameter Van der Meer used $H_s/\Delta d_{n50}$ (see also equation (8.7)), the wave-period appears in the surf-similarity parameter, ξ_m (m means related to the average wave period), while there is a discontinuity in the stability relations between surging and plunging breakers. In the following the influence of the parameters in equations (8.8) will be demonstrated with some computations and a comparison with the Hudson-formula. A standard case will be used:

$H_s = 2$ m	$T_m = 6$ s	$\Delta = 1.65$
$\cot \alpha = 3$	$N = 3000$	$S = 2$
$P = 0.5$	$d_{n50} = 0.6$ m (300 - 1000 kg)	$K_D = 3.5$

Wave period

The influence of the wave period is incorporated in the breaker parameter, ξ. Figure 8-7a shows the relation between the stability parameter, $H_s / \Delta d_{n50}$, and ξ (with all other parameters as in the standard case). Near $\xi = 3$, $H_s / \Delta d$ has its minimum. This is in the transition zone between surging and plunging breakers, which appears to give the most severe attack on the slope, see also chapter 7. Note that typical values of $H_s/\Delta d$ lie around 2 for both formulae. Below $\xi = 1$, the Van der Meer curve is dashed, since the empirical relation was not tested for those values, see also Figure 8-13. In Figure 8-7b the period is the independent parameter. Note that periods smaller than 4 s have not been drawn, since a maximum wave steepness ($s = H_s / L_0$) $= 0.06$ has been assumed, which, with a wave height of 2 m, leads to a minimum wave period of $\approx 3\sqrt{H_s}$. In the Hudson-formula, T does not play a role.

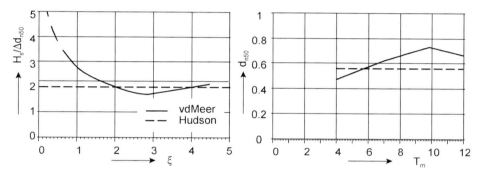

Figure 8-7 Stability parameter as a function of ξ and diameter as function of T

Permeability

A permeability parameter P has been introduced and the value for different structures has been established by curve-fitting the results. Figure 8-8a gives the values for various situations. A homogeneous structure (no core) gives $P \approx 0.6$, a rock armour-layer with a permeable core ($d_A / d_F \approx 3$): $P \approx 0.5$, an armour layer with filter ($d_A / d_F \approx 2$) on a permeable core ($d_F / d_C \approx 4$): $P \approx 0.4$ and an "impermeable" core: $P \approx 0.1$. Impermeable is again a relative notion, wave penetration in clay or even sand is almost negligible, so, in these stability relations the slope is considered impermeable. Attempts have been made to derive P from porous flow calculations, see Van Gent, 1993.

Figure 8-8b shows the influence of P on the necessary diameter. For permeable slopes, the results from Hudson or Van der Meer are almost identical, but for impermeable slopes, like dikes or revetments, the difference is considerable (0.75 m versus 0.55 m, which is a 2.5 times heavier stone).

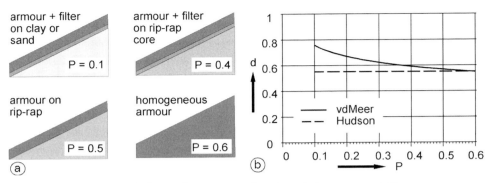

Figure 8-8 Stone diameter as a function of slope permeability

Number of waves

N is the number of waves. Figure 8-9 gives the increase of the damage with the number of waves during a test. With $N = 7500$, the damage can be considered to have reached an equilibrium. When only storms of short duration occur and intensive maintenance will be done, a smaller N can be chosen, leading to a smaller d, which is possibly cheaper. 3000 waves with an average period of 6 s, represent a storm of 5 hours.

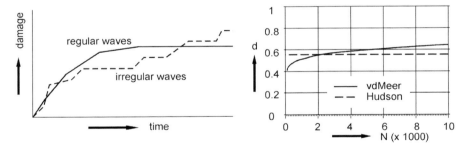

Figure 8-9 Damage versus time and stone diameter versus number of waves

Damage level

The damage level has been defined in a more manageable way, see Figure 8-10a: $S = A_e / d^2$. This is an erosion area divided by the square of the stone diameter. In a strip with a width d perpendicular to the page, S is more or less equal to the number of removed stones. The advantage of S is the use of an area which can be objectively measured by soundings. For the threshold of damage, $S = 2-3$, can be used. When the armour layer is locally completely removed and the filter layer becomes exposed, the damage can be defined as failure of the structure. Depending on the slope, the matching S is about 10.

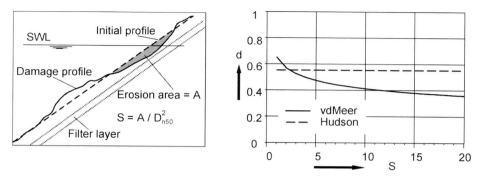

Figure 8-10 Stone diameter as a function of the damage level

Wave height and slope angle α

Figure 8-11, finally, shows the influence of the wave height and of the slope angle. For these two parameters, Hudson and Van der Meer show the same tendency.

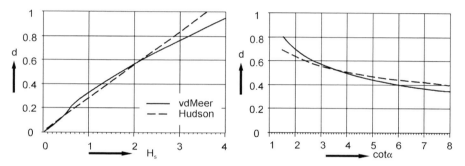

Figure 8-11 Stone diameter as a function of wave height and slope angle

Damage development

Another advantage of the Van der Meer formulae, is the possibility to take the damage development into acccount. This can be important for repair and maintenance policies, but also for the construction of a protection. S can be computed explicitly by rewriting equation (8.8). Figure 8-12 shows the damage as a function of wave height for stones with $d_{n50} = 0.25$ m, cot $\alpha = 2$ and $P = 0.6$. This could be the core of a breakwater, which is exposed to waves during construction. By building the breakwater in the "moderate" season, the waves are usually lower than the design waves for the breakwater. Figure 8-12a shows the influence of the wave period on the damage, for a duration of 1000 waves and Figure 8-12b shows the influence of the number of waves with a period of 6 s. Note that a period of 6 s gives more damage than a period of 3 or 9 s, which again has to do with the transition between plunging and surging breakers. The number of waves is not so important. With these figures and wave or wind statistics from the area, an expectation of the damage can be determined.

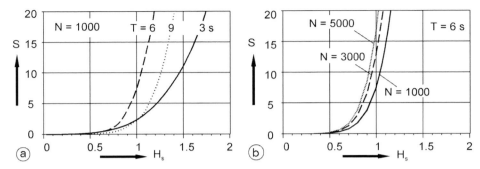

Figure 8-12 Damage as a function of wave height, wave period and storm duration

Example 8-2

A dike has to be protected with rock. The slope is 1:3 and H_s for the design waves = 1.9 m with a wave steepness, based on the average wave period, s_m = 0.04. What stone class is needed?

Starting point is the Van der Meer equation. For a dike revetment, in equation 8.8, P = 0.1. We consider 7000 waves (N = 7000) and S = 2 (only a little damage). The surf similarity parameter, ξ = 0.33/0.2 = 1.7. The transition between the plunging and surging (equation 8.8), ξ_{trans} = $(6.2*0.1^{0.31}*\sqrt{0.33})^{1/(0.1+0.5)}$ = 2.5. Hence, the plunging part of the equations has to be used. This leads to: $H_s/\Delta d_{n50}$ = $6.2*0.1^{0.18}*(2/\sqrt{7000})^{0.2}*\xi^{-0.5}$ = 1.5. With a value for Δ = 1.6 this leads to d_{n50} = 0.8 m or stone class 1000-3000 kg.

Using Hudson's equation we would have found: $H_s/\Delta d_{n50}$ = $(3.5*3)^{0.33}$ = 2.2 and a d_{n50} = 0.55 or stone class 300-1000 kg which would be an underestimation of the necessary rock.

Mild slopes

The relations by Hudson and Van der Meer were based on experiments with slopes in the range ~ 1:1.5 to 1:6. Very mild slopes (cot α > ~ 10) with rock protection do exist, e.g. when a pipeline outfall on a beach is protected. In that case the breaking type is no longer plunging but spilling. Schiereck et al., 1996, did some research for this case and found that the results are even more favourable than the Van der meer equations indicate, see Figure 8-13. In a spilling breaker the wave energy absorption is distributed quite evenly along the slope, see chapter 7, indeed resulting in a higher stability number. So, as a first approximation, the Van der Meer equations can be used even for spilling breakers (with the wave height in deep water as input value).

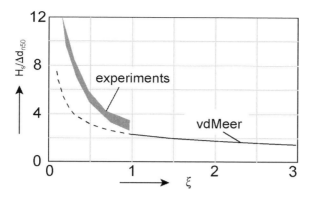

Figure 8-13 Comparison Van der Meer with mild slope experiments

8.3.3 Low crests

The Van der Meer and Hudson formulae were derived for non overtopped slopes. When slopes are overtopped, there is a certain wave transmission, see chapter 7. That means that not all energy will be destroyed on the slope and thus the stability of the armour stones will increase. For crests *above* the still water level Van der Meer (see Pilarczyk, 1990) found a reduction factor for the d_{n50} as derived from equation (8.8):

$$\text{Reduction } d_{n50} = \frac{1}{1.25 - 4.8 \dfrac{R_c}{H_s} \sqrt{\dfrac{s_{0p}}{2\pi}}} \qquad (8.10)$$

where R_c is the crest height with respect to SWL, see Figure 8-14a and s_{0p} is the (deep water) wave steepness related to the peak period (T_p); the minimum value of the reduction factor is 0.8 and the maximum is 1. As a result of the wave transmission, however, armouring needs to be heavier on the other side of the breakwater and the question is whether the total damage will reduce, see also Burger, 1995. One approach for low dams is to apply the same armour units on both sides.

For crests *below* the still water level Van der Meer formulated a relation for the stability (see Pilarczyk,1990) which can be rewritten as follows:

$$\frac{H_s}{\Delta d_{n50}} = -7 \ln\left(\frac{1}{2.1 + 0.1S} \frac{h_c}{h}\right) \sqrt[3]{s_p} \qquad (8.11)$$

Figure 8-14b shows some results. s_p in equation (8.11) is the *local* wave steepness instead of the deep water steepness. These formulae have been established separately and should be used with caution. Further research is needed in order to establish an overall relation for the stability increase of low crested dams.

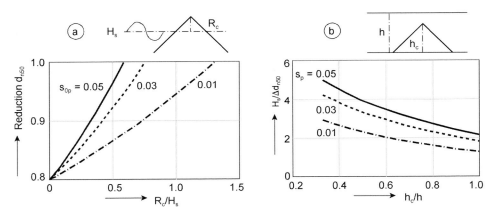

Figure 8-14 Stone stability for low crested dams

8.3.4 Toes

The toe's main function is to support the armour layer of a slope. The stones in a toe can often be smaller than those in the armour layer, since they are less heavily attacked by the waves. The waves do not break at all on deep lying toes.

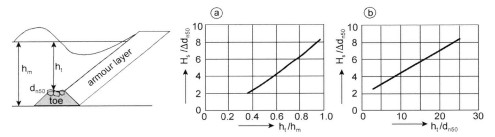

Figure 8-15 Stability of stones in toe structure

Figure 8-15a gives the results of experiments on toe stability, see CUR/CIRIA, 1991. The line in the figure represents the relation between the stability parameter and the relative toe depth:

$$\frac{H_s}{\Delta d_{n50}} = 8.7 \left(\frac{h_t}{h_m} \right)^{1.4}$$ (8.12)

This equation shows an increase in stability for relatively deep lying toes. The validity of this equation is restricted to situations with little damage ($S \approx 2$) and relatively deep toes ($h_t/h_m > 0.4$).

Figure 8-15b shows the results of research by Gerding (see d'Angremond et al., 1996). These experiments can be described with:

$$\frac{H_s}{\Delta d_{n50}} = 1.1 \left(0.24 \frac{h_t}{d_{n50}} + 1.6 \right) \tag{8.13}$$

In this expression, iteration is necessary, as d_{n50} appears at both sides of the equation. The stone size in the armour layer can be seen as the maximum stone size in a toe structure, whatever the results derived from these equations.

All of these experiments were done for breakwaters with relatively steep slopes. Toes of vertical walls lie in the antinode of a standing wave where, theoretically, only the pressure fluctuates. Wider toes may extend to the node where the velocities are at their maximum. For non-breaking waves, it is recommended to follow the procedure with the modified Shields-diagram of section 8.3.1, see also d'Angremond et al., 1996.

8.3.5 Heads

The head of a breakwater or coastal groyne is usually the most exposed part of such a structure. No systematic test results are available. The Shore Protection Manual (SPM, 1984) recommends constructing a head with stones which are twice as heavy as in the trunk or, alternatively, a head with a slope which is twice as gentle as the trunk's slope.

8.4 Stability of coherent material

8.4.1 Placed-block revetments

In revetments, particularly sea defences, placed elements, mostly made of concrete are very important. In the last twenty years, much research has been done in The Netherlands regarding the stability of placed blocks on a slope under wave attack. Therefore, this construction type will be discussed in more detail.

Placed blocks can have many shapes. Their coherence varies, blocks may be: pinched, connected with cables or geotextile, or interlocked. Another variation in the shape can be the upper side of the blocks which can be designed to reduce the wave run-up, but for this chapter this variation is not very important. Figure 8-16a gives some examples. The transition between the blocks and the underlying soil is another variable. Figure 8-16b shows some possibilities (many other combinations are possible). When the blocks are not exactly equal in height, a layer to embed the blocks is necessary to correct the height differences. This is the case when natural material like basalt is used. The extra layer is sometimes combined with a filter. Sometimes the blocks are placed directly on clay; this requires a high construction quality standard, as irregularities of the surface decrease the strength of the revetment.

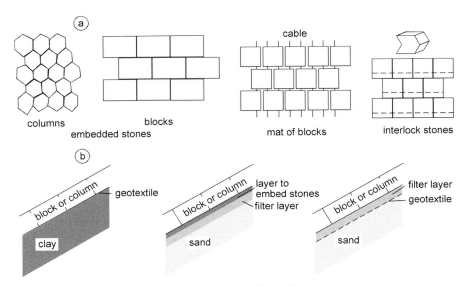

columns   blocks
          embedded stones

mat of blocks   interlock stones

Figure 8-16 Block types and filters in revetments

Stability mechanism

In section 8.2 we saw that the characteristic length of the loading phenomenon was of the same order of magnitude as the leakage length: $L \approx \Lambda$. For loose grains and asphalt the orders of magnitude are completely different and the relation between external and internal pressures becomes very simple. When $L \approx \Lambda$, the situation is more complex and much energy has been put into fathoming the secrets of placed block stability. Figure 8-17 shows the phenomena that might play a role in the stability of placed blocks during a wave cycle. From tests and calculations it was reasoned that phenomena b and c are dominant in the process. In fact, the wave action on and under the blocks can not be separated and porous flow phenomena have to be taken into account.

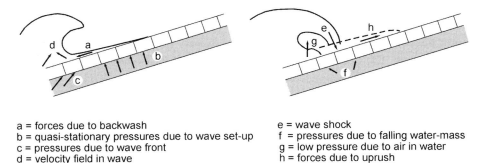

a = forces due to backwash
b = quasi-stationary pressures due to wave set-up
c = pressures due to wave front
d = velocity field in wave

e = wave shock
f = pressures due to falling water-mass
g = low pressure due to air in water
h = forces due to uprush

Figure 8-17 Possible loading mechanisms in block revetment

Figure 8-18a shows which forces determine the situation at maximum downrush: the pressure <u>on</u> the blocks is low in front of the wave, while <u>under</u> the blocks it is high, due to the water pressure in the filter layer caused by the propagating wave and due to the relatively high phreatic level in the slope. This causes uplift forces on the blocks.

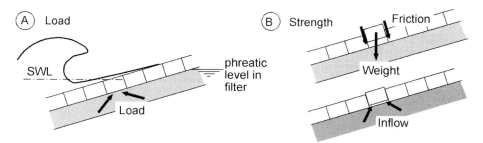

Figure 8-18 Load and strength of block revetments

Whether a block is pushed out or not depends on the strength of the revetment, in which two factors play an important role: in the first place, of course, the coherence of the blocks, which, in the case of placed blocks, is the friction between them and, even more importantly, the clamp phenomenon which will be discussed later on. The second factor is the flow towards a stone when it is pushed out. With a relatively small permeability of the filter layer, the block is sucked onto the slope because only very little water can flow into the growing hole leading to a sudden decrease of the pressure under the block, see Figure 8-18b. Both load and strength will be discussed in more detail.

Load

For the magnitude of the uplift force, the relation between the permeability of the top layer and that of the filter layer, expressed in the leakage length, Λ, is very important, see also section 8.2. To demonstrate the relative importance of permeabilities, the porous flow in the revetment is again simplified: the flow through the filter layer is assumed to be parallel to the slope while the flow through the top layer is supposed to be perpendicular to it (x is the coordinate along the slope), see Figure 8-19. The flow in the filter layer can be expressed as:

$$v_F = -k_F \frac{d\phi_F}{dx} \tag{8.14}$$

and through the top layer:

$$v_T = k_T \frac{(\phi_F - \phi_T)}{d_T} \tag{8.15}$$

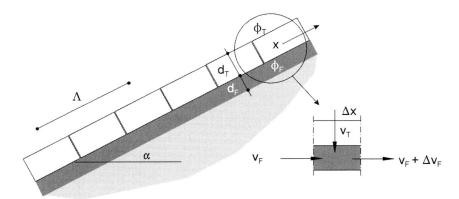

Figure 8-19 Flow through block revetment and leakage length

Based on continuity, $\Delta v_F \cdot d_F = v_T \cdot \Delta x$, see Figure 8-19, hence $v_T \approx d_F \cdot dv_F / dx$, from which follows:

$$\frac{d^2\phi_F}{dx^2} = \frac{-k_T(\phi_F - \phi_T)}{k_F \, d_T \, d_F} = -\frac{(\phi_F - \phi_T)}{\Lambda^2} \quad \rightarrow \quad \phi_F - \phi_T = -\Lambda^2 \frac{d^2\phi_F}{dx^2} \tag{8.16}$$

in which ϕ_T and ϕ_F are the piezometric head ($\phi = p / \rho g + z$) on the top layer and in the filter layer, k_T and k_F are the permeability of the top layer and filter layer and d_T and d_F are the thickness of the top layer and filter layer, respectively. From this equation it can be seen that the head difference over the top layer depends directly on Λ, defined in equation (8.3). A relatively thick and permeable filter layer and/or a relatively thick and impermeable top layer give a large Λ and hence, a large head difference over the top layer.

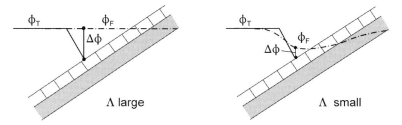

Figure 8-20 Head difference over block for large and small leakage length

This equation can be solved analytically if boundary conditions are highly schematized and if flow in the filter layer is assumed to be laminar (or more precisely: the relation between velocity and pressure is presumed to be linear). Its description is beyond the scope of this book; the reader is referred to CUR/TAW, 1992 or to Bezuyen et al., 1990. Figure 8-20 shows the situation with a very high value of Λ and with a very small value. *So, a large Λ, the leakage length, is unfavourable for the stability of the blocks* and it is clear that a permeable top layer and an "impermeable" filter layer lead to the most stable structure. This means that filter layers should be kept as thin as possible!

Strength

The first element of the resistance against uplift is, of course, weight. Friction between the blocks is next and Figure 8-21a shows another potentially important mechanism in a placed block revetment, see also Suiker, 1995. When a block is lifted by the pressure from below, it will lift other blocks as well, due to friction. As a result of their geometry, this will lead to large normal forces in the blocks, which are passed on to other blocks and finally to some bordering structure.

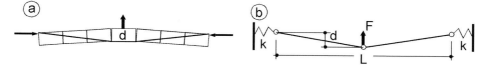

Figure 8-21 Clamping mechanism

The mechanism also works when two beams are hinged to a spring support, between two walls at a distance which is slightly smaller than the total beam length, see Figure 8-21b. From structural mechanics it is known that the maximum force, *F*, will be:

$$F_{\max} = \frac{16}{9} \sqrt{3} \frac{k d^3}{L^2} \tag{8.17}$$

This force is mobilized when a block is raised and there is little or no room between two blocks. The clamping will be less when slots between the blocks are present. The blocks will be placed as close to each other as possible, but slots will always be present. Taking slots into account in equation (8.17) it is found that the strength of a block revetment can be several times higher than the proper weight of the blocks, but the resulting strength depends on the slot width and the assumed value of k, as they are both uncertain.

Stability

A simple approach to determine the stability of placed blocks is as follows, see also Gerressen, 1997. The head difference across the blocks is caused by the run-down of

the waves, see chapter 7. Figure 8-22 gives the pressures along the slope.
The most unfavourable block is situated just ahead of the wave front. When the expression for the run-down from chapter 7 is used and when the only strength comes from the proper block weight, the stability is given by:

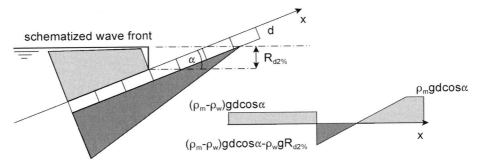

Figure 8-22 Stability of block revetment

$$(\rho_m - \rho_w)gd\cos\alpha - \rho_w g 0.33\xi H_s = 0 \quad \rightarrow \quad \frac{H_s}{\Delta d} = 3\frac{\cos\alpha}{\xi} \tag{8.18}$$

Of course, the constant 3 in equation (8.18) is only a first guess. Λ will influence the load (see Figure 8-20) and friction and clamping will influence the strength.

Results

It is obvious that, with so many assumptions and so many factors, it is impossible to give more than proportionalities in the relation between all of the involved parameters. It was found:

$$\frac{H_s}{\Delta d} \propto \left(\frac{d}{\Lambda \xi_p}\right)^{0.67} \tag{8.19}$$

For various structure types, experiments were carried out to establish the constants of proportionality, leading to the graphs presented in Figure 8-23a, which serve as an indication of the stability. There is also an "analytical" method which takes more detail into account. For more information the reader is referred to CUR/TAW, 1992.
A closer look at equation (8.19) and Figure 8-23, reveals:
- The simple stability computation using equation (8.18) is too pessimistic, but the trend is similar to equation (8.19)'s trend,
- a large Λ, leads to a low $H_s/\Delta d$, which is in line with equation (8.16) where a large Λ gives a large head-difference $(\phi_F - \phi_T)$ and, hence, lower stability,

- the same holds for ξ, which is the same trend as found for loose rock, see equation (8.8)a and Figure 8-23b,

- d plays a complicated role in this equation. It is part of Λ, indicating that a thick top layer gives a large head-difference (equation(8.16)), which is unfavourable. But a thick top layer also means more weight and hence more strength. Assuming all other parameters in equation (8.19) are constants, it appears that: $H \propto d^{4/3}$, so a thick top layer is over-all favourable for stability.

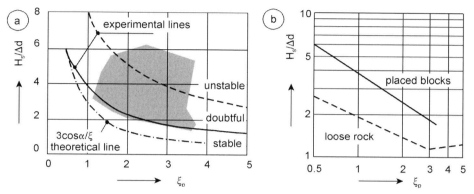

Figure 8-23 Test results for placed blocks on filter layer and comparison with loose rock

In this book no discrimination is made between the combinations of different kinds of blocks (pitched, interlocked etc.) and transitions (filter layers, geotextiles, directly on clay etc.). Compared with rip-rap, placed blocks look superior and even better results than the line presented in Figure 8-23 are possible. However, much depends on the quality control during construction and thereafter. A permeable top layer is favourable, but it can become impermeable, as a result of dirt, vegetation, shells etc. We have also seen that a thin and "impermeable" filter layer, creates a more stable situation. That is why the idea of blocks placed directly on clay is a very attractive one from a theoretical point of view. However, everything again depends on the quality of the construction, which is difficult to assure. Biological activities in the clay after construction may also have an unfavourable effect on the stability.

Example 8-3

The same dike as in Example 8-2 has to be protected with concrete blocks. The slope is 1:3 and H_s for the design waves = 1.9 m with a wave steepness, based on the average wave period, $s_m = 0.04$. When the density of the concrete is 2600 kg/m3, what is the necessary thickness?

The stability relation for block revetments in Figure 8-23 is given for ξ_p and not for ξ_m. With a peak period that is about 20 % higher than the average period ξ_p will be about 1.9. $H_s/\Delta d$ from Figure 8-23 then becomes ≈ 2.4 In sea water, Δ becomes ≈ 1.5, which leads to blocks with a thickness of about 0.5 m.

8.4.2 Impervious layers

Uplift

Impervious layers can be made of asphalt or concrete. The differences with blocks
are that the protection forms a whole instead of consisting of separate elements and
that the fluctuating wave pressures can penetrate only a limited distance from the
edge of the protection. Usually there is no filter layer (the protection itself is
sandtight).

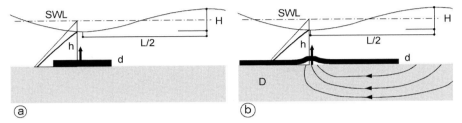

Figure 8-24 Impervious layers in waves

In the case of a protection which is shorter than a wave length, the equilibrium is
simply given by:

$$\left(\rho_m - \rho_w\right)g\,d > \rho_w\,g\,\frac{H}{2}\,\frac{1}{\cosh\left(\dfrac{2\pi}{L}h\right)} \tag{8.20}$$

In certain places, the wave pressure on top of long protections is half the waveheight
below the subsoil pressure, see Figure 8-24b. In those places, the pore pressure tries
to lift the protection. This can only be effectuated if water flows into the hole
between bottom and protection. TAW, 1984, has shown that this is hardly possible
and that this is not an important stability factor for impervious revetments. Threats
are wave impacts and porous flow due to waterlevel differences over the protection,
see chapter 5.

Wave impacts

In chapter 7 the special character of wave shocks compared with the cyclic quasi-
static pressures was discussed. These shocks induce stresses in the material which
should not exceed the failure stress, see Figure 8-25.
The response of the layer depends, among other things, on the stiffness of the layer
and the subsoil. The protection layer (asphalt or concrete) has an elasticity E (Pa) and
the subsoil has a stiffness c (Pa/m). The wave impact causes a stress, σ, in the layer.
Figure 8-25 shows the influence of the stiffness ratio, E/c, based on some assumed
values of wave impact and layer properties. A stiff layer on a soft subsoil gives the

highest stresses. So, a concrete layer on soft clay is not a logical choice when wave impacts are to be expected.

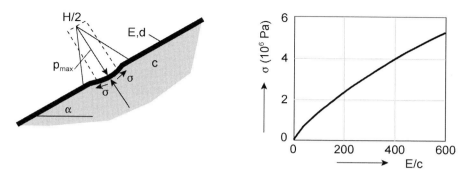

Figure 8-25 Wave impact on slope and influence material properties

To calculate the stresses, the load and response mechanism is schematized as a cushioned-spring system, see TAW, 2000. The results are:

$$\sigma = \frac{3p_{max}}{\beta^3 H d^2}\left[1 - \exp(\frac{-\beta H}{2})\left(\cos\frac{\beta H}{2} + \sin\frac{\beta H}{2}\right)\right] \quad with \ \beta = 4\sqrt{\frac{2.65c}{Ed^3}} \qquad (8.21)$$

in which p_{max} is the value for the wave pressure as mentioned in section 7.4.3 and H is the individual wave height. Typical values for the parameters involved can be found in annex A, Materials. Fatigue plays an important role in the critical stress, σ_c. Compared with a onetime load, σ_c can be 2 to 10 times lower. d has to be determined iteratively from equation (8.21), which is something that has to be done with a computer. In Figure 8-26 the results are given for asphalt concrete ($E = 10 \cdot 10^9$ Pa) on rather well compacted sand ($c = 100 \cdot 10^6$ Pa/m) and on clay ($c = 30 \cdot 10^6$ Pa/m) and a storm duration of 10-20 hours with wave impacts.

This picture shows that for mild sand slopes (< 1:3), the thickness of an asphalt concrete protection is relatively thin: < 0.2 m, even for very high waves. In that case the quasi-stationary pressures due to a high phreatic level inside determine the thickness of the layer, see chapter 5. On clay, however, wave impacts can make thick layers necessary.

Filters

A filter under an impervious protection layer is not necessary for sandtightness. For the stability of the layer as a whole it is even undesirable. The leakage length, Λ, is already infinitely high if the top layer is impervious and a filter layer does not improve the situation. When the weight of the layer is less than the excessive wave pressure, the strength of the layer must take care of the equilibrium. The situation is similar to the one in Figure 8-24b and what was said there, is also valid here. A filter can be applied for drainage, to prevent excessive pressures against the layer due to a high phreatic level inside the slope, see Figure 8-27. The effect of a filter on pressures has to be determined with a porous flow model, see chapter 5.

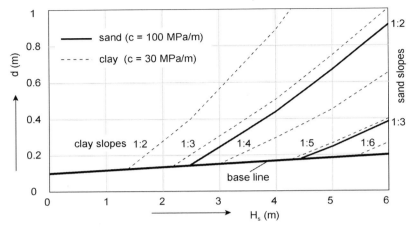

Figure 8-26 Necessary thickness of asphalt concrete on sand or clay

Figure 8-27 Influence filter under impervious layer on phreatic level

8.5 Summary

Firstly some simple relations for erosion of unprotected slopes and of bottoms in front of walls have been presented.

In this chapter the main item is stability in waves. It appears that most relations for flow situations can also be used for non-breaking waves, as long as the increased shear stress under waves is taken into account. In breaking waves, only empirical relations, like the one by Hudson:

$$M = \frac{\rho_s H_{sc}^3}{K_D \Delta^3 \cot\alpha} \quad (\text{or: } \frac{H_{sc}}{\Delta d} = \sqrt[3]{K_D \cot\alpha})$$

or Van der Meer:

$$\frac{H_{sc}}{\Delta d} = 6.2\, P^{0.18} \left(\frac{S_d}{\sqrt{N}} \right)^{0.2} \xi^{-0.5} \qquad \text{(plunging breakers)}$$

$$\frac{H_{sc}}{\Delta d} = 1.0\, P^{-0.13} \left(\frac{S_d}{\sqrt{N}} \right)^{0.2} \xi^P \sqrt{\cot \alpha} \qquad \text{(surging breakers)}$$

can be used.

The latter is to be preferred, since more parameters are included. The permeability of a slope is one of the most important differences between the two formulae. With Van der Meer's formulae it is also possible to calculate the damage as a function of wave parameters. **Note**: *The subscript **c** indicates 'critical' values in the stability parameter $H_{sc}/\Delta d_{n50}$, thus distinguishing between stability and mobility parameters, but is not used consequently.*

Typical values of the stability parameter in breaking waves, $H_{sc}/\Delta d_{n50}$, lie around 2 for rather steep, continuous slopes. Higher values are valid for mild slopes. For toes, this value can also be higher, depending on the relative depth of the toe. Low crests can lead to an increase of the stability parameter, while at the head of a dam or breakwater twice the mass computed for the trunk needs to be used.

For placed-block revetments, the mechanisms are explained and an empirical stability relation is presented. It can be concluded that block revetments of good quality, can stand greater wave loads than loose rocks. Typical values of the stability parameter, $H_{sc}/\Delta d_{n50}$, are about 1.5 - 2 times the value for loose rock.

Waves will usually not be able to lift impervious-layer revetments of some size. Wave impacts can influence the thickness of the layer. A filter is not necessary, except possibly for the drainage of excessive pressures due to tides or surges.

9 SHIPS
Loads, stability and erosion

Ships in canal (North Brabant), courtesy Bas Klimbie

9.1 Introduction

This chapter treats the protections of waterways' beds and banks of waterways against loads caused by ships. Ships cause an interesting wave pattern and currents, involving many elements of the previous chapters. A summary of these phenomena will be given, with a focus on the resulting loads. For a more extensive treatment of some aspects the reader is referred to e.g Stoker, 1957 or RWS/DHL, 1988

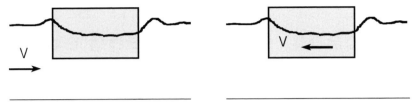

Figure 9-1 Flow around fixed object versus moving object in stagnant water

From a hydrodynamical point-of-view, a moving ship is similar to flow around a body (see Figure 9-1). Figure 9-2 gives an overview of the relevant phenomena. The waterlevel depression along the ship's hull and the so-called return flow, which are well-known from observations of ships in canals, are also present when water flows around a body. This water-level depression is the *primary wave* with a wave length of about the ship's length. From the point of view of an observer on the bank, the primary wave starts with the front wave, followed by the depression and ending with the stern wave. Within the primary wave this stern wave usually results in the most severe attack on the banks.

The much shorter waves that originate from the hull and which are visible on aerial photographs, are the so-called *secondary waves*. Both types of waves behave like "normal" water waves, which means that the relations for wavelength, celerity etc. in chapter 7 are valid. In practice, the primary wave can be long or short, depending on the same depth-length relations as mentioned in chapter 7; secondary waves are practically always short, except for secondary waves produced by high-speed boats. In relatively narrow navigation channels the primary waves are important, otherwise only the secondary waves play a role.

The currents caused by the ship's propeller, the so called *propeller wash*, have the characteristics of a jet. This jet is particularly important when ships manoeuvre near a berthing place or a jetty. When sailing, the jet flow is partly neutralized by the speed of the ship.

Many investigations into ship motion have been carried out in the past, but most of them were for shipbuilders. The results of the investigations by Delft Hydraulics Laboratory for the Dutch Rijkswaterstaat in the seventies are an important source of information. Although this research was set up primarily for inland waterways, the results, especially the literature survey, have a wider range of applications. For more details, see RWS/DHL, 1988.

Most damage on revetments is not caused by the largest ships, which often sail relatively slowly and relatively far from the banks. Small service crafts, tugs etc, sailing full power near the banks, actually do most harm.

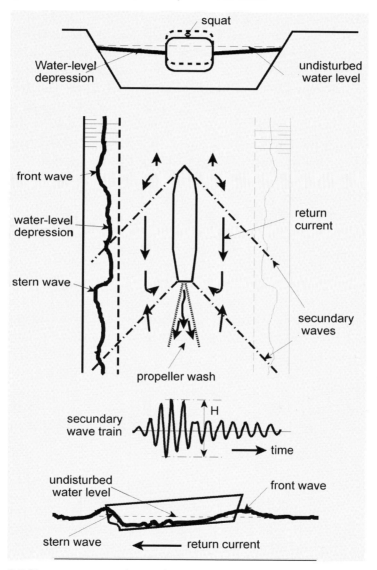

Figure 9-2 Phenomena around a moving ship in a waterway (from RWS/DHL 1988)

9.2 Loads

9.2.1 Limit speed

All mentioned hydraulic aspects are related to the ship's speed. For design purposes, this speed can be determined in several ways. When speed data of the relevant type of ship are available, it is preferable to use these. If not, the speed can be calculated from the engine power and the ship's resistance (which consists of both skin friction and form drag, see chapter 2). These calculations will not be treated here. For a first estimate, the physical properties of waves can be used. Regardless of its engine power, a ship can not sail faster than its own primary wave. This limit speed is analogous to the sound barrier for airplanes. Self-propelled ships can pass this barrier if they are able to plane on the water (e.g. speed boats). The speed of displacement-type ships can not exceed the celerity of their own generated waves, unless they are being towed by another (longer) ship. Assuming that the maximum wave length caused by a ship is equal to the ship's length, the limit speed can be approximated with the wave celerity in the linear wave equations by eliminating the wave period from the equations:

$$\left.\begin{array}{l} c = \dfrac{gT}{2\pi}\tanh\dfrac{2\pi h}{L} \\[2mm] L = cT \end{array}\right\} \qquad V_l = c = \sqrt{\dfrac{gL}{2\pi}\tanh\dfrac{2\pi h}{L}} \qquad\qquad (9.1)$$

In deep water, V_l is only proportional to the square root of the length of the ship. In shallow water, V_l is only proportional to the waterdepth and the length of the ship does not play a role.

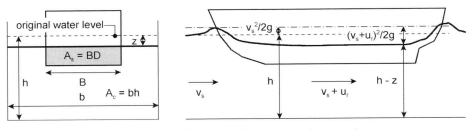

Figure 9-3 Definitions in 1-dimensional approach

If the cross-section area of the ship ($A_s = B \cdot D$) is not negligible with respect to that of the waterway ($A_c = b \cdot h$), the above approach is no longer valid. In a relatively narrow and shallow canal, the watermovement around the ship can be schematized as a 1-dimensional flow situation (see Figure 9-3), where the ship lies still and the water flows with v_s. This was done by Schijf, 1949. In that case, the flow can be approximated with the Bernoulli-equation, neglecting energy losses and the continuity equation. This method also includes the following assumptions: the ship sails in the axis of the canal, prismatic cross-section of ship's hull, ship is horizontal, channel cross-section is prismatic, return flow and waterlevel depression are constant over the remaining channel's cross-section and the depression of the ship is equal to the water-level depression. The equations then become:

Bernoulli :
$$h + \frac{v_s^{\,2}}{2g} = h - z + \frac{(v_s + u_r)^2}{2g}$$

(9.2)

continuity:
$$b\,h\,v_s = (b\,h - B\,D - b\,z)(v_s + u_r) = Q$$

The limit (maximum) speed is reached when the return flow becomes critical or, in other words, the derivative of the return flow with regard to the water-level depression, z, is zero:

$$\frac{dQ}{dz} = \frac{d(v_s + u)(Ac - As - bz)}{dz} = 0$$

(9.3)

In combination with the Bernoulli-equation, it is found:

$$\frac{A_s}{A_c} + \frac{V_l^2}{2gh} - \frac{3}{2}\frac{V_l^{2/3}}{(gh)^{1/3}} = 1$$

(9.4)

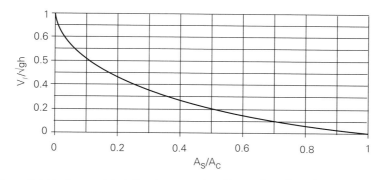

Figure 9-4 Limit speed (relative to water depth) as a function of blockage A_s/A_c

Figure 9-4 represents the solution of equation (9.4). When A_s is negligible compared with A_c, it shows that the limit speed is again the same as follows from the linear wave theory: $V_l = \sqrt{gh}$. If the ship blocks the canal completely, of course, the limit speed is 0.

Figure 9-5, finally, combines the results for the limit speed according to linear wave theory and for the 1-dimensional approach. In deep water, without blockage, the limit speed solely depends on the length of the ship. With different degrees of blockage, this limit is reached in a greater water depth. These lines result from completely different assumptions, so the transition between the solid line and the other lines may not be represented correctly.

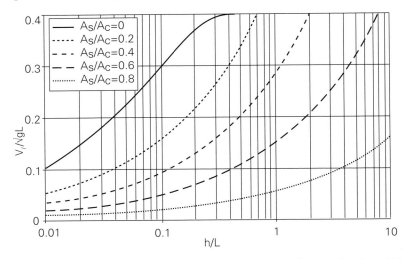

Figure 9-5 Limit speed (relative to ship's length) as a function of water depth and blockage

For the purpose of bank design, a speed of 90% of V_l is recommended. A higher speed is technically and economically not attractive and will not be used by any captain. A lower speed, even with traffic regulations, is not recommended for design purposes, because there will always be captains that will not adhere to the speed limit.

9.2.2 Primary waves

The water-level depression (z) and the return current (u_r) can be derived from equation (9.2). Written in a dimensionless form:

$$\frac{v_s^2}{gh} = \frac{2\,z/h}{\left(1 - A_s/A_c - z/h\right)^{-2} - 1}$$

$$\frac{u_r}{\sqrt{gh}} = \left[\frac{1}{1 - A_s/A_c - z/h} - 1\right]\frac{v_s}{\sqrt{gh}}$$

(9.5)

The water-level depression as a function of the blockage (A_s/A_c) and v_s can now be read from Figure 9-6, which is a graphical presentation of the solution of the first part of equation (9.5). A speed of 90% of V_l is recommended for design purposes (the solid line in the figure). Figure 9-7 gives the return current in the same way.

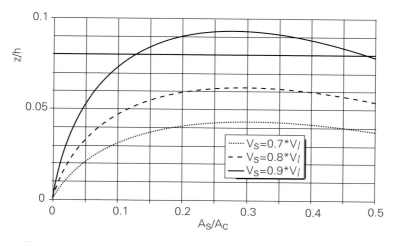

Figure 9-6 Relative water-level depression as a function of blockage

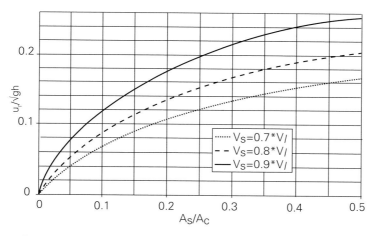

Figure 9-7 Relative return flow velocity as a function of blockage

Practical complications

The equations for z and u_r were derived for idealized situations, assuming a rectangular cross-section of the canal, the position of the ship in the canal centre-line, etc. For design purposes, the following corrections can be made:

Canal profile

When the profile is not rectangular, but e.g. trapezoidal, the width, b, is equal to the width at the water surface, but the depth, h, to be used is A_c / b, see Figure 9-8.

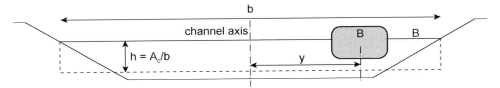

Figure 9-8 Deviations from ideal case in 1-dimensional approach

Flow velocity

If there is a flow in the channel, v_s and V_l have to be relative to the flow velocity (so, $0.9V_l$ keeps the same value). From this, the water-level depression directly follows. The (positive or negative) flow velocity has to be added to the return flow, u_r.

Eccentric position ship

If a ship is sailing at a distance y from the axis (see Figure 9-8), the water-level depression and the return flow can be corrected as follows:

$$z_{ecc} = \left(1 + \frac{2y}{b}\right) z \qquad\qquad u_{r-ecc} = \left(1 + \frac{y}{b}\right) u_r \qquad\qquad (9.6)$$

As the minimum distance to the bank, the width B can be used (see Figure 9-8).

Multiple ships

Frequently there are situations where there is more than one ship in a cross-section of a waterway, e.g. when two ships encounter or when a ship overtakes another one. Only the last situation causes a higher load than a single ship. A first approximation is then to add the cross-section areas of the two ships and to make the calculations for one large ship.

Finally, for the height of the stern wave, which forms usually the most important load on a bank, the following approximation can be made:

$$z_{max} = 1.5 z_{ecc} \qquad\qquad (9.7)$$

9.2.3 Secondary waves

The primary wave more or less has the characteristics of a negative solitary wave; the secondary waves are formed by a number of periodic waves, a wave train (see Figure 9-2). Secondary waves are caused by the pressure pattern due to the discontinuities in the hull profile. These discontinuities are found at the bow and at the stern both of which emit waves. The bow is usually dominant. The wave emission from the bow can be thought of as a succession of travelling disturbances (see Figure 9-9).

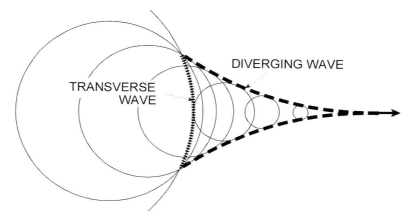

Figure 9-9 Origin of diverging and transverse waves

Each disturbance creates a circular wave, like the waves caused by a pebble thrown into a pond. These circles are enveloped by the so-called diverging wave. Behind the ship, the remnants of the circles can be seen as transverse waves. The transverse waves travel in the same direction and with the same speed as the ship, while the diverging waves travel more slowly. The velocity difference is accounted for by the angle between the waves and the sailing line $(c = v_s \cos\phi)$.

Figure 9-10 shows the secondary waves behind a ship with the diverging and transverse waves. The first type diminishes with the square root of the distance perpendicular to the sailing line and the second type diminishes with the square of this distance. The diverging waves are dominant in this pattern (for $Fr = v_s/\sqrt{gh} <$ 0.75). Where transverse and diverging waves meet, they interfere forming cusps, which are dominant for the stability of revetments. For moderate speed ($Fr < 0.75$), the cusp locus line is at an angle of about 20° with the sailing line and the direction of propagation of the cusps is at an angle of about 35° with the sailing line, hence the angle of approach for a bank parallel to the sailing line is 55°. For higher speeds the cusps propagate more in the direction of the sailing line. For $Fr = 1$, transverse and diverging waves coincide and for $Fr > 1$ transverse waves can no longer exist. This is interesting for high-speed ferries and other very fast sailing ships, but self-propelled displacement ships will never reach this limit. Assuming a practical limit of $0.9V_l$, the value of Fr remains below 0.75 already for small blocking percentages, as can be seen from Figure 9-4. This means that the pattern of Figure 9-10 is generally applicable for waterways with self-propelled displacement ships.

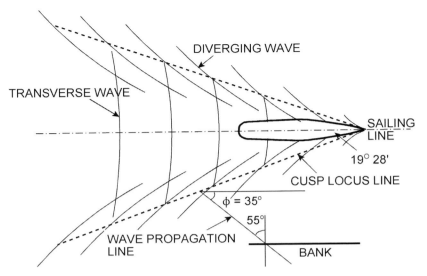

Figure 9-10 Secondary wave pattern

Unfortunately, much less is known when it comes to the wave *heights*, which have to be determined experimentally. More details are given in RWS/DHL, 1988 and Sorensen, 1973. In this book "wave heights" of secondary waves refer to the height of the interference peaks, the cusps, as they are the dominant load. The wave height greatly increases with the speed, reaching a theoretical maximum at $Fr = 1$. The sharp increase of the wave height with the speed can be explained as follows: the limit speed is similar to the sound barrier for airplanes. The energy from the engine can no longer be transformed into speed, but the remainder must go somewhere. Hence, much fuel is used to make waves. This explains why ships usually sail well under their limit speed.

In relatively deep water, the wave height of the cusps diminishes with the cubic root of the distance to the sailing line, which is again a good approximation for moderate Fr-numbers (< 0.75).

Figure 9-11 gives experimental data from several sources, both from small-scale model tests and full-scale tests. These data can be described with the following relation:

$$\frac{H}{h} = \zeta \left(\frac{s}{h}\right)^{-1/3} Fr^4 \qquad (9.8)$$

The coefficient of proportionality, ζ, represents the ship's geometry. It was found that the ship's length and its block coefficient hardly influence the secondary waves. The draught, D, and the shape of the bow dominate ζ. This is in line with the fact that the discontinuities in the hull of the ship are responsible for the emission of pressure waves. $\zeta = 1.2$ gives a reasonable upper limit of the experimental data (see Figure 9-11). When more accurate data are needed, model tests can be carried out for a specific type of ship (see also RWS/DHL, 1988).

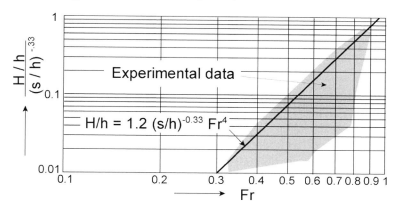

Figure 9-11 Secondary wave height measurements (from many sources)

The wave period also reaches its maximum at $Fr = 1$. The wave period and the wavelength can be calculated as follows. As reasoned before, for $Fr < 0.75$ the secondary waves are deep water waves. In that case (with $c = v_s \cos \phi$ and $\phi = 35°$):

$$c = \frac{gT}{2\pi} \quad \rightarrow \quad T = \cos 35° v_s \frac{2\pi}{g} \approx 0.82 v_s \frac{2\pi}{g} \tag{9.9}$$

Model and prototype-measurements correspond well with this formula.

9.2.4 Propeller wash

The flow behind a ship's propeller is very similar to flow in a circular jet, so it can be expected that the same proportionalities are valid. This is not necessarily the case for the numerical values, since there are also differences. For example, the water in the jet is already turbulent because of the propeller blades; this will make the flow establishment region shorter than in a free jet. Another difference is the water surface, which will influence the divergence of the jet.

Intermezzo 9-1 High-speed ferries

Recently, in several places in the world, high speed ferries have come into operation. These ships sail at supercritical speeds ($Fr > 1$). Some phenomena then differ from the ones that were previously described above. These phenomena are illustrated in the figure below.

The angle between the wave propagation line and the sailing line is 35^0 for deep water waves ($Fr < 0.75$). When the speed of the ship increases, transverse waves and diverging waves will coincide and the angle becomes 0^0: the waves will travel in the same direction as the ship. For $Fr > 1$, transverse waves no longer exist and the angle increases again, as the wave celerity, c, has reached a maximum value ($\sqrt{gh}$), hence $\cos\phi = c/v_s$ decreases. The wave height and the period both have a maximum around $Fr = 1$. The maximum load on banks is therefore reached when the ship accelerates or decelerates and $Fr \approx 1$. For cruising speed, when $Fr > 1$, the waves and the wake of a high speed ferry usually do not cause any damage.

Due to a completely different shape of the hull of high-speed ships, the relations as given in Figure 9-11 are not valid. In practice, the wave height is of the same order of magnitude as with displacement ships sailing at moderate speed ($Fr < 0.75$). The most striking difference is the wave propagation angle and the wave period. The wave period reaches typical values of 10 s compared with 5 s for displacement ships. This means that the behaviour of these waves on banks differs considerably. The wave propagation angle is similar to the stern wave in a primary wave. It therefore seems justified to treat these waves in the same manner when it comes to stability relations. At this moment general inform-ation on wave heights by these ships is lacking, so measurements will be needed in cases where high speed ships play a role.

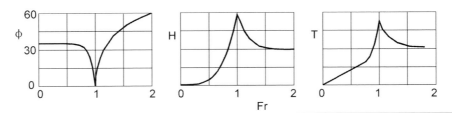

Figure 9-12 shows the turbulent velocity fluctuations in a free circular jet, compared with the fluctuations in a propeller wash. The relative fluctuations in the fully developed jet lie around 30% for both jets, but with the propeller this value is reached much earlier. It can therefore be expected that the propeller jet will diverge more than a free jet.

The jet properties can be described with expressions analogous to the relations for the jets in chapter 2:

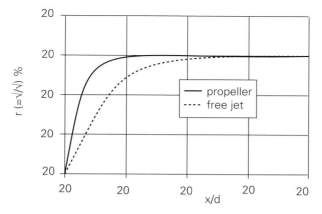

Figure 9-12 Turbulence in propeller wash and free circular jet (from RWS/DHL,1988)

$$
\left.
\begin{aligned}
u_m &= \frac{2.8u_0}{x/d} \\
b &= 0.21x \\
u &= u_m e^{-0.69\left(\frac{r}{b}\right)^2}
\end{aligned}
\right\}
\qquad
u = \frac{2.8u_0}{x/d}\, e^{-15.7\left(\frac{r}{x}\right)^2}
\tag{9.10}
$$

These expressions and Figure 9-13 show that a propeller jet indeed diverges faster than a free circular jet. The width, b, is about twice the value in a free circular jet, see also chapter 2.

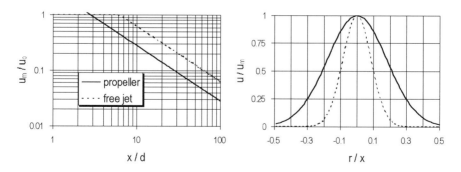

Figure 9-13 Velocity distributions in propeller wash and free jets

The values in these formulae can be estimated with: $d = 0.7 \cdot$diameter for a normal propeller and $d = $ diameter for a propeller in a jet tube. When the diameter of the propeller is not known, it can be estimated at about 70% of the ship's draught when unloaded, but usually such data is available. The outflow velocity u_0 can be estimated with:

$$u_0 = 1.15\left(\frac{P}{\rho d^2}\right)^{1/3} \tag{9.11}$$

where P is the power of the engine (in W).

When the location of the maximum or the distribution of the velocity on the bottom is not important, the maximum velocity on the bottom can be determined by differentiating equation (9.10) to x. This gives $x = 5.6\ r$ for the location of the maximum velocity. This value in equation (9.10) gives:

$$u_{b-\max} = 0.3 u_0 \frac{d}{z_b} \tag{9.12}$$

where z_b ($= r$ in equation (9.10)) is the vertical distance between the propeller axis and the bottom.

Figure 9-14 shows the relation between equation (9.10) and equation (9.12) for distances of 1 m and 10 m below a propeller of 1 m diameter and an engine power of 10^7 Watt.

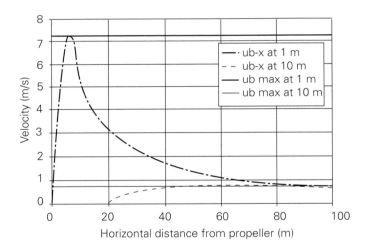

Figure 9-14 Velocities behind propeller for various cases

The velocities behind a ship's propeller are important in case the revetment or the bottom can be attacked by this flow when the ship is stationary or is manoeuvring near the bank. Once the ship is moving, this load becomes less important as the velocities in the jet are compensated by the speed of the ship (in the ideal case the ship has a velocity equal to u_0 while the jet remains motionless; compare a rocket at full speed). In that case, an indication of the velocity at the bottom can be found by reducing the values for u found with equation (9.12) with the speed of the ship.

RWS/DHL,1988 recommends reduction with half the speed of the ship, which can be seen as a conservative design approach:

$$u_b = u_{b-\max} - 0.5 V_S \tag{9.13}$$

9.3 Stability

9.3.1 Primary waves

Banks are loaded by primary waves in two ways: by the return flow, u_r, and by the stern wave, z_{max}. With regard to a stone on the bank, the return flow first accelerates and then decelerates again. In chapter 3 we have seen two possible stability approaches, Shields and Izbash, the first of which is hard to apply, since the transition from velocity to shear stress is very difficult. Therefore an Izbash-type relation is used, while the coefficient in the expression is deduced from the outcome of experiments. In RWS/DHL, 1988, a coefficient, almost twice the value of the coefficient in uniform flow, is recommended (see also chapter 3):

$$\Delta d_{n50} = 1.2 \frac{u_r^2}{2g} \frac{1}{\sqrt{1 - \dfrac{\sin^2 \alpha}{\sin^2 \phi}}} \tag{9.14}$$

For stability in the stern wave, RWS/DHL, 1988 recommends, based on experimental data:

$$\frac{z_{\max}}{\Delta d_{n50}} = 1.8 \cot \alpha^{0.33} \tag{9.15}$$

This equation is practically identical to Hudson's formula given in chapter 8. The stern wave usually leads to larger stone sizes than the return flow. The slope as a whole has to be protected against damages caused by the return flow. The upper part also has to be protected against the stern wave (and the secondary waves, see next section). The height of this part should be $(2-3)*z_{\max}$ around the water level.

9.3.2 Secondary waves

Again, only limited experimental data are available for a design formula. RWS/DHL, 1988 recommends a formula which is similar to the equation by Van der Meer. "Similar" does not mean the same, since the wave height used in Van der Meer is H_s. This means that there are many waves higher than H_s (see chapter 7). A secondary wave train contains a limited number of waves and H in equation (9.8) represents the maximum wave height (see also Figure 9-2). When the plunging part of the Van der Meer equation, see chapter 8, is used with $P = 0.1$ (revetment), $S = 2$ (little damage) and $N = 7000$ (ship loads are usually very frequent, hence the maximum number of waves), it appears that instead of H_s, $1.4*H_s$ has to be used to obtain the same results presented in RWS/DHL, 1988. The stability equation finally becomes:

$$\frac{H\sqrt{\cos 55°}}{\Delta d_{n50}} = 2.7\,\xi^{-0.5} \quad \rightarrow \quad \frac{H}{\Delta d_{n50}} = 3.6\,\xi^{-0.5} \tag{9.16}$$

where $55°$ is the angle of approach of the waves to the bank. The protection against secondary waves should reach from about $(2-3)*H$ below water level to about $(2-3)*H$ above water level.

Example 9-1

A ship with a width of 10 m and a depth of 3 m is sailing in a canal, 40 m wide and 5 m deep. The banks have a slope of 1:3 (the depth is the corrected depth according to Figure 9-8). What stone class is needed to protect the banks?

Firstly, we have to find the limit speed for this ship in this canal. $A_s/A_c = 10*3/40*5 = 0.15$. Either iterating with equation 9.4 or reading from Figure 9-4 we find $V_l \approx 0.55*\sqrt{gh} \approx 3.8$ m/s. The design speed will then be about $0.9*3.8 \approx 3.4$ m/s. Iterating with equation 9.5 or directly from Figure 9-6 we find $z \approx 0.083*h \approx 0.42$ m. We assume further that the ship is sailing 10 m from the bank, hence, $y = 5$ m (see Figure 9-8) leading to a stern wave (equation 9.6 and 9.7) $z_{max} \approx 1.5*(1+2*5/40)*0.42 = 0.78$ m. The return current is found in a similar way: $ur \approx 0.15*\sqrt{gh} \approx 1.04$ m/s. Due to the eccentricity this becomes $(1+5/40)*1.04 \approx 1.17$ m/s. The secondary waves can be calculated from equation 9.8: $H = 1.2*h*(s/h)^{-0.33}*v_s^4/(gh)^2 = 1.2*5*(10/5)^{0.33}*3.4^4/(10*5)^2 \approx 0.27$ m. The wave period becomes 1.8 s (equation 9.9).

The next step is to determine the stone classes that match these loads. For the stern wave in the canal we need (see equation 9.15): $d_{n50} = 0.78/(1.65*1.8*3^{0.33}) \approx 0.18$ m corresponding with 10-60 kg. For the return flow we find (equation 9.14 with $\tan\phi = 1$): $d_{n50} = 1.2*1.17^2 / (1.65*2*9.81*\sqrt{(1-0.31^2)}) \approx 0.06$ m. For the secondary waves (equation 9.16) the load leads to $d_{n50} = H*\sqrt{\xi} / (1.65*3.6) \approx 0.06$ m. It is obvious that the stern wave in this case dominates the load. When the waterway becomes much wider, the limit speed will be higher, the water-level depression will be smaller and (due to the higher limit speed) the secondary waves will be higher. On inland waterways, the primary wave is often dominant, while in estuaries and lakes the secondary waves can be dominant.

9.3.3 Propeller wash

The flow in a jet is very turbulent, see Figure 9-12. Values for the relative turbulence are three times higher than in uniform flow (compare Figure 2-5 with Figure 9-12). For stability in the jet flow of a propeller, an Izbash type of equation is used. Experiments in the framework of RWS/DHL, 1988 led to:

$$\Delta d_{n50} = 2.5 \frac{u_b^2}{2g} \frac{1}{\sqrt{1 - \dfrac{\sin^2 \alpha}{\sin^2 \phi}}} \qquad (9.17)$$

Note that this gives diameters that are about 4 times larger than for uniform flow! RWS/DHL,1988 indicates that this equation is based on the assumption that some transport is acceptable (in terms of the Shields-parameter, $\psi \approx 0.045$). When hardly any transport is accepted ($\psi = 0.03$), the coefficient in equation (9.17) should be 5 instead of 2.5, giving a diameter almost 8 times that for uniform flow! This will lead to very large stones and one should question whether such a no-damage criterion is realistic, since the propeller load moves and is only short-lived at a certain location. Of course the slope correction factor becomes 1 for a bottom protection.

Example 9-2

The same ship as in example Example 9-1 (width is 10 m, depth is 3 m) is moored at a jetty. The engine has a power of 1000 kW and a propeller with a diameter of 1.4 m. The propeller axis is situated about 1.5 m above the bottom. Which stone class is needed to protect the bottom at this jetty?

The effective (jet) diameter for a normal propeller is about 70 % of the real diameter $\approx$ 1 m. With equation 9.11 we find for the outflow velocity $u_0 = 1.15*(10^6/1000*1^2)^{0.33} \approx 11.2$ m/s. With equation 9.12 this leads to a velocity at the bottom, $u_b = 0.3*11.2*1/1.5 \approx 2.25$ m/s. With equation 9.17 this leads to a stone $d_{n50} = 2.5*2.25^2/(1.65*2*9.8) \approx 0.4$ m corresponding to stone class 60-300 kg.

9.4 Erosion

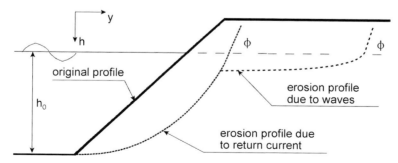

Figure 9-15 Erosion profile in navigation channel due to return current and waves

For erosion due to ships hardly any data is available. For waves, the approach in chapter 8 could be used, where the slope erosion was determined as a function of incident wave height and sand grain diameter. For erosion due to to return currents, the approach by Ven Te Chow,1959 for irrigation canals, which assumes an equal distribution of the shear stress along the bank, can be used. This leads to (see also Figure 9-15):

$$h = h_0 \cos\left(\tan(\phi)\frac{y}{h_0}\right) \tag{9.18}$$

There is, however, only qualitative evidence that these relations are also valid in navigation channels.

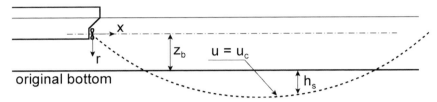

Figure 9-16 Bottom erosion due to propeller wash

For bottom erosion in a propeller wash, equation (9.10) can be used to determine the scouring depth by equating u to u_c of the bottom material. In this equation, u, is the average velocity in the jet. The extra turbulence is taken into account by applying the multiplier that comes from the stability relation in a propeller wash, see section 9.3.3. A comparison between equation (3.4) and equation (9.14) shows that the necessary diameter in a jet is about 4 times greater than in normal flow. A 4 times larger diameter means a 2 times larger velocity load, hence the factor in the jet equation (9.10) becomes 5.6 instead of 2.8. Every location in the jet where $u = u_c$ can now be computed from:

$$h_s = x\sqrt{\frac{-\ln\left(\dfrac{u_c x}{5.6 u_0 d}\right)}{15.7}} - z_b \tag{9.19}$$

where $h_s + z_b = r$ in equation (9.10). Figure 9-16 clarifies the approach. Again, there is no experimental evidence to confirm this equation. It should therefore be used carefully as an indication of the necessity of a protection.

9.5 Summary

This chapter deals with the design of protections against ship loads in waterways. The loads due to ships show many similarities with loads presented in the previous chapters on flow and waves. The loads in this chapter are divided into three types: the primary wave, related to the length of the ship, the secondary waves, related to the shape of the bow of the ship, and the propeller wash, related to the propulsion of the ship.

The limit speed of a ship is important for the design of protections. It depends on the length of the ship, the waterdepth and the blockage of the waterway. Figure 9-5 gives an overview of these relations. For the design of protections it is recommended to assume a value of 90% of the limit speed.

The primary wave is dominant in relatively small waterways where the blockage of the cross-section by ships is not negligible (which can already be the case when the blockage is about 10%). Within the primary wave, the stern wave usually gives the most severe attack on the banks of the waterways. Expressions are given to compute the height of the stern wave from the characteristics of ship and waterway, see section 9.2.2. Equation (9.15) gives the relation between the stern wave and the necessary stone dimensions.

The secondary waves are dominant when blockage does not play a role. Their magnitude depends much more on the shape of the bow than on the draught or length of the ship. Equation (9.8) gives an upper limit of the wave height that can be expected. The angle of approach to the banks is more or less constant (55°) for displacement ships. Equation (9.16) gives a simple relation for the stability of stones in secondary waves.

The propeller wash plays an important role at locations where ships manoeuvre at low velocities. The flow in that situation can be approximated with the same kind of relation as used for a circular jet. Equation (9.17) gives the relation between the velocity at the bottom (from equation (9.12) or (9.13)) and the stone size.

At present, very little is known about erosion due to ship motion. Section 9.4 gives some advice on how to get a first impression of the possible erosion on unprotected banks and bottoms.

10 DIMENSIONS

Dice blocks on breakwater Hook of Holland (courtesy Rijkswaterstaat)

10.1 General

In the previous chapters, many relations between load and strength have been presented. This chapter focuses on how to deal with the variations in loads as encountered in nature and with the strengths of structures which also show variations and which decreases after construction. Maintenance and repair are therefore important to maintain the strength of a structure in time.

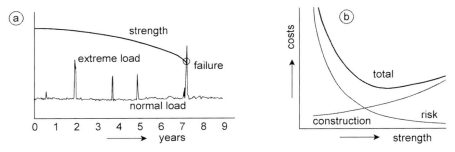

Figure 10-1 Failure, risk and costs

Figure 10-1a shows the eventual destiny of any structure without maintenance. The strength decreases due to wear and tear, erosion or fatigue. The load has a random nature and there is a probability that, at some moment, the load will exceed the strength and that the structure will collapse, crack or fail. Figure 10-1b shows the idea behind an economic design: a very strong structure runs little risk of failure but is expensive, while a less strong structure is cheaper but the risk is high. Somewhere in between, there is an optimum. Although very logical and simple, it is hard to quantify this into workable numbers.

The risk of failure can be expressed in general terms:

$$\text{Risk} = \text{probability} * \text{consequence} \tag{10.1}$$

in which the probability (a number between 0 and 1) can be expressed as probability per year and the consequence can be expressed as loss of money and/or human lifes, hence the unit of risk is loss per year. To keep the risk acceptably low, an event with a large consequence should have a low probability. Section 10.2 gives more details.

The consequence is related to the choice of the limit state in the failure analysis. As already said in chapter 1, a distinction is made between the Serviceability Limit State (SLS) and the Ultimate Limit State (ULS). The ULS defines collapse or such deformation that the structure as a whole can no longer perform its main task. It is usually related to extreme load conditions. The SLS defines a partly or temporarily unusable state of structure. In the case of a harbour, the SLS could represent such wave penetration that transshipment of goods is impossible, while the ULS could be the collapse of a breakwater. The choice of the SLS is less clear e.g. for a bottom or dike protection. In Vrijling et al.,1992, the SLS is seen as the deterioration of strength under persistent loads. In the light of equation (10.1) it is obvious that the probability of reaching a ULS should be much lower than reaching an SLS. In the example of the harbour P_{SLS} = 1/year and P_{ULS} = 0.01/year. Section 10.3 goes into more details concerning maintenance policies and the relation between the SLS and maintenance.

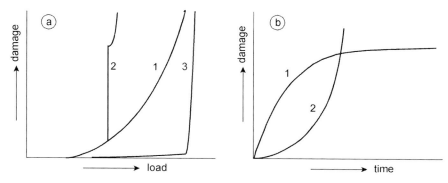

Figure 10-2 Differences in structural behaviour

When an acceptable probability needs to be determined, the behaviour of the structure and the material is important. Figure 10-2a shows the damage as a function of load, given a certain loading period. Line 1 represents the "normal" behaviour of rock under waves or flow. Line 2 is valid for a relatively thin layer of the same rock with a filter layer of much finer material underneath. Besides removal of the top layer, this leads to a large increase of damage and further deterioration depending on what is behind the filter. Line 3 may hold for a block revetment with clamping forces. Up to a certain load, the blocks are lifted somewhat, but there is no real damage. Beyond a certain level, the load causes a sudden instability and a complete row of blocks can be removed.

Figure 10-2b is similar to Figure 10-2a, but shows the damage as a function of time given a certain load. Line 1 is the behaviour found for rock under waves or current. In the Van der Meer formulas, the number of waves influences on the damage (S ∝ √N) more or less according to line 1. In the Van der Meer relations, the number of waves is limited to a maximum of 7000 since in experiments the damage hardly increases beyond that number. Line 2 may be valid for an asphalt revetment which is lifted by water pressure. After some time, due to the plastic behaviour of asphalt, the deformation suddenly increases and even rupture of the material is possible.

In a probabilistic analysis, some of the above-mentioned properties can be included in the distributions of strength and load. In many cases, however, these peculiarities are handled with "engineering judgment", a strange mixture of experience, common sense and intuition, often leading to keeping a safe margin. A designer must always foresee the consequences of a failure. Thinking about failure mechanisms is therefore important, as will be treated in Section 10.4.

10.2 Probabilistics

10.2.1 Introduction

It is impossible or uneconomical to make a structure so strong that it will never fail. A hydraulic engineer's task is therefore to design structures with an acceptable risk of failure. In terms of load and strength, this means that the probability that the load exceeds the strength should remain below a certain value. This value depends on the consequences, see again equation (10.1). It is possible to express load and strength in a limit state function Z:

$$Z = \text{Strength - Load} = R - S = R(x_1, x_2, \ldots x_m) - S(x_{m+1}, \ldots x_n) \tag{10.2}$$

where R is the strength and S the load and $x_1 \ldots x_m \ldots x_n$ represent all random variables involved in strength and load (like wave height, wave period, stone diameter, slope angle etc.). S as a symbol for load and not for strength does not seem logical, but this is according to international agreement. R and S are acronyms related to the French words *Résistance* and *Sollicitation* ("asking"). We will stay in line with this agreement, despite the confusion at first glance.

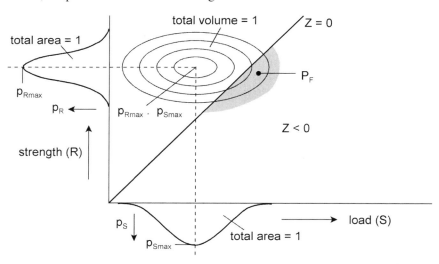

Figure 10-3 Probability mountain

Figure 10-3 gives the probability density distributions of load and strength (p_S and p_R, respectively). When $R = S$ ($Z = 0$), the structure (or part of it, depending on the scope of Z) reaches the limit state as defined with Z. A three-dimensional representation of the probability distributions is most appropriate, as, theoretically, for every load S there is a $R < S$ and, the other way around, for every strength R there is a $S > R$. So, the combination of probabilities of $Z < 0$ determines the total probability of failure: P_F.

P_F is the volume of the part of the "probability mountain" (of which the contour lines have been drawn in the figure) where $Z < 0$. This volume is given by:

$$P_F = P(Z < 0) = \iint\limits_{Z(x)<0} ... \int p_{\underline{x}}(x) \, dx_1 .... dx_n \tag{10.3}$$

where $x_1 ... x_n$ again represent all parameters involved. Every parameter has its own probability distribution and the determination of P_F can lead to very complex and labour-intensive computations. Finding reliable probability functions for every parameter is also a difficult and time consuming process.

Intermezzo 10-1

Some history

Dikes in The Netherlands are much older than probabilistic design methods. The height of a dike was always related to the "highest water level ever", the locally registered maximum water level in history. Whenever the maximum registered water level was exceeded a new "highest water level ever" could be defined. This went on for many centuries and people lived with these ways of life. In a modern society, where man tries to minimize risks, this is highly unsatisfactory. In 1939 the knowledge of statistics led to the awareness that higher water levels would always be possible, but with a lower probability (see Wemelsfelder, 1939).

However, it took another disaster in 1953 to affect decision making. No maximum wind velocity or rainfall is known in nature, hence, there is no maximum water level due to storm surges or river discharges either. A probabilistic approach is the only way to deal with loads in nature. In structural engineering the situation is somewhat different. The maximum load on a bridge can be influenced by regulations e.g. by means of the maximum permissible loading capacity of trucks. But even then uncertainties remain. What is the probability of an overloaded truck crossing the bridge or what will the future intensity of heavy loaded trucks be in relation to fatigue of the bridge material?

Four different levels of the probabilistic design approach can be discerned:
- *Level III: fully probabilistic approach*

  P_F is determined by means of numerical integration of the probability functions in equation (10.3) or by means of randomly drawn realisations of these functions ("Monte Carlo" approach), see section 10.2.3.
- *Level II: approximate probabilistic approach*

  P_F is approached by means of simplified functions, see section 10.2.4.

- *Level I: quasi-probabilistic approach*
  For every parameter involved in load or strength a partial safety factor is used, e.g. one derived from a level II computation, see section 10.2.5.
- *Level 0: deterministic approach*
  Some maximum load and minimum strength is taken based on experience and/or intuition (see Intermezzo 10-1) and one overall "safety factor" is applied. It will be clear that this is not a probabilistic approach at all. An overall safety coefficient, γ usually says nothing about the safety, see Figure 10-4, where γ is the same for both cases, but P_F is completely different. Good engineering judgement requires a larger γ when the variation is larger.

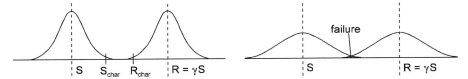

Figure 10-4 Equal overall safety coefficient with different probability distributions

Another approach can be to work with characteristic values for load and strength, with a small chance that these values are too low or too high, see Figure 10-4. But even then, if the distribution of the extreme values is not known, the real meaning of γ is uncertain.

10.2.2 Comparison of methods

To illustrate the use of a traditional deterministic approach compared with the various levels of probabilistic approach, the following example are used, see Figure 10-5. A building is situated in the vicinity of the coastline and is protected by a rip-rap revetment, sufficiently high and including filter layers and a toe protection.

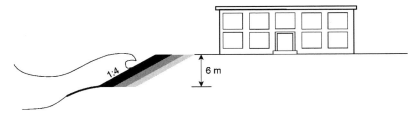

Figure 10-5 Example structure for comparison of probabilistic and deterministic approach

The idea is the following. Firstly, a design is made with a "traditional" deterministic approach (level 0). A characteristic wave height is chosen from the available wave observations and the threshold of motion is taken as a characteristic strength,. Subsequently, a fully probabilistic method (level III) is used to show what is a more realistic approach. Next, the results of level II and level I methods are shown to illustrate the merits of these methods. Finally, a level 0 approach is applied again, using the results of the probabilistic methods, showing that common sense always pays. The example is simple and is meant as an educational tool, not as a practical application.

Ten years of wave observations are available, see Table 10-1. Only waves > 0.5 m have been processed, as they are representative for "storms", lasting several hours. In these ten years, the highest recorded wave height was 1.62 m.

Table 10-1

Wave height interval (m)	Occurrences in 10 years	Exceedances in 10 years	Return period (yrs)
0.51-0.6	48	98	0.1
0.61-0.7	29	50	0.2
0.71-0.8	21		
0.81-0.9	6	20	0.5
0.91-1.0	4		
1.01-1.1	3	10	1
1.11-1.2	2		
1.21-1.3	2	5	2
1.31-1.4	1		
1.41-1.5	1	2	5
1.51-1.6	0		
1.61-1.7	1	1	10

Deterministic approach

With the highest recorded wave height, the stone size for the top layer is calculated with the Van der Meer formula, see chapter 8:

$$d_{n50} = \frac{H_{sc}\xi^{0.5}}{\Delta 6.2 P^{0.18}\left(\dfrac{S}{\sqrt{N}}\right)^{0.2}} \qquad (10.4)$$

Swell does not play a role at this coast, so for the wave steepness a value of 0.05 can be used. With a slope angle of 1:4 this means that only the plunging part of the formulas has to be used. The permeability, P, for a revetment on sand ≈ 0.1, $\Delta \approx 1.6$ and the number of waves, N, is 7000 (maximum). The damage number, $S = 2$ is chosen, as it is representative for the threshold of motion. This leads to $d_{n50} = 0.56$ m and a choice of rock class 300-1000 kg ($d_{n50} \approx 0.6$ m).

This can be seen as an example of a classical deterministic approach. Now, several probabilistic methods will be used to establish the risk of failure of this structure.

Probabilistic approach

Probabilistic calculations will be done with the VaP package from ETH Zürich (see Petschacher, 1994). In a probabilistic approach, a limit-state function has to be defined. For this example the Van der Meer relation is rewritten:

$$Z = 6.2\, P^{0.18} \left(\frac{S}{\sqrt{N}} \right)^{0.2} \xi^{-0.5} - \frac{H_{sc}}{\Delta d_{n50}} \tag{10.5}$$

Note that it is not strictly necessary to separate strength and load factors. If this Z-function < 0, the structure fails. With the values from the deterministic approach we find $Z = 0.15$, slightly positive since we used a larger stone (0.6 instead of 0.56 m). Probability distributions for all parameters are used in the computation of the total probability. So, firstly, these distributions have to be estimated. The wave height distribution is determined from the available wave observations, see Table 10-1 and Figure 10-6.

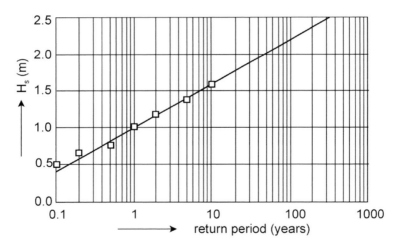

Figure 10-6 Long-term wave height distribution

In chapter 7 (Appendix 7.7.2) wave statistics were given for an irregular wave field. Such a wave field could be represented by the significant wave height, H_s, while the wave heights in a registration were described with a Rayleigh-distribution. So, H_s represents the wave condition at a certain moment, or better for a short period of one or more hours and the Rayleigh-distribution can be seen as the short term wave height distribution. All registered values of H_s give a distribution for the long term. This distribution is normally described with a Weibull-distribution, but often an exponential distribution gives reasonable results. Figure 10-6 shows an exponential distribution. This distribution has to be described mathematically to be used as input for the VaP-model.

The general expressions for the exponential distribution are:

$$f(x) = \lambda \exp(-\lambda(x - \varepsilon)) \rightarrow F(x) = 1 - \exp(-\lambda(x - \varepsilon)) \rightarrow (1 - F(x))^{-1} = \exp(\lambda(x - \varepsilon))$$

probability   density   probability   X < x   return period

(10.6)

Note: the function is defined only for $x \geq \varepsilon$ since negative probabilities are impossible, see second equation in (10.6).

Figure 10-6 gives the return period of the wave heights, so the parameters λ and ε have to be derived from the third equation of (10.6). This can be done by taking two values of the line in Figure 10-6, e.g. for return periods of 1 and 10 years:

$$\left. \begin{array}{llll} 1 = e^{\lambda(1-\varepsilon)} & \rightarrow & \ln 1 = 0 = & \lambda - \lambda\varepsilon \\ 10 = e^{\lambda(1.6-\varepsilon)} & \rightarrow & \ln 10 = 2.3 = 1.6\lambda - \lambda\varepsilon \end{array} \right\} \rightarrow \varepsilon = 1,\ \lambda = 3.83 \qquad (10.7)$$

For d_{n50}, Δ, $\tan\alpha$ and s, a normal distribution is assumed. The mean values for these parameters are equal to the ones used in the deterministic approach: 0.6 m, 1.6, 0.25 and 0.05, respectively. The standard deviations are estimated as: 0.05 m, 0.1, 0.0125 and 0.01, respectively. P is assumed to have a log-normal distribution, to avoid errors caused by negative values of P in the calculation. The mean value is 0.1 and the standard deviation is 0.05. N is given a deterministic value of 7000.

Note: The normal distribution for d_{n50} is not a distribution curve within a stone class (e.g. a sieve curve). It represents the deviations in characteristic diameter for a whole mass of stones. Compare the Rayleigh-distribution within a wave record (characterized by H_s) and the long-term distribution of H_s

10.2.3 Level III

Numerical integration of Z is one of the available level III methods but will not be used here. Another method is the Monte Carlo method. The basis of this method is quite simple, see Figure 10-7. For all parameters, a random number is drawn, taking into account the probability distribution. This means that a value with a high probability density will appear more often. So, after many draws, the histogram of a normally distributed parameter will show the well-known Gauss-shape.

When all parameters have a value, the resulting value for Z is computed from equation (10.5). This whole procedure is repeated N times after which P_F simply is N_F/N (N_F being the number of times that $Z < 0$). The procedure is simple but the number of repetitions is very high, which makes the Monte Carlo method a computer job par excellence.

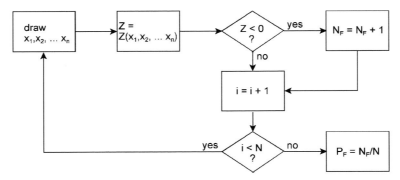

Figure 10-7 Procedure of Monte Carlo method

Computations with S = 2

Figure 10-8 shows two realisations of the Monte Carlo procedure for $S = 2$, resulting in $P_F = 0.095$ and 0.091, respectively. The number of draws, N, determines the accuracy of the method. A rough estimate of the necessary value of N is: $N > 400/P_F$, so N has to be checked after P_F has been determined. An average, after three calculations, $P_F = 0.094$, which is the probability per year since the wave heights have been introduced as numbers of exceedance per year, see Figure 10-6. This value seems logical: the deterministic design was made with a 1/10 year wave (which has a probability of exceedance in one year ≈ 0.1) and a slightly larger stone was chosen ($d_{n50} = 0.6$ instead of 0.56 m).

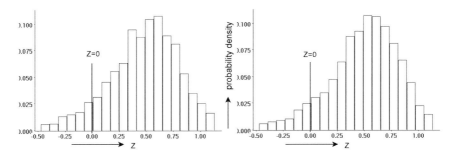

Figure 10-8 Realisations of Monte Carlo simulation

$S = 2$ was used in the deterministic approach. This can be seen as a Serviceability Limit State (SLS), beyond which some repair can be necessary. So, there is a 10% chance per year that some damage to the revetment will occur. This result is not really astonishing. But how safe is our building really? To judge the safety, it is necessary to determine the probability of damage to the building. This requires determination of the Ultimate Limit State (ULS).

Computations with S = 10

The ULS will be approximated here simply with $S = 10$. The reason for this is as follows: $S = 10$ indicates the damage when the top layer (which is about $2 \cdot d_{n50}$ thick) has been removed completely. Once the top layer removed, the filter layers, which have much less resistance against wave loads, will be removed as well. Underneath the revetment there is only sand. Equation 8.2 gives an idea of how far the coast will erode. With a wave of 1.6 m and fine sand, this will be about 40 m. This means collapse of the building near the shore. Implicitely it is assumed that the storm lasts for several hours, which is reasonable. So, although roughly, $S = 10$ can indeed be seen as total failure.

The Monte Carlo simulation for $S = 10$, with all other parameters equal to those in the previous case, gives $P_F = 0.011$. So there is a 1% chance per year of total collapse of the building. Is this a problem? The lifetime of a building is normally many decades, say 50 years. The probability of collapse is then given by:

$$P_F \ in \ 50 \, years = 1 - \left(1 - P_F \ / \ year\right)^{50} = 0.42 \tag{10.8}$$

So, there is almost a 50% chance that this building will be destroyed during its lifetime! It should also be noted that even the best maintenance policy can not prevent this, since a storm that causes a damage level of 10 will also do so when the armour layer is still completely intact. The capitalized risk is:

$$R = \sum_{n=1}^{50} P_F D \frac{1}{\left(1+r\right)^{50}} = P_F D \frac{1 - \left(\dfrac{1}{1+r}\right)^{50}}{r} \tag{10.9}$$

in which D is the total damage and r is the interest rate. D, (including the economic activities related to the building) when $S = 10$ is set to $10 \cdot 10^6$ € and r is assumed to be 5%. The capitalized risk is then $0.011 \cdot 10 \cdot 10^6 \cdot 18.25 = 2 \cdot 10^6$ €.

The final answer to this dilemma has to come from econometric considerations. These considerations are presented very simply. Revetments with various strength will be compared with a focus on costs and risk. For the involved risk, Table 10-2 gives the results of Monte Carlo computations and equation (10.9):

Table 10-2

Armour layer (kg)	d_{n50} (m)	P_F per year (-)	P_F per 50 years (-)	Risk (10^6 €)
60 - 300	0.4	0.189	0.999	34.5
300 - 1000	0.6	0.011	0.42	2.0
1000 – 3000	0.85	0.001	0.049	0.18
3000 – 6000	1.1	0.00017	0.0085	0.03

This has to be compared with the costs of the different revetments. Again, some simple assumptions will be made. The length of bank necessary to protect the building is assumed to be 200 m. The evolved length of the protection along the slope of the bank directly follows from the geometry of Figure 10-5: $\sqrt{(6^2 + 24^2)} \approx 25$ m. The layer is taken $2 \cdot d_{n50}$ thick. Only the costs of the top layer are different and one extra filter layer for the largest stone classes is needed. The costs of the other activities to construct the revetment (excavating, creating slope, filter layers, toe protection) are assumed to be $1 \cdot 10^6$ € for all revetments. Table 10-3 gives the costs:

Table 10-3

Armour layer (kg)	Cost per m³ (€)	Volume (m³)	Costs extra filter layer (10^6 €)	Costs (incl. extra filter) (10^6 €)	Total costs revetment (10^6 €)
60 - 300	20	4000	0	0.08	1.08
300 - 1000	24	6000	0	0.14	1.14
1000 – 3000	30	9000	0.02	0.27	1.27
3000 – 6000	36	11500	0.02	0.42	1.42

The difference in costs is small: the initial costs to construct the revetment ($1 \cdot 10^6$ €) are dominant. Comparison of costs and risks now shows the following picture, see Figure 10-9. It is obvious that stone class 1000-3000 kg instead of 300-1000 kg is a good choice, since the risk decreases with a factor 10 for just a small amount. Another solution could be a thicker top layer, permitting a larger S before the structure is endangered.

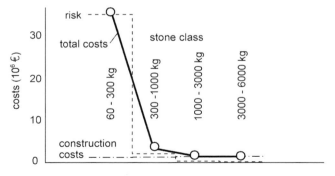

Figure 10-9 Risk and costs for various top layers

This choice becomes completely different when there is no expensive building directly on the shore. For example, if there is a protection, only to prevent further meandering of an estuary channel, the cost-risk ratio is completely different and the deterministic approach of section 10.2.2 is perfectly adequate (see also the evaluation in section 10.2.6).

10.2.4 *Level II*

The probabilistic level II approach is a collective term for approximate solutions of the failure probability by means of linearisation around a well-chosen point, the so-called *design point*. The limit-state function, Z, is described with a normal distribution just like all parameters that make up Z. This means that a deviating distribution of a parameter will be replaced by a normal distribution, which has the same value and slope in the design point as the original probability function. The failure probability, finally, is determined from the properties of the normally distributed Z-function, μ_Z and σ_Z via, see also Figure 10-10:

$$\beta = \frac{\mu_Z}{\sigma_Z} \tag{10.10}$$

The failure probability P_F and β are directly related in the normal distribution and can be found in standard tables. So, if β can be derived from the known parameter distributions, P_F can also be known.

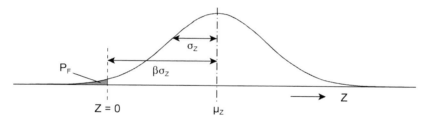

Figure 10-10 Failure probability in level II approach

The *design point* is the point on the line $Z = 0$ where the probability density of the combination of load and strength has its maximum. When a structure fails, load and strength will probably have values near the design point values. The distance between the average value of a parameter and its design value is an indication of its importance with regard to σ_Z and, hence, to the failure probability of the structure. This importance is indicated with α_i for each parameter. Appendix 10.6 can serve as an explanation if the procedure is not clear.

The VaP results for the original Z-function, equation 10.5, are:

Table 10-4

$\beta = 2.33$, $P_F = 0.0099$			
Parameter	α-value	α-value^2	Design value
H_s	0.90	0.81	2.054 m
d_{n50}	-0.24	0.06	0.571 m
P	-0.25	0.06	0.068 -
Δ	-0.18	0.03	1.557 -
$tan\alpha$	0.07	0.01	0.252 -
s	-0.15	0.02	0.046 -
		$\Sigma = 1$	

The influence of the wave height variation on the failure probability is clearly dominant. This is often the case; the load variations are more important than the strength variations.

Comparing the failure probability from the level II-analysis with the results of the Monte Carlo approach of section 10.2.3 gives an idea of the reliability of the level II method as a whole. $P_F \approx 0.011$ is found with the Monte Carlo approach compared with $P_F \approx 0.01$ from the level II approach. A difference of 10% is acceptable, since computations like this only serve to obtain an indication of the failure probability.

The advantage of a level II method is the resulting α_i-values, which indicate the relative importance of a parameter in the total failure probability. The combination of a large σ_i and a large power in the Z-function lead to a high α-value for a parameter. So, with good engineering judgement, a large α-value does not come as a surprise and an engineer intuitively chooses a conservative value for such a parameter.

10.2.5 Level I

A level I approach adds nothing new to what has been said above. It is actually an application of the results of a higher-level method, especially level II methods. The approach requires a design value and a safety coefficient to be established for every parameter. The safety coefficients are derived from a level II computation, using the α-values and β-value. The β-value stands for the required safety and the α-values stand for the relative importance of each parameter. The partial safety coefficient for each parameter is then given by:

$$\gamma_i = \frac{\mu_i - \alpha_i \beta \sigma_i}{\mu_i}$$
(10.11)

α-values are negative for loads and positive for strength, leading to $\gamma > 1$ for loads and < 1 for strengths when used as a multiplier in both cases. Other definitions are possible and can be found in literature. The safety factors in equation (10.11) are defined with regard to the average values. When using other characteristic values, the safety factor changes correspondingly. This approach can be seen as the application of engineering judgement as mentioned in the previous section.

10.2.6 Evaluation

In section 10.2.2 we started with a deterministic approach and compared the result with various levels of probabilistic methods. A deterministic method results in a certain strength (in the example the necessary stone class), given a certain load (in the example the wave height). A probabilistic approach results in a failure probability, given the distributions of load and strength. So, a probabilistic method never leads directly to dimensions for a structure. Given an acceptable failure probability, the dimensions have to be determined iteratively.

The risk analysis in section 10.2.3 showed that the deterministic approach led to an unacceptable high risk in the case of a building near the shore line. Would it have been possible to avoid this high risk without sophisticated probability models?

The deterministic approach was based on a wave height with a return period of 10 years and a low damage level, $S = 2$. From the risk analysis it became clear that the real issue is an acceptable low probability of failure of the building, equivalent to $S = 10$. Comparing the cost of a revetment, which is in the order of magnitude of 0.5 mln €, with the risk of collapse of a building of 5 mln €, the failure probability should be less than 10% during the lifetime of the building, which is about 50 years, say 5%. A convenient formula to deal with this issue is the Poisson-equation, which is an approximation of equation (10.8):

$$P = 1 - \exp(-fT) \tag{10.12}$$

in which: P = probability of occurence of an event one or more times in period T
   T = considered period in years
   f = average frequency of the event per year

For $P = 0.05$ and $T = 50$ years we find $f = 1/1000$. In other words: a wave height with a return period of 1000 years should be used. From Figure 10-6 we find: $H_s \approx 2.8$ m. With S = 10 in equation 10.4, this gives: $d_{n50} \approx 0.7$ m. This would lead to a stone class 1000-3000 kg instead of 300-1000 kg. The same procedure could be applied to the maintenance. In that case a 10% chance of failure (SLS) or, in other words, the need for maintenance, is reasonable. This would lead to the same calculation as carried out in the deterministic approach of section 10.2.2 (once in 10 years wave height and $S = 2$). In this case, however, the ULS is dominant. In the case where there is not an expensive building on the shore, maintenance is the only criterion.

The Poisson equation (10.12) combined with common sense as outlined above can be seen as a semi-probabilistic approach which can serve as a very useful tool in a preliminary design.

Of course, to establish the dimensions of a complex and important structure, like a storm surge barrier to protect a large and densely populated area, a probabilistic approach is an important design tool. Since probabilistic computations can only be done with some given structure, deterministic calculations are always carried out first and the final dimensions result from iteration between both approaches. A level III method is the best way to determine the failure probability. Level II methods are useful to get an insight into the relative importance of the various parameters involved.

Equation (10.12) can also be used to show other elementary statistical aspects:

- A common mistake is to design a structure with an envisaged lifetime of 5 years (e.g. a temporary situation during construction) with a load condition with a frequency of 1 per 5 years. Equation (10.12) shows that the probability that this load is reached or exceeded is 1-exp(-1) $\approx$ 0.63. So, there is a 63 % chance that conditions will be worse than assumed, which is usually not acceptable.

- In the case of a sea defence structure, to protect a low-lying area, it is impossible to define the lifetime of the structure since it is supposed to protect "forever". Assuming an acceptable probability of 1% in a human life (e.g. 75 years), instead of using the lifetime of the structure, equation (10.12) gives the frequency for the design conditions: $f \approx 1/7500$ year which is quite a normal number for dangerous flooding hazards.

10.3 Maintenance

10.3.1 Introduction

Maintenance is essential for every structure. Throwaway products may also penetrate into engineering, but they will always be limited to parts of structures. Even then, the replacement of these parts is maintenance. Maintenance is primarily focused on the strength of a structure but monitoring of boundary conditions (e.g. sea level rise) can also be part of a maintenance program. Maintenance is therefore part of the management of a structure and consists mainly of inspection and repair. The total management policy links design, maintenance and the risk of failure of a structure, see also section 0 and Figure 10-1. A picture similar to Figure 10-1 can be drawn for an optimum maintenance strategy.

10.3.2 Maintenance policies

Several maintenance policies are possible, depending on the predictability of the decrease of strength in time, the costs and possibilities of inspection and repair, and the consequences of failure. The different policies are, see also Vrijling et al.,1992 and CUR, 1991:

Failure-based maintenance

Figure 10-11 shows the concept of failure-based maintenance. No action is taken until a structure, or part of it, fails. After that it is repaired or replaced by a new one. An example from every day life is a light bulb in a living room which is only replaced after it stops functioning. This is an efficient maintenance policy since the complete lifespan is used, but it is only permissible for non-essential parts with low risk. In hydraulic engineering this is no common practice.

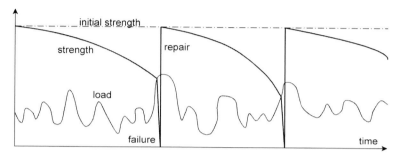

Figure 10-11 Failure-based maintenance

Time-based maintenance

If the deterioration can be predicted reasonably well, a time-based maintenance policy can be applied, see Figure 10-12. This is e.g. the case when the deterioration is governed by wear and tear due to the weather. An example from every day life is the painting of window frames of a house. An example from hydraulic engineering is painting a steel gate of a storm surge barrier.

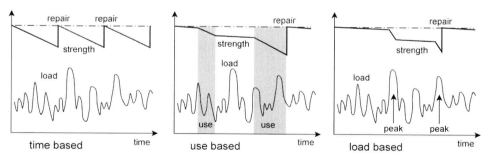

Figure 10-12 Time-, use- or load-based maintenance

Use-based maintenance

Use-based maintenance depends on the usage of the structure. This is the case when the wear and tear is mainly a function of use intensity. An example from every day life is the overhaul of a car engine after so many km's. An example from hydraulic engineering is the overhaul and repair, if necessary, of the machinery of a gate of a storm surge barrier.

Load-based maintenance

When the deterioration is mainly governed by extreme loads, repair is carried out after the occurrence of an extreme load or after cumulation of loads. An example from everyday life is the replacement of safety belts in a car after a collision.

State-based maintenance

This is probably the most common maintenance policy in hydraulic engineering. An example from everyday life is the repair of a roof which is inspected every now and then and treated when its state gives rise to repair. Examples from hydraulic engineering are revetments.

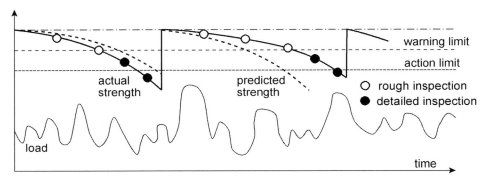

Figure 10-13 State-based maintenance

Figure 10-13 gives the procedure for state-based maintenance. The structure is inspected roughly at fixed intervals, based on the expected and predicted decrease of strength. After some warning limit state is reached, the inspections become more frequent and more detailed. When the action limit state is reached, repairs have to be carried out.

Figure 10-14 shows how to arrive at the best maintenance policy. In this picture monitoring means keeping track of time or of usage or of the load that the structure has been subjected to. Inspection has to be done on the spot.

In practice, a mixture of the various policies is often used. E.g. in addition to a time-based policy, inspection can lead to postponing repair if the state of the structure is better than expected. In general, one can say that inspection reduces the uncertainties in the parameters.

Which maintenance policy is most appropriate, depends on the consequences of failure and the predictability of the strength as a function of time, see Figure 10-14.

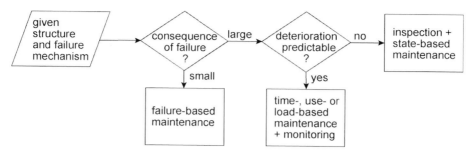

Figure 10-14 Choice of maintenance policy

10.3.3 Probabilistic approach of inspection

An outlet sluice is constructed with a bottom protection to avoid damage due to scour. The apron is just long enough to prevent collapse of the sluice structure if a slide occurs in the scour hole. Before that happens, the slope of the scour hole should be covered with slag material or gravel. If the slope slides, the scour depth is an estimated 8 m (the critical scour depth) and the slope angle after sliding is estimated to be 1:6, leading to a necessary apron length of say 50 m, see Figure 10-15a and chapter 4 for further details.

Note: *Of course, the stability of a scour hole depends heavily on the geotechnical properties of the soil and the upstream slope of the scour hole. All of these factors have been simplified in this example into one parameter, the critical scour depth: h_{sc}.* The scour depth as a function of time is determined with the Breusers-equation, see chapter 4:

$$h_s(t) = \frac{\left(\alpha\bar{u} - \bar{u}_c\right)^{1.7} h_0^{0.2}}{10\,\Delta^{0.7}}\, t^{0.4} \tag{10.13}$$

This function can be rewritten as a limit state function:

$$Z = h_{sc}(t) \quad - \quad \frac{\left(\alpha\bar{u} - \bar{u}_c\right)^{1.7} h_0^{0.2} t^{0.4}}{10\,\Delta^{0.7}} \tag{10.14}$$

The following values are assumed in these equations (all parameters with normal distributions):

Table 10-5

Parameter	α	$\bar{u}$	$\bar{u}_c$	h_0	Δ	h_{sc}
Mean (μ)	2.5	1 m/s	0.5 m/s	10 m	1.65	8 m
Deviation (σ)	1	0.1 m/s	0.05 m/s	0.25 m	0.05	2 m

The most uncertain parameters are α and h_{sc}, hence they get a large deviation. Figure 10-15b shows the development of the scour depth as a function of time according to equation (10.13), using the average value of Table 10-5 ($\alpha = 2.5$).
Note: α is used as a turbulence parameter in equation (10.13) and in the probabilistic approach it is used as an influence parameter.

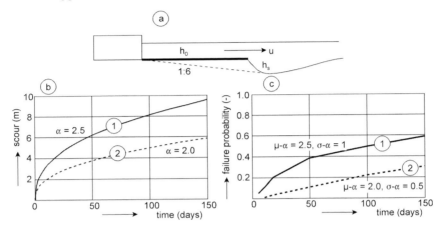

Figure 10-15 Failure probability scour hole

The question now is how soon and how often should the scour depth be sounded in order to know when the scour slope has to be covered, e.g. when the hole is 5-6 m deep. A level II analysis (with VaP) is carried out with the other values in Table 10-5 with time, t, (effective flow time in days) as a deterministic parameter with different values. Figure 10-15c shows the results (upper line). According to equation (10.13) the critical scour depth of 8 m will be reached after 95 days (Figure 10-15b). Figure 10-15c shows that there is a 50% probability that the depth is indeed 8 m after 95 days. This is, of course, logical: with all parameters normally distributed, there is 50 % chance that the value is higher and 50% chance that it is lower. No matter how large the deviations, the Gauss-curve is centred around the expectation value.
But a probability of 50% is very high and the risk is unacceptable. So, the reponsible manager will not postpone inspection till the expected critical depth is reached according to equation (10.13). When a low probability is considered acceptable, e.g. 5%, Figure 10-15c shows that inspection should be carried out after 5 days of effective flowing time! Sounding is not very expensive and this can be seen as a test run of the new sluice.
During the first 20 days of use of the sluice, more soundings can be done and the result could be that the scour development is not along line 1 in Figure 10-15b, but along line 2. This means that the estimated α-value (determined either by rough estimates as illustrated in chapter 4, or by considering the results of model tests) was too pessimistic. With the results of the soundings, a new value for α can be

determined (assuming that the values for the other parameters are correct). The α-value for line 2 is 2 for which, again, a probabilistic calculation can be carried out. The results are directly valid for the considered structure, so a lower deviation can also be assumed, e.g. 0.5 instead of 1, as was the case in Table 10-5. Line 2 in Figure 10-15c shows the results for the new probability failure. The next inspection could be after 50 days of flow. If the scour process is still according to line 2 in Figure 10-15b, the slope could be covered after 100 to 150 days flow time.

Note: When applying a lower α-value based on soundings, one should be sure of the composition of the soil. If the upper layer consists of a more resistent layer with some cohesion, scour can accelerate again after breaking through this layer.

This example demonstrates the use of inspection in maintenance. The costs are not prohibitive and it reduces the uncertainty about the strength, hence it increases the predictability of the behaviour of the structure.

10.4 Failure mechanisms

10.4.1 Introduction

Chapter 1 already stated that it is paramount for a designer to have an idea of the different failure mechanisms of a hydraulic structure. It is repeated once again that *most structures fail, not because the incoming wave height has been underestimated with 10 or 20 %, but because a failure mechanism has been neglected.* The contents of chapter 1, concerning fault trees, will be further elaborated on relating them to probabilistics and illustrating them with some examples.

10.4.2 Systems

A systems approach to the design of a hydraulic structure is a rather abstract notion, but it can serve to illustrate some important elements of a design. Two important concepts in structural systems are series and parallel systems, see Figure 10-16. In a series system, every broken element means total failure. In a parallel system, the function of a broken element can be transferred to other elements. An example of a series system amongst protections is the various possible slip circles in a slope, see chapter 5. The weakest circle determines the safety. An example of a parallel system is a single slip circle in non-homogenous soil. Strong soil layers can compensate for weak parts in the soil, as all layers contribute to the total resistance.

A series system consisting of two or more elements, has a total probability of failure between:

$$\max P_{Fi} \quad \leq \quad P_{F-tot} \quad \leq \quad \sum P_{Fi} \tag{10.15}$$

The lower boundary is valid if there is full correlation between the elements, or in other words, if failure of each element is coupled to failure of the other elements. The upper boundary is valid if there is no correlation at all between failure of the elements. In that case the total probability is the sum of all partial probabilities.

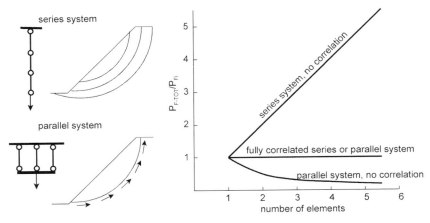

Figure 10-16 Series and parallel systems

For a parallel system, the total probability is given by:

$$P_{F1}.P_{F2}. \quad .. \quad .P_{Fn} \quad \leq \quad P_{F-tot} \quad \leq \quad \max P_{Fi} \tag{10.16}$$

Now the lower boundary (a multiplication of all of the probabilities) is valid without any correlation and the upper boundary represents failure with full correlation.

The first thing that is striking is that there is no difference between the total probability of failure of series and parallel systems if they have fully correlated elements. Figure 10-16 shows the extremes of equations (10.15) and (10.16) for elements with equal P_{Fi}. For partial correlations between the elements, the total probability lies between the extremes in this figure.

For a practical application, much statistical data is needed for all parts of a structure. This is beyond the scope of this book, the reader is referred to CUR, 1995. A qualitative conclusion is that series systems should be avoided wherever possible and that the system should contain enough redundancy. An example is a dike with a protection layer of blocks and a body of sand or clay. With a sand body, severe damage of the protection layer will quickly lead to breach of the dike, while a clay body can resist the wave load in a storm for many hours.

10.4.3 Fault trees

In chapter 1 a rough fault tree for a revetment was presented. Here we will go into more detail. The series and parallel systems of section 10.4.2 also play a role in fault trees. A series system is represented in a fault tree by a so-called 'OR'-gate, indicating that failure of one of the elements leads to failure of the system under consideration. An 'AND'-gate represents a parallel system: failure of both elements is necessary for failure of the system.

Figure 10-17 shows a (simplified) fault tree for a caisson on a sill as part of a closure dam. Water flows through the open caissons. The flow can cause erosion of the top layer of the sill and the head difference can cause erosion of the subsoil through the filter. Both directly cause collapse of the caisson, hence an OR-gate in the fault tree. Both phenomena are independent, hence not correlated and the probability of collapse of the caisson is the sum of the probabilities of both. Scour can cause a slide when there is insufficient inspection. If a slide occurs in loosely packed sand, this can lead to liquefaction and subsequently to a flow slide, see chapter 5. A flow slide and a short bottom protection can lead to collapse of the caisson. All of these combinations are AND-gates because both conditions are necessary to induce the next step.

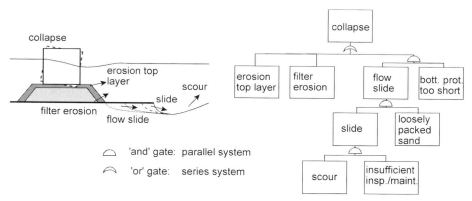

Figure 10-17 Series and parallel systems in fault tree

The fault tree for a revetment given in chapter 1 is reproduced in some more detail in Figure 10-18. Now (fictitious) probabilities (per year) have been added. Most elements form series systems where the total probability is the sum of the partial probabilities. When a partial probability is an order of magnitude smaller than the dominant mechanism, it can be neglected, see the initial geotechnical instabilities compared with the toe erosion.

Toe erosion and instability of the top layer have a probability equal to the hydraulic conditions that cause them. This is the case when the design has been based on a deterministic approach, see also section 10.2.2. The combination of wave conditions and subsidence of the soil form a parallel system, giving a probability of wave overtopping equal to the multiplication of both partial probabilities. In the series system of local instability this appears negligible compared with the other mechanisms.

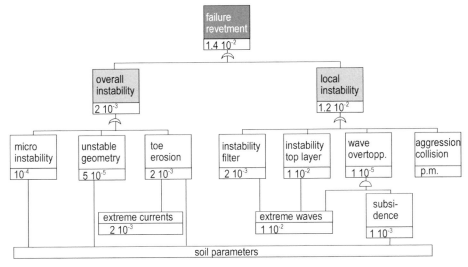

Figure 10-18 Fault tree revetment with probabilities

The use of fault trees lies mainly in the awareness of weak spots of a structure. It is extremely difficult to establish a reliable absolute value for the total probability. It is often advantageous to go through a fault tree bottom-up and top-down and, if necessary, several times. When going bottom-up, the conclusion can be that a probability somewhere in the system is too high, leading to an adaptation of a detail in the design. When going top-down, one can start with an acceptable probability of the top event and see what has to be done to each mechanism to reduce the probability to this value. In the fault tree in Figure 10-17 it is obvious that erosion of the sill is more dangerous than scour. So, the probability of erosion of the sill has to be kept low.

10.4.4 Examples

This section gives some examples from the hydraulic engineering practice concerning cases where one or more failure mechanisms have been neglected, leading to failure of a structure with large consequences. Starting point is, again, the fault tree of Figure 10-18.

Wave overtopping

If wave overtopping is underestimated, the result can be as shown in Figure 10-19, an example from Australia. The top layer seems strong enough and filter layers have also been included in the design. However, due to overtopping, wave action eroded the unprotected soil on the crest of the revetment, leading to its total collapse.

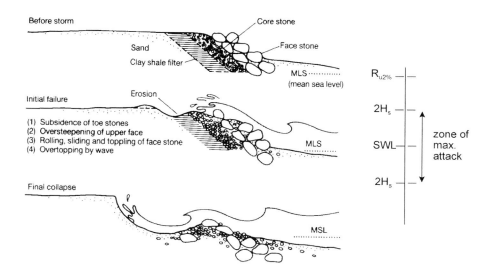

Figure 10-19 Collapse due to wave overtopping

Toe erosion

A bridge with an abutment was built across a river in Bangladesh, see Figure 10-20. Due to the meandering of the river, the location of the abutment suffered severe erosion. The revetment of the abutment consisted of a concrete slab and a sheet pile. This sheet pile had been driven only a few meters into the subsoil and had not been secured against erosion of the foreland. The sheetpiles were undermined and the stiff concrete slab cracked leading to the final collapse of the abutment. Repair work could be carried out before the bridge itself was damaged.

The abutment, situated in the outer bend of the meandering river, should have been designed to be able to withstand the erosion process at the toe of the revetment, which had a probability of occurence of about 100%. Either a much longer sheetpile or an extensive toe protection in the initial design would have been the result.

Wave overtopping and micro-stability

In 1953 many dikes in the south western part of The Netherlands collapsed. The crest level of these dikes was too low and the inner slopes were too steep (1:1.5 - 1:2). The water on the crest penetrated into the slope and the resulting groundwater flow, combined with the steep slope, led to an unstable situation, see also chapter 5.

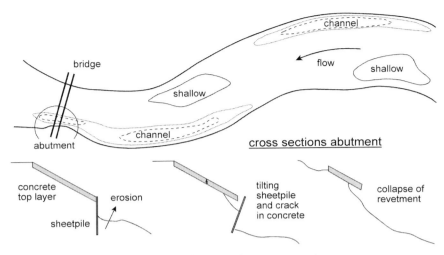

Figure 10-20 Collapse due to toe erosion

How useful are fault trees?

The ultimate question is whether all this could have been prevented if a fault tree had been used. Of course, the use of the fault tree in Figure 10-18, would have led to awareness of the erosion at the toe of the abutment in Figure 10-20 and of the wave overtopping of the revetment in Figure 10-19. These phenomena are already mentioned in the tree, so one is forced to think about it. But how can all failure mechanisms be included if one has to make a fault tree for a new case, as, by definition, you omit what you overlook? In the case of the dikes of Figure 10-21, the probability of high water levels was insufficiently recognized, see Intermezzo 10-1, and the instability of the inner slopes was insufficiently understood at that time. Elements (dolos) of large breakwaters have been known to break due to high stresses in the concrete, a mechanism that had not been included in any fault tree.

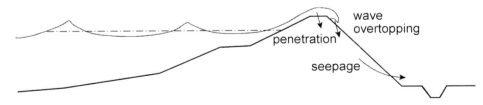

Figure 10-21 Collapse of dike due to micro-instability

So, a fault tree can be useful but it offers no guarantee that nothing has been overlooked or underestimated. The general message in this book is that insight and knowledge of the phenomena is essential. Without that, a fault tree is useless. A brainstorm session with people of various backgrounds can be very helpful to avoid omitting mechanisms in a fault tree. The first part of the brainstorm session can just be summing up possible failure mechanisms. The second part can be trying to establish the probability of every mechanism, in some cases only in a qualitative way. The use of a fault tree is then a powerful tool to avoid blunders and to recognize weak spots in a design.

10.5 Summary

Starting with the general notion:

Risk = probability * consequence

the probabilistic approach of a design is treated. A reliability function is defined:

Z = strength – load

from which the failure probability follows:

$P_F = P(Z < 0)$

This approach can be elaborated on various levels:
Level III: *a full (numerical) integration of Z, or a Monte Carlo approach, where the various parameters that make up Z get a value according to their probability distribution. In the Monte Carlo approach, a structure is built and tested, as it were, say 10000 times, the number of times the structure fails gives the failure probability.*
Level II: *Z, and all underlying probability distributions are assumed to have a normal distribution again resulting in the establishment of P_F and influence factors of all parameters involved (α).*
Level I: *starting with a desired P_F and the influence factors from a level II approach, a partial safety coefficient for each parameter is defined.*
Level 0: *a semi-probabilistic or deterministic approach based on experience in which design values for loads and strengths are selected. The Poisson distribution:*

$P = 1 - exp\,(-fT)$

can be a useful tool when a design value for the load is to be established.
Several ways of maintenance are mentioned. For protections, a state-based maintenance approach is usually the most appropriate. Inspection is an essential part of a state-based maintenance approach.
An overview of all relevant failure mechanisms is essential to be able to create a "fool proof" design. This overview can best be presented by combining the mechanisms in a fault tree. A brainstorm session with several experts can help to avoid overlooking a relevant mechanism. This is important, since most structures do

not fail because of an underestimation of the load, but due to an important failure mechanism which has been overlooked.

10.6 Appendix: Probabilistic approach Level II

In this appendix, the algorithm to come to a failure probability will be given by means of a simple example. The following iterative procedure is to find β, α_i etc.:

1. Estimate μ_Z from Z and the estimated values of the various parameters in the design point:

$$\mu_Z = Z\left(x_1^*, x_2^*, .. x_i^*, ..., x_n^*\right) + \sum_{i=1}^{n} \frac{\partial Z}{\partial x_i}\left(\mu_{x_i} - x_i^*\right) \tag{10.17}$$

As a first estimate for x_i^*, μ_{x_i} can be used.

2. Determine σ_Z from the contribution of the various parameters to the variation of Z:

$$\sigma_Z = \left[\sum_{i=1}^{n}\left(\frac{\partial Z}{\partial x_i}\right)^2 \sigma_{x_i}^2\right]^{1/2} \tag{10.18}$$

3. Calculate β from equation (10.10)

4. Determine the α-values (which indicate the influence on P_F) for each parameter from:

$$\alpha_i = \frac{\partial Z}{\partial x_i} \frac{\sigma_{x_i}}{\sigma_Z} \tag{10.19}$$

This shows that both an important role of x_i in Z (indicated by $\partial Z/\partial x_i$) and a large uncertainty about the value of x_i (indicated by σ_{xi}) are responsible for a large α-value.

5. Make a new estimate for the design-point values with:

$$x_i^* = \mu_{x_i} - \alpha_i \beta \sigma_{x_i} \tag{10.20}$$

6. Find a substituting normal distribution for not-normally distributed parameters for the design-point value of the parameter.

These steps are to be repeated until the design-point values have converged with sufficient accuracy. With the final value of β, the failure probability, P_F, is derived from the standard normal distribution.

For a limit-state function with many parameters, a computer is the most suitable instrument to do the job. However, in order to enlarge the insight into this method, an example for the revetment in section 10.2.2 will be shown with further simplifications. The limit-state function for the revetment will be reduced to just two variables, which are representative for load and strength. This has three advantages: the joint probability of load and strength can be easily visualized, contributing to a better understanding, the iteration can be shown with a manual computation and an "exact" solution by means of numerical integration (level III) can be obtained.

Simplified Z-function

The Z-function according to equation (10.5) is simplified further. Only the most important parameters concerning load (H_s) and strength (d_{n50}) are defined as random variables. Again, the wave heights are assumed to have an exponential distribution (10.6). The iteration to find the design point, needs to be started with the average value, μ, and standard deviation, σ, of the parameters. For an exponential distribution, μ and σ are given by:

$$\mu_H = \varepsilon + \frac{1}{\lambda} = 1.26 \qquad \sigma_H = \frac{1}{\lambda} = 0.26 \tag{10.21}$$

see also equation (10.6) and mathematical textbooks.

For the stones the same, normally distributed diameter is chosen ($\mu_d = 0.6$ m, $\sigma_d = 0.05$ m). The other parameters will be defined as deterministic parameters, using the values in section 10.2.2 with S = 10. The Z-function (10.5) then reduces to:

$$Z = 2.533 - \frac{H_s}{1.6\,d_{n50}} \tag{10.22}$$

The first step in the iteration is to find μ_Z with μ_{xi} as first estimate for the design point values. Note that now the second member on the right-hand side of equation (10.17) is 0. We find $\mu_Z = 1.22$ m.

σ_Z is determined from equation (10.18) again with μ_{Hs} en μ_{dn50} as first estimates:

$$\frac{\partial Z}{\partial H_s} = \frac{-1}{1.6\,d_{n50}} \qquad \frac{\partial Z}{\partial d_{n50}} = \frac{H_s}{1.6\,d_{n50}^2} \tag{10.23}$$

leading to -1.042 m^{-1} and 2.188 m^{-1} respectively. From equation (10.18) we then find $\sigma_Z = 0.292$ and from equation (10.19): $\alpha_{Hs} = -0.927$ and $\alpha_{dn50} = 0.374$. Equation (10.10) gives $\beta = 4.179$ and equation (10.20): $H_s^* = 2.267$ m and $d_{n50}^* = 0.522$ m.

The next step in the iteration procedure is to replace the exponential distribution for H_s with a normal distribution which has the same probability density and probability in the design point, H_s^*. This is done by equating:

$$3.83 \exp\left(-3.83\left(H_s^* - 1\right)\right) = \frac{\exp\left(-0.5\left(\frac{H_s^* - \mu_{H_s\,normal}}{\sigma_{H_s\,normal}}\right)^2\right)}{\sigma_{H_s\,normal}\sqrt{2\pi}} \tag{10.24}$$

and:

$$1 - \exp\left(-3.83\left(H_s^* - 1\right)\right) = \int_{-\infty}^{H_s^*} \frac{\exp\left(-0.5\left(\frac{H_s^* - \mu_{H_s\,normal}}{\sigma_{H_s\,normal}}\right)^2\right)}{\sigma_{H_s\,normal}\sqrt{2\pi}} dH_s^* \tag{10.25}$$

The left-hand side of equation (10.25) is equal to 0.9922. This probability is equivalent to a β-value for the normal distribution of 2.42, which can be found in standard tables and which is equal to $(H_s^* - \mu_{Hs\text{-}normal}) / \sigma_{Hs\text{-}normal}$. From the equations (10.24) and (10.25), with the H_s^*-value from the first iteration, we then find the two unknowns: $\mu_{Hs\text{-}normal} = 0.495$ and $\sigma_{Hs\text{-}normal} = 0.735$.

The iteration is now repeated with the value for H_s^* from the first iteration and using $\mu_{Hs\text{-}normal}$ and $\sigma_{Hs\text{-}normal}$ in equations (10.17) and (10.18) leading to a new value for H_s^*, and subsequently with equations (10.24) and (10.25) leading to new values for $\mu_{Hs\text{-}normal}$ and $\sigma_{Hs\text{-}normal}$.

Note that at the start the second member on the right-hand side of equation (10.17) was 0, while in the end the first member will be zero, since the design point is situated on the $Z = 0$ line ($Z[x_i^*] = 0$). The second part then represents the average value of Z as "predicted" from the design point: $\partial Z/\partial_{xi} \cdot (\mu_{xi} - x_i^*)$.

Finally this leads to:

Table 10-6

Parameter	α-value^2	μ (normal)	σ (normal)	Design value
H_s	0.93	0.508 m	0.728 m	2.293 m
d_{n50}	0.07	0.6 m	0.05 m	0.566 m
	$\Sigma=1$			

$\beta = 2.55$, $P_F = 0.0054$

It is clear that the variations of the wave height are dominant in the results (compare α-values). Note that α-values for loads are negative and for strength positive.

Since equation (10.22), the simplified Z-function, has only two (independent) random variables, the probability density can be computed from:

$$p(Z) = p(H_s) \cdot p(d_{n50}) = \frac{3.83 \, \exp(-3.83(H_s - 1))}{0.05\sqrt{2\pi}} \exp\left(-0.5\left(\frac{d_{n50} - 0.6}{0.05}\right)^2\right) \quad (10.26)$$

From this function, the failure probability can be found directly by numerically integrating the probability density, in accordance with equation (10.3). This has to be done for all values $H_s > 1$ m (for which the wave height distribution is valid) and knowing that (from equation (10.22)) $Z < 0$ for $H_s > 4.05 d_{n50}$:

$$P_F = \iint\limits_{H_s > 4.05 d_{n50}} \frac{3.83 \exp(-3.83(H_s - 1))}{0.05\sqrt{2\pi}} \exp\left(-0.5\left(\frac{d_{n50} - 0.6}{0.05}\right)^2\right) dd_{n50} \, dH_s$$

$$(10.27)$$

This integration gives: $P_F = 0.0057$ which is the "exact" solution for this simplified case (Level III, numerical integration). This number, compared with Table 10-6, is only slightly different, thus an approximation with a level II approach gives reasonable results.

The final results can also be presented graphically, further clarifying the approach and the meaning of the various notions in the equations. Figure 10-22 shows lines of equal probability density from equation (10.26) together with lines of equal Z from equation (10.22). The lines of equal probability density are the contourlines of the probability mountain (see also Figure 10-3). The design point indeed appears to be situated on the line $Z = 0$.

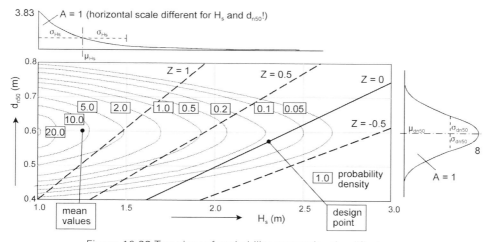

Figure 10-22 Top view of probability mountain, simplified case

As said before (in section 10.2.1), the volume of the part of this mountain where $Z < 0$, divided by the total volume gives the failure probability. This part was calculated directly with equation (10.27).

Figure 10-23 is a 3-dimensional representation of Figure 10-22. Figure 10-23a gives a view of the Z-plane. $\partial Z/\partial H_s$ and $\partial Z/\partial d_{n50}$, as used in equations (10.17), (10.18) and (10.19), are the slopes in the d_{n50}- and H_s-directions at the design point. Figure 10-23b represents the design point on the flanks of the probability mountain showing that the probability density indeed has its maximum in the design point.

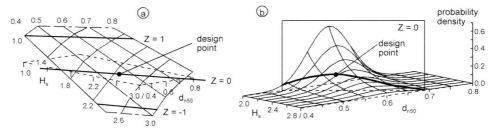

Figure 10-23 3-dimensional view of Z-function and probability mountain with design point

Figure 10-24, finally, shows the equivalent normal distribution of the wave heights as derived from equations (10.24) and (10.25). Again, the intersection and 230the tangent of the two distributions are at the design point.

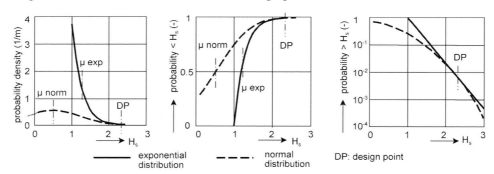

Figure 10-24 Exponential distribution and substitute normal distribution

Only two variables were included in the limit state function to make the example as clear as possible. The manual computations as executed above can serve as an algorithm also for more than two variables. This algorithm results in a simple computer program for a level II approach.

11 PROTECTIONS

Brushwood for the construction of a mattress

11.1 Introduction

If erosion occurs somewhere, five main options are available:

1 Do nothing

If erosion causes no problems, this is the favourite option, see also chapter 1.

2 Take away the cause of the problems

This is theoretically always the best thing to do, but since you need to know what the cause is and you have to be able to do something about it, it is not always feasible. A solution that comes close is to build groynes along a coast where the erosion is caused by longshore transport. If the erosion is caused by cross-shore transport, groynes are useless!

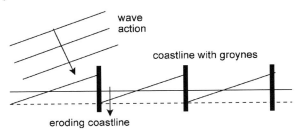

Figure 11-1 Groynes along coast with longshore transport

3 Supply sediment

This may seem a poor solution, since it cures nothing and it may have to go on forever, but it is often a good measure. It is very flexible and fits very well in environment strategies, see also chapter 12. Whether this is the right option depends mainly on costs and risks, since sediment is usually supplied with intervals of several years. For coasts and rivers this "soft" solution can be an attractive alternative. Further treatment of this option is part of morphological studies and is beyond the scope of this book.

4 Reduce the loads

This can be done by constructing a breakwater in front of an eroding coast. Again the effects on erosion are part of morphological studies and will not be treated here. Only the design of such a breakwater is a subject in this book.

5 Increase the strength

This the "hard" solution. Although this book mainly focusses on this type of solution, it should be considered in a certain perspective. It is not a very flexible option and it can sometimes lead to more erosion in adjacent areas. In many cases however, especially when there is a lot of pressure to preserve the available space, it is the only solution.

11.2 Bed protections

11.2.1 General

The main solutions as presented in the previous section are not all possible for bed protections. When the cause of erosion is the disturbance of the flow by some structure, and of course this structure was designed on purpose, taking away the cause is not an option. Sediment supply to the bottom is often not feasible, load reduction can be realised by streamlining the outflow of a sluice, leading to a lower velocity and/or less turbulence. Mostly, however, the increase of strength by means of a bottom protection is the most suitable measure. From chapter 4 we already know that the main function of a bottom protection is not to prevent scour, but to keep the scour hole at such a distance from the structure, that the risk of falling into its own hole is minimized.

11.2.2 Loose rock

General

A bottom protection made of loose rock is relatively simple to realise. Rock or gravel is easy to obtain and, as long as the accuracy demands are not too high, the necessary equipment is available almost everywhere. If no parallel filter gradient is active in the bottom protection, (when there is only turbulent open channel flow on top of the protection) one or two layers is usually enough to prevent loss of bottom material through the protection layer.

When the necessary diameters of the top layer and the original bed material are not too large (e.g. gravel or light rock on sand) one layer can be sufficient, provided the layer is thick enough. The equation by Wörmann, 1989, see equation 6.3, indicates a reasonable thickness in relation to the diameter ratio. The minimum thickness of a layer is usually $2d_{n50}$. For small stones like fine gravel, this leads to a minimum thickness which is impossible to realise. In that case, a minimum thickness of a few dm is used, see also chapter 13.

When a filter layer is applied between the top layer and the original soil, a material with a wide gradation is preferred, like slag. The use of a single geotextile as a filter layer in a bed protection is difficult when the protection has to be made under water, since it is very hard to unfold the textile in that case. A combination with a fascine mattress is a common alternative.

Rotterdam Waterway

An example of a very complex bed protection consisting of loose grains, is the storm surge barrier in the Rotterdam Waterway. Figure 11-2 shows the barrier which consists of two convex steel gates with 300 m long arms. Each gate rotates around a spherical hinge and floats into position. The gates are then lowered on top of concrete blocks. During this closure, high velocities occur between the gates and the concrete blocks, due to the growing head difference across the barrier, which causes a severe load on the bed. When the gates are closed, the head difference across the barrier causes a large parallel gradient inside the sill.

Figure 11-2 Overview Storm surge barrier Rotterdam Waterway

Figure 11-3 gives the cross-section of the barrier which shows the complete bed protection with the stone classes of the top layer. At the sea side, the maximum stone weight is 300-1000 kg and at the riverside 3000-6000 kg. This is because the critical situation is closing the gates during a strong flood flow. In that case, the flow accelerates at the sea side while it decelerates downstream of the gate. Chapter 2 and 3 explain that deceleration is associated with a high degree of turbulence, hence the top layers at the sea side and at the river side differ. When the barrier is open, normal ebb and flood flow occur with much lower velocities. Farther away from the sill, the stone sizes in the top layer decreases.

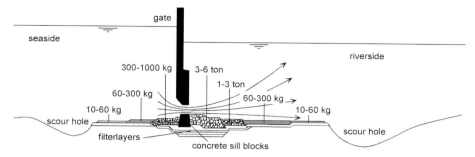

Figure 11-3 Cross-section bed protection barrier Rotterdam Waterway

Figure 11-4 shows the composition of the sill. The design conditions for the filter below the sill are now governed by the situation with closed gates when a large head difference is present acroos the barrier. The original bed material consists of fine sand and silt. The first filter layer is sand ranging from 0.5 to 5 mm. The other layers below the top layer are gravel 3.5-35 mm, rock 30-140 mm and rock 10-60 kg. This is a good example of a geometrically closed filter.

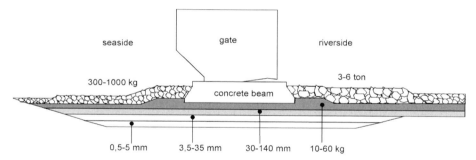

Figure 11-4 Cross-section sill and filter Rotterdam Waterway

The parts of the bed protection without an active filter gradient, have a 0.4 m thick layer of phosphorous slags (40-160 mm) directly on the original bed, mostly with only one layer on top of that, see Figure 11-5.

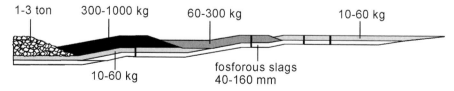

Figure 11-5 Cross-section bed protection river-side

Figure 11-6, finally, shows the complexity of the geometry of the bottom protection and the variety in top layer stone sizes.

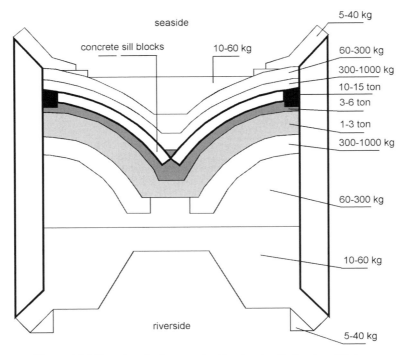

Figure 11-6 Top view bottom protection barrier Rotterdam Waterway

11.2.3 Fascine mattresses

General

The traditional bottom protection in the Netherlands is a fascine mattress which consists of willow faggots (bundles of twigs) and is covered with stones. When made elsewhere, willow can be replaced by bamboo shoots or any other local vegetation that is flexible and strong enough. When the mattresses stay well under water, they can last for more than a hundred years.

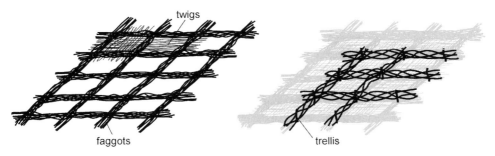

Figure 11-7 Fascine mattress

Figure 11-7 shows a classical fascine mattress. The faggots serve as the backbone of the structure while the bottom twigs carry the stones while still floating (see also chapter 13). A minimum coverage of stones is applied to sink the mattress. Depending on the external load, the mattress is further covered with stones by dumping after the mattress has been sunk.

Filter function

Willow twigs are too open to serve as a filter. In classsical mattresses the filter function is performed by a reed mat under the mattress. The critical gradient of such a structure (parallel to the bottom) is about 3-4 %, so, with higher demands another solution has to be found. In that case, a geotextile is often applied, see Figure 11-8.

Figure 11-8 Geotextile under mattress

11.2.4 Composite mattresses

General

When large areas have to be protected and/or the filter demands are high, the construction of loose rock protections can become time consuming and expensive. The same is, more or less, true for the use of fascine mattresses. For the Deltaproject, very large areas of bottom had to be protected (in total several km^2) and much effort has been put into the development of alternative protections. As an example the bottom protection of the Eastern Scheldt storm surge barrier is treated.

Eastern Scheldt

Figure 11-9 shows the Eastern Scheldt storm surge barrier and its bottom protection. Again, different functions can be discerned:

1	pier	8	sill beam	15	sand filling of pier base slab
2	quarry sto ne dam for land abutment construction	9	road	16	sill beam stops/bearings
		10	road box girder and machinery for gate operation	17	upper mattress
3	beam supporting operating equipment			18	grout filling
4	hydraulic ciliders	11	power supply unit	19	block mattress
5	capping unit	12	sand filling of sill beam	20	bottom mattress
6	upper beam	13	top layer of sill	21	compacted sand under the bed of the Eastern Scheldt
7	gate	14	core of sill	22	gravel bag

Figure 11-9 Overview Eastern Scheldt storm surge barrier (drawing by Rudolf Das, courtesy Rijkswaterstaat)

1. keeping enough distance between the scour holes on both sides of the barrier,
2. preventing erosion under the barrier when the gates are closed and a large head difference exists across the barrier with a large gradient in the filter,
3. protecting the bottom near the gates when large velocities occur through the gates while closing or when all gates are closed except one (the probability of one gate not being closed on time is rather high with 66 gates!).

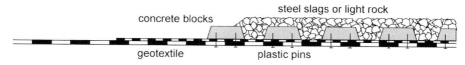

Figure 11-10 Blockmat Eastern Scheldt

Figure 11-10 shows the composition of concrete block mats as applied on both sides of the barrier. The mat only has to withstand the turbulent flow. The filter gradient parallel to the bottom is equal to the gradient of the free surface, hence in the order of magnitude of less than 1 %. A simple geotextile on the bottom is enough in that case. The concrete blocks are attached to the geotextile with plastic pins. The mats were manufactured mechanically in a specially built factory. The concrete blocks sink the mat and keep it stable on the bottom. After sinking, the mat is covered with steel slags for more stability in the turbulent currents of the mat as a whole. The dimensions of the mats for the storm surge barrier were 30×200 m^2. The sinking of the mats required special equipment, see chapter 13.

Under the barrier a large gradient parallel to the bed is active. Head differences of more than 5 m are possible while the width of the concrete sill on which the gates close is about 20 m (see Figure 11-12), leading to gradients of about 25 %. Given the consequences if the barrier subsides (distortion of the barrier frames and gates) for the essential function of the barrier structure as a safeguard against flooding, the failure probability of the filter must be very low. Hence, a geometrically closed filter seems appropriate. The construction of such a filter with several layers of relatively fine material in flowing water of more than 25 m deep is a Herculean task and, given the length of the barrier (4000 m), would be very expensive. A sophisticated geotextile mat could serve the purpose, but no guarantee could be given that the geotextile would last the expected lifetime of the barrier (100 – 200 years). The solution that was decided upon is a combination of a geometrically closed filter and geotextiles, see Figure 11-11. The geotextile wraps the filter grains and thus enables to bring the thin layers of fine material to the bottom in such a way that the integrity of the filter structure remains intact. So, when the geotextile eventually deteriorates, the geometrically closed granular filter is right in place.

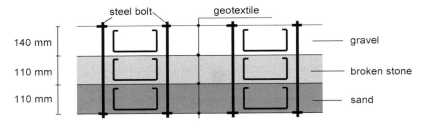

Figure 11-11 Filtermat under barrier Eastern Scheldt

When the gates are closed, large velocities occur near the sill. In order to protect the foundation of the barrier, large rocks and concrete blocks are placed, see Figure 11-12.

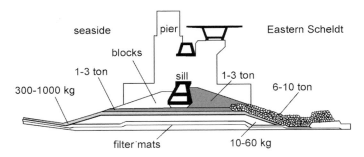

Figure 11-12 Cross section barrier foundation Eastern Scheldt

Figure 11-13 shows the bottom protection of a part of the barrier in the Eastern Scheldt.

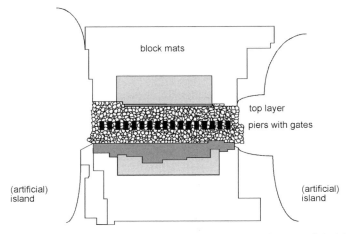

Figure 11-13 Top view bottom protection barrier Eastern Scheldt

11.2.5 Evaluation

In the previous sections, several alternative bottom protections have been reviewed. What to choose when? The first striking difference is the choice for a granular filter in the Rotterdam Waterway and for composite mattresses in the Eastern Scheldt, while at first sight both structures are very much alike: both are large scale storm surge barriers with large head differences etc. The choice was based on very practical grounds. Both are large structures, but there is large and extra large. The barrier in Rotterdam is about 400 m long while the total length of the Eastern Scheldt barrier is about 4000 m (divided into three parts). This scale difference, and the circular plan of the Rotterdam barrier, see Figure 11-6, made it economically not feasible to build a special factory and pontoon to sink the protection in the Rotterdam case, while for the Eastern Scheldt this was a cost-effective solution. Moreover, the intensive shipping in Rotterdam made it impossible to work with an anchored pontoon.

For small scale projects, one should not think of mechanically fabricated bottom protections. Loose rock and/or fascine mattresses will do for many projects. A reason to consider the use of a geotextile in a bottom protection can be that it offers better protection against a sliding scour hole, see Figure 11-14. If a protection consists of loose rock, the scour process can go on after sliding, since the filter structure has been disrupted. A coherent protection can "hang" on the slope while the filter function remains largely intact. In Figure 11-14, a concrete block mat has been drawn, but more or less the same holds for a geotextile under a fascine mattress.

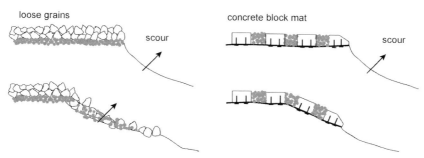

Figure 11-14 Sliding of scour hole with loose and coherent protection

11.2.6 Piers

Although the previous sections were classified along the different types of protection, not discriminating between the structures to be protected, this section deals exclusively with the protection of (bridge) piers. Bottom protection behind sluices and barriers should be long enough to keep the scour hole at distance of the structure to be protected. A longer bottom protection usually does not significantly reduce the scour depth (see chapter 4), but it reduces the danger of damage due to a slide or flow slide. The bottom protection around (bridge) piers is different in this respect. The disturbance of the flow around the pier is of a very local nature: a few diameters from the pier, the flow has its normal value again. This means that a larger protection indeed reduces the scour depth. A circular protection of 3 – 4 times the pier diameter is usually enough, see Figure 11-15a.

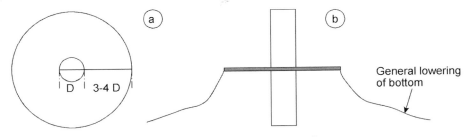

Figure 11-15 Bridge pier protection

When the area is morphologically unstable one should take this into account or otherwise the protected bridge pier will act as an extra obstruction to the flow, see Figure 11-15b, aggravating the erosion and possibly finally leading to collapse of the pier.

11.3 Bank protections

11.3.1 Revetments

Revetments can protect a river or canal bank. Only the most frequently utilised materials will be treated here; often the locally availibility of materials determines the choice. Aesthetic aspects can also play a role. Vast lengths of revetments with rock possibly do not look good, but then again, the same holds for concrete blocks or asphalt or even grass. Monotony can be prevented by diversifying materials or shapes, see also chapter 12 concerning the environment.

Some general aspects of revetments, like transitions and toe protections, which are also valid for shore protections, are discussed in section 11.5

11.3.2 Loose rock

Loose rock can be used in almost any case. When accessibility is an issue, rock has some disadvantages. But, being easily available and applicable without specialized equipment or personnel, it will always be an important material. Figure 11-23 shows loose rock as part of a simple revetment, only consisting of a geotextile with light rock (10 – 60 kg) on top.

Figure 11-16 Simple bank protection with geotextile

A fascine mattress can also be applied as a bank protection. The hedges ("trellis") in Figure 11-7 serve to give the stones some more stability on the slope.

11.3.3 Composite mattresses

Figure 11-17 Block mat as bank protection

Composite mattresses can be applied when the accessibility of the bank is part of the terms of reference. Also, aesthetic aspects can play a role. Application is usually only viable for long stretches of bank. For a very small scale project, the special equipment necessary to place the mats might lead to high costs. Figure 11-17 shows a block mat on a canal bank.

11.3.4 Rigid structures

Rigid structures are usually considered when a steep slope is wanted due to limited available space. In that case, the structure has the characteristics of an earth retaining structure. Sheet pile walls of steel, concrete or wood are examples of revetments e.g. along canals, see Figure 11-18.

Figure 11-18 Sheet pile wall along canal (CUR 200, 1999)

11.3.5 Groynes

River groynes can serve to maintain a certain depth in the river (for navigation purposes) by forcing the water to flow between the groynes during normal or low discharges or to keep the high flow velocities away from the river banks or a combination of both. Figure 11-20 and Figure 11-19 show an example of a groyne along the rivers in the Netherlands which consists of placed blocks on the higher parts and fascine mattresses with loose rock under water. The crest of the groyne is situated around mean water level. When the water level is higher, the river gets more space in order not to hamper the run-off of the river flood. The distance between the groynes should not be more than 5 times the length, but 3-4 times is better.

Figure 11-19 River groyne in The Netherlands (river IJssel)

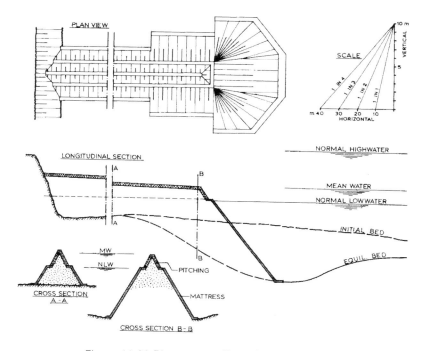

Figure 11-20 River groyne (from Jansen et al,1979)

Structures such as the one in Figure 11-20 are expensive, particularly as usually many groynes are necessary along a river stretch. Figure 11-21 shows a cheaper alternative to protect the bank. By varying the openings between the piles, the flow gradient along the bank can be adjusted within certain limits. To maintain a navigation channel this type of groyne is less suitable.

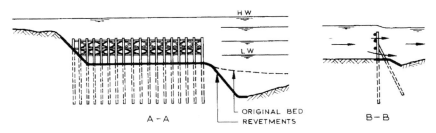

Figure 11-21 Pile screen (from Jansen et al,1979)

Figure 11-22 Pile screens in Jamuna river Bangladesh

11.4 Shore protection

11.4.1 Revetments and dikes

Loose rock

In shore protections, loose rock can also be used in (dike) revetments. In the Netherlands, loose rock on dikes is, traditionally, only applied below mean water level. Accessibility and aesthetics are the main reasons. Figure 11-23 shows an example from Surinam, where rock is applied up to the crest of the dike.

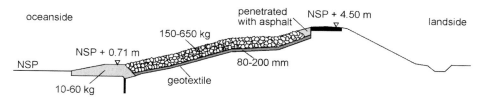

Figure 11-23 Loose rock dike revetment (Nickerie, Surinam, Leo Philipse)

Placed blocks

Placed blocks (or columns or other shapes) can yield good (dike) revetments, provided the elements are placed with skill and care. Figure 11-24 gives an example of a dike revetment with placed blocks.

Figure 11-24 Placed block revetment (Vietnam)

Asphalt

When the wave attack is severe, an asphalt revetment can be the answer. The construction demands special equipment and regular inspection for cracks is necessary. But a well designed, constructed and maintained asphalt revetment is very strong and reliable. Figure 11-25 shows an example. A filter under an asphalt revetment is not necessary.

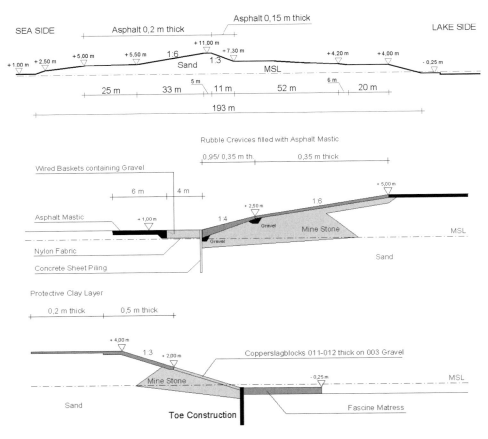

Figure 11-25 Dike with asphalt revetment (Brouwersdam)

Grass

When wave attack does not play an important role (waves up to ~ 0.5 m), a grass revetment or another vegetation *with good cover*, can be an attractive solution also from an economical point of view. Grass can stand water for many hours but can not live under water permanently, so for the lower parts of a revetment, another material always has to be applied.

Figure 11-26 Dike with grass revetment

Rigid structures

Seawalls are usually made of concrete or masonry, see Figure 11-27.

Figure 11-27 Seawall Flushing (photo Wil Riemens, Middelburg)

11.4.2 Groynes and breakwaters

The function of groynes is usually maintaining a channel or a coastline. Breakwaters serve as a protection against waves for a harbour or a coast. Breakwaters are mostly larger than groynes, but the differences are not strictly defined.

Coastal groynes can serve to reduce the longshore sediment transport and/or to keep the current velocities away from the shore. In the first case it is not necessary to make the groyne higher than, say 0.5 to 1 m, above the wanted beach level. Figure 11-28 gives an example of a groyne of this type. Since, in most seas, the sea level goes up and down with the tides, the zone of wave attack on this type of groyne will also go up and down. A light stone class, penetrated with asphalt, on a fascine mattress makes a simple and strong groyne.

Figure 11-28 Simple, low coastal groyne

When the groyne also keeps the current velocities away from the shore during high tides, the wave attack will be concentrated on the head of the groyne. In that case there is hardly any difference with a breakwater. Figure 11-29 shows an example of such a groyne. A bottom protection will be necessary to keep enough distance between the scour hole (caused by the current around the head) and the structure.

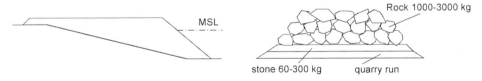

Figure 11-29 Simple, high coastal groyne

11.4.3 Breakwaters

A breakwater can simply be seen as a large groyne. It is then usually cost effective to compose the cross-section of several stone sizes. Figure 11-30 and Figure 11-31 show an example.

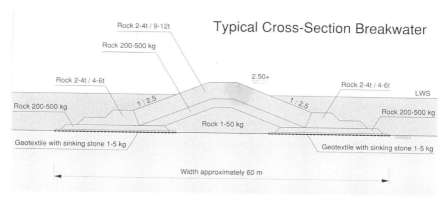

Figure 11-30 Cross-section breakwater (van Oord ACZ)

Figure 11-31 3-dimensional view breakwater cross-section (van Oord ACZ)

Figure 11-32 shows an example of offshore breakwaters as a coastal protection. The effect of the wave load reduction is clearly visible in the position of the shore line.

Figure 11-32 Offshore breakwaters

11.5 General aspects: revetments

This section treats considerations which are relevant for canal or river banks as well as shore protections.

11.5.1 Choice

The choice of the type of revetment depends on the availibility of materials, the available space, the other functions of the revetment, the loads, harmonisation with the landscape etc. Table 11-1 gives a checklist of criteria which can determine the choice.

Table 11-1

Type Criterion	Loose rock	Placed blocks	Asphalt	Grass	Rigid
Heavy loads	+	+	++	-	++
Costs	depends	depends	depends	depends	depends
Flexibility for subsidence	++	+	+	++	-
Space required	0	0	0	-	++
Construction /maintenance	++	0	-	+	0
Landscape	depends	depends	depends	depends	depends
Accessibility	-	+	+	+	-

Figure 11-33 gives an idea of appropriate sea defence structures for a wide range of depths and wave heights. Concrete armour units are available in many shapes, such as tetrapods, doloss etc. Treatment is beyond the scope of this book, see e.g. d'Angremond, 2001.

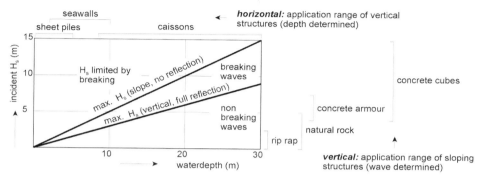

Figure 11-33 Conceptual design of protection (VanderWeiden,1989)

11.5.2 Transitions

General

Transitions in revetments are inevitable since different locations on the slope often require different solutions. But at the same time, transitions are vulnerable elements, damage often starts in these places. This is common knowledge in everyday life: problems resulting from rain and storm never occur in the middle of a wall but always at the transition between wall and roof or windows.

A transition can decrease the strength and/or be subjected to an increased load. Figure 11-34a shows the transition between a concrete block revetment and a loose rock protection. The board is meant to keep the blocks in place. Due to subsidence of the soil, a split can occur resulting in a decreased strength (unprotected soil under the blocks). The board may also lead to an increased load: the wave pressures against the blocks are higher due to locally insufficient drainage, see also chapter 8.

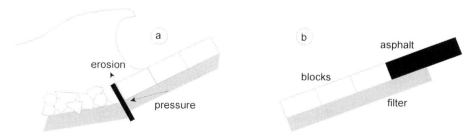

Figure 11-34 Transitions – general aspects

The ideal transition is just as strong and flexible as the adjoining layers. This is very hard to realise, therefore the following aspects should be considered:

- **Care:** possibly the most important aspect. With extra attention during construction, inspection and maintenance many problems can be avoided.
- **Permeability:** when there is a difference in permeability of the revetment on both sides of the transition, the transition should be dimensioned with the difference in mind, see below.
- **Overlap:** a split down to the sand should be avoided, there should always be some overlap between layers, see Figure 11-34b.

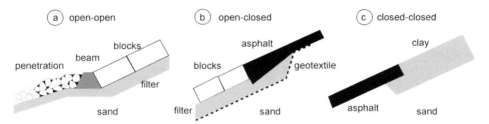

Figure 11-35 Transitions between open and closed revetments

Open-open

Figure 11-35a shows a transition between two open revetments with different permeabilities: concrete blocks and loose rock. The very permeable rock can cause extra pressure under the blocks. By penetrating the first meter or so of the rock with asphalt, the wave pressures will penetrate less easily. The support of the blocks above the transition is guaranteed by the concrete beam, which has a special shape.

Open-closed

Between the blocks and asphalt in Figure 11-35b there is an overlap in the filter layer in order to avoid a clear cut seam. Via the blocks, an extra pressure under the asphalt can build up. The asphalt layer is thicker at the transition to withstand this pressure.

Closed-closed

A filter is not necessary in this case, but a seam down to the sand has to be avoided in order to prevent a leak with pressure from inside which will cause erosion, see Figure 11-35c.

Many other transitions can be encountered, e.g. between blocks and grass. If cattle grazes on the slope, the animals can damage the transition. Partially open blocks, which allow grass to grow through, can serve as a gradual transition.

11.5.3 Toes

Toes are a special kind of transition, i.e. from a slope to a horizontal plane. The same aspects as in the previous section play a role, but the support function of a toe is an extra aspect to reckon with.

A slope causes a horizontal force at the toe. In a hard bottom this force can easily be withstood, see Figure 11-36a, but in soft soil it causes deformation of the toe and possibly erosion or other damage to the revetement, see Figure 11-36b. In that case, a toe protection against erosion, which is necessary anyway, can be constructed with stones, which in turn are able to withstand the horizontal force.

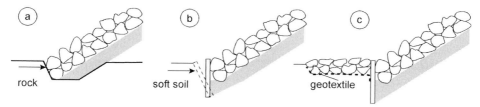

Figure 11-36 Horizontal support of slope

In order to avoid pressure build-up at the toe, a permeable structure can be used at the toe of a block revetment, see Figure 11-37.

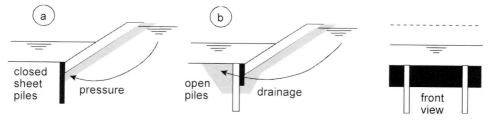

Figure 11-37 Drainage at toe

Eroding toes

If the toe of a revetment is subject to erosion, e.g. in the case of an outer bend of a meandering river, several measures can be taken, see Figure 11-38. In case (a), the bank is excavated to the expected erosion depth and the revetment is extended during construction. This is an appropriate, but often expensive solution.

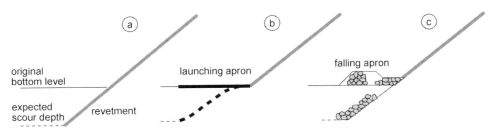

Figure 11-38 Toe protections in case of expected erosion

In case (b), a mattress or a block mat or any other coherent structure is applied. When the toe erosion starts, the mattress will hang down in the erosion hole. This is sometimes called a "launching apron".

A special type of toe protection shown in Figure 11-38c is the falling apron, see Van der Wal, 2000. It is a layer of loose rock without mattress or geotextile. With toe erosion, the loose stones "fall" and cover the slopes of the scour hole. The stones in the falling apron:

- should be large enough to withstand the possible flow forces
- should be fairly uniform in size: concrete blocks form a fine falling apron. They are used to create an "imperfect filter" and to allow erosion of the bottom material through the stones. This is essential in a falling apron: a perfect filter would hinder erosion through the stones, resulting in an uncontrolled drop of the apron after some time.
- should have a high density (> 2500 kg/m^3). A high density contributes to a better falling process.

The necessary volume of a falling apron can be estimated with:

$$V = (5-7)h_s d_{n50} \ \ \text{m}^3/\text{m} \tag{11.1}$$

The falling apron should be monitored regularly in order to check the erosion and falling process.

Figure 11-39 Model test with falling apron (CNR,1994)

12 ENVIRONMENT

Vegetated canal banks (North Brabant). Courtesy CUR.

12.1 Introduction

12.1.1 General

Beds, banks and shores are the borders between soil and water. Until some decades ago, hydraulic engineers were only interested in shaping these boundaries to fight erosion or to create transshipment possibilities, but growing attention in society to environmental aspects has lead to an other approach. The transition between land and water plays an important role in nature and landscape and this awareness also has consequences for hydraulic engineers. Nature friendly protections have become an issue worldwide and mono-functional protections are becoming obsolete. Section 12.1.2 contains more background information concerning ecological aspects.

Another environmental aspect is the choice of materials. To prevent pollution, it is no longer allowed to use certain materials which contain heavy metals or toxic substances. The choice of material also depends on the laws which are in force in the area where the project is being executed. In general it can be said that many waste products are suspicious. Using lead slag is plainly wrong, as it emits too much heavy metal, while phosphorous slag can be used in a salt water environment but not in fresh water. More information can be found in (CUR 200 (1999)).

Ecology and hydraulic engineering may seem diametrically opposed, but that is not necessarily so. A navigation canal serves as an example. Tradition or the available space and funds often decided that a canal's banks should be constructed with sheet piles. The ecological functions of a canal with sheet-pile banks are seriously hindered since the vertical banks offer no room at all for flora and fauna. But navigation on such a canal can also be unpleasant. Due to the vertical walls, the ship's wave action persists for a long time and when there is much traffic, the canal water sloshes which particularly hinders small ships. So, there is not necessarily a contradiction between these interests. Another example is the presence of vegetation as natural protection. Leaving this vegetation unspoilt can save a lot of money.

Another approach to protections is also a challenge for hydraulic engineers. A new set of boundary conditions enters the field of protection design, making it more interesting. Also in research, a new field of attention has come into the picture since the effect of vegetation on load and strength has led to fusions like eco-engineering or engineering biology. Sections 2, 3 and 4 of this chapter contain examples of this new approach.

12.1.2 Ecology

General

Ecology is the science that deals with the mutual relations between living organisms and their relations with the environment. This science has become more and more important during the last decades. One does not need to be an environmentalist to know that economic development has several side effects which can form a threat to mankind in the long run. Man, being a living organism, is also part of ecosystems. Banks and shores are part of larger watersystems, like rivers or lakes. Healthy functioning of such a system is a prerequisite for all other interests related to that watersystem. If a watersystem is not a stable ecosystem, the supply of water as raw material for drinking water, for fisheries or recreation etcetera is also endangered.

Reducing pollution is a necessary condition but it is not sufficient to achieve a healthily functioning ecosystem. E.g. introducing salmon back into the Rhine requires more than reducing industrial discharges, it is also necessary to create banks and other quiet places where fish can spawn. So, a little knowledge of ecology can do no harm to a hydraulic engineer.

Ecological infrastructure

On a planning level one can discern several elements in an ecological infrastructure, see Figure 12-1:

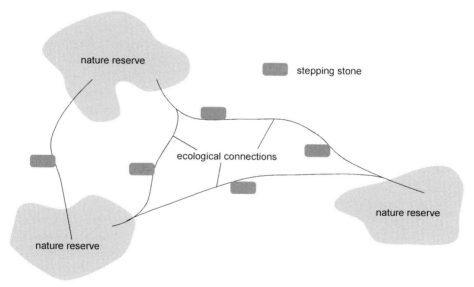

Figure 12-1 Ecological infrastructure

- Nature reserves
  Areas where certain species can live in more or less ideal circumstances. These areas are large enough to aid as long term conservation of certain populations.
- Ecological connections
  These connections can serve certain species or are of more general use. When the connections do not function satisfactorily, fragmentation will follow, especially

for species which are less mobile, have a low rate of reproduction or live in small numbers.

- Stepping stones
  If the distance between nature reserves is too large, a stepping stone in between can be of use for the dispersion of species. These areas are suitable as a temporary living area, but they are too small to allow a population to stay healthy in the long run.

Banks play a role in ecological connections, but can also serve as stepping stone for certain species.

Figure 12-2 River as ecological connection with stepping stones

Dynamics

Like people, plants and animals have their preference for a certain degree of dynamics. Nature offers a high variety of circumstances, from quiet ponds to rapids and surf zones. Figure 12-3 shows the relation between dynamics and the well being of organisms.

Many species can live in a moderately dynamic area but the number of organisms per species will be small. In a highly dynamic environment, only few species can live but the number of organisms per species will be high. If an area is too dynamic, nothing can live. An example of the latter is a beach, where waves break and the environment demands so much of organisms that it is not suitable for plants or animals as a habitat. A little less dynamic are tidal flats, where mangrove trees can stand the waves and shellfish can live under the surface. Every part of the flats is dry or wet during a different percentage of time, depending on the height with regard to the tides. And in every part only a few species can meet the environmental demands. Per square meter, however, numerous organisms of a certain species can be found. Another example can be found in dunes, where the slopes that are exposed to the sun have few species, while the other side of the dune, in the shadow, shows a much greater variation.

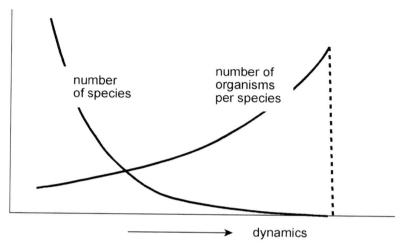

Figure 12-3 Species and dynamics

With regards to banks and shores, it is clear that a steep, stony slope exposed to wave action is not a suitable habitat for most organisms and measures to reduce the dynamics are necessary to create better living conditions.

Banks and ecosystems

Banks are part of larger ecosystems. They form the transition from water to land and vice versa and show characteristics of both. For many species, both aquatic and terrestrial, banks are vital. Fish use shallow parts to spawn or to hide between plants. Insects live on aquatic plants, birds look for food or rest while mammals come to drink or feed. Figure 12-4 shows some relations for a fresh water bank.

Figure 12-4 Ecotopes in a bank

A gradual transition offers more opportunities for species to survive and to migrate. A canal with sheet pile banks is impossible to cross for mammals, like deer. Figure 12-5 shows possible solutions to overcome this problem.

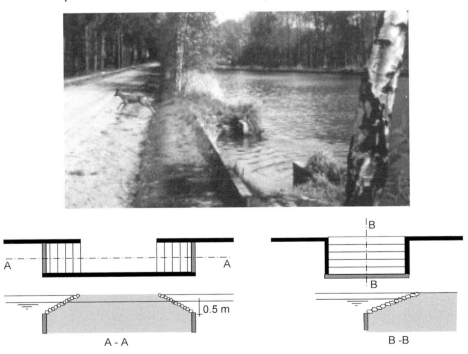

Figure 12-5 Crossing location for mammals

When dealing with a design for any bank, one should be aware of what is possible in a certain area. Natural banks in other comparable areas can serve as a reference. The other functions of the bank, like recreation, also have to be taken into account. A stepwise approach can help to draw up a final plan in which future maintenance and monitoring should also be included. The relation between construction and maintenance costs can be different for nature friendly banks, which can actually be cheaper.

Figure 12-6 shows the various steps in the realisation of a nature friendly bank project. It starts with a vision of what is wanted and what is possible and advances with the usual steps. In the first stage of the project, some thought is already given to maintenance, monitoring and evaluation. This is even more necessary than in "normal" bank protection projects since the functioning of a nature friendly bank requires attention throughout its life but particularly in the first years.

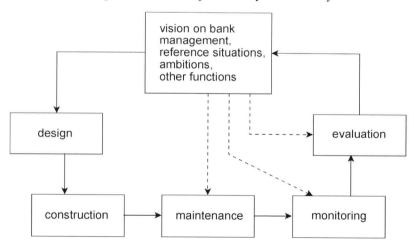

Figure 12-6 Planning and execution steps

12.1.3 Load and strength

During extreme conditions any protection can be damaged requiring maintenance. Chapter 1 stated that erosion only needs to be prevented if it harms certain interests. Gain and loss are normal phenomena for a bank or shore in a natural situation and something similar can be said for nature friendly protections. If vegetation serves as protection, some loss can occur which is acceptable as long as the protected interests are not endangered and the vegetation can recover after the extreme load.

Vegetation as a protection can both reduce loads and increase strength. Vegetation has a relatively large resistance to waves and currents, thus reducing the loads. Roots can increase the strength by protecting the grains on a micro scale or by reinforcing them. Figure 12-7 shows vegetation with wave load. The outside plants are "front soldiers" and have to withstand a higher load than the inside plants. At the front, due to the high velocities, scour can also occur if the roots are not able to retain the soil. The effect can be that the outside plants are damaged or disappear. As long as the number of soldiers is large enough, the battle can still be won. When there is enough time to recover before the next extreme event, the protection will function well. If not, some extra protection, e.g. a stone dam, will be needed.

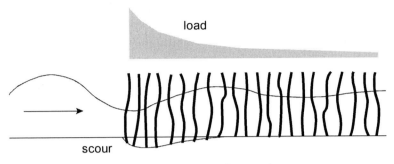

Figure 12-7 Load and strength

12.2 Bed protections

12.2.1 General

Bed protections are situated completely under water, so from an ecological point of view, bed protections seem to be of little interest. However, in contrast with stone revetments, bottom protections of loose rock can even ecologically improve the original situation. Shells and other organisms can easily attach themselves to or hide in between large stones, creating new ecological opportunities.

12.2.2 Fascine mattresses

Chapter 11 mentioned that geotextile is necessary for the filter function of a fascine mattress, since twigs or reed have a low critical gradient (only a few percent parallel to the bottom). Some research has been done to find a natural material to replace this geotextile (Lemmens, 1996).

Figure 12-8 shows the test set-up to investigate the critical gradient for various natural fabrics. A geotextile was used as a reference in order to compare the results with existing solutions.

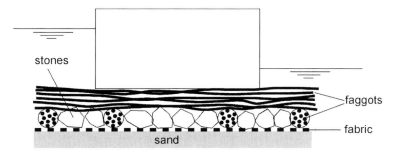

Figure 12-8 Filter tests for nature friendly fascine mattress

Nr.	Type of fabric	Critical gradient (%)
1	None (classical fascine mattress)	7
2	Synthetic geotextile (not penetrable for roots)	59
3	Synthetic geotextile (penetrable for roots)	16
4	Reed (classical filter cloth in fascine mattresses)	18
5	Coconut fibre cloth	12
6	Coconut fibre cloth filled with 4 cm of woodpulp	14
7	Jute cloth	26
8	Jute cloth filled with 4 cm of woodpulp	56

It appears that it is possible to find substitutes for geotextiles. A reed mat (nr 4) has the same critical gradient as a geotextile that can be penetrated by roots of plants (nr 3), which is favourable when it is used not as a bed protection but on a slope. Jute with woodpulp (nr 8) can stand the same gradient as high quality geotextile (nr 2) which can not be penetrated by roots and which can be used in cases where high quality demands hold for bottom protections.

Fabrics like jute, coconut or other natural cloth can be found in many places. The problem is the durability which is usually not more than a few months when applied on a slope in the air-water-sunshine zone. Under water it can possibly last much longer but experiments need to be done to find out.

12.3 Bank protections

12.3.1 General

As already indicated in section 12.1.2, a gradual transition from water to land is one of the keywords in a nature friendly design. This means a gentle slope in the first place, so, nature friendly bank protections need space! In small waters it can be enough to create a gentle slope and wait for the vegetation to grow or to help somewhat by sowing or planting. In rivers or canals this will usually not be enough. With increasing loads one can think of reinforced vegetation as a protection or load reduction in front of the bank. Some examples are given in the following sections.

12.3.2 Vegetation

Loads

Vegetation, such as reed, rush, willow trees or other aquatic plants can reduce both waves and current velocities. The orbital wave motion flows around reed stalks and bends them to and fro, making wave reduction a very complex process. Completely stiff stalks are most effective as wave reductors. Bending in the wave direction makes reed less effective, but vibrations perpendicular to the wave direction again cause a greater reduction due to a larger resistance to the orbital motion (Klok, 1996). Finally, the internal friction inside the stalks also absorbs some energy. With all these factors it is not yet possible to determine the wave reduction analytically or even numerically. Empirical research has produced the following result for wave transmission (CUR, 1999):

$$K_T = \frac{H_T}{H_I} = 1 - \left[1 - \exp\left(-0.001 N^{0.8} \frac{B}{\cos\beta} \right) \right] \tag{12.1}$$

in which: N is the number of stalks per m², B is the width of the vegetation and $\cos\beta$ is the angle of incidence of the waves (0° for waves perpendicular to the bank, ~ 55° for ship waves), see Figure 12-9a which shows the wave transmission for various values of N. The values found with equation (12.1) are valid for reed stalks in the growing season. In other seasons, depending on the length and appearance of the reed, the wave transmission can reach much higher values and there is hardly any wave reduction left. So, the performance of reed as wave reductor depends on the season, hence only if the load is also seasonal (e.g. recreational navigation) this is a suitable solution.

With regard to what has been said in section 12.1.3 concerning load and strength, equation (12.1) is valid for waterdepths up to 1 m and waves up to 0.4 m. For trees, the same equation can be applied with a higher, but yet unknown maximum wave height.

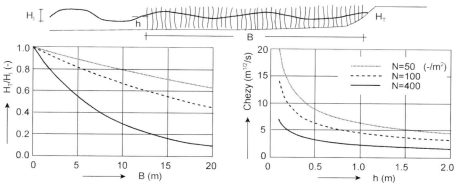

Figure 12-9 Wave damping and roughness in reed

Vegetation also reduces the current velocity. The equivalent Chezy-value for vegetation has been found to be (CUR, 1999):

$$C_{veg} = \sqrt{\frac{g}{0.5NDh}}$$ (12.2)

in which D is the diameter of the stalks and h is the waterdepth. Figure 12-9b shows the results for various values of N. Note that the C-values are very low compared with values found in open channel flow. This C-value has no relation with the logarithmic velocity profile in uniform flow but is just an algebraic parameter to be used in relations like:

$$u_{veg} = u_{open} \frac{C_{veg}\sqrt{h_{veg}}}{C_{open}\sqrt{h_{open}}}$$ (12.3)

as was treated in chapter 2 and in which the subscript $_{open}$ stands for the values outside the vegetation.

Strength

In section 12.1.3, load and strength were discussed. From the previous examples it has become clear that vegetation can function as a load reductor. Increase of strength is not self-evident: a tree or a plant, standing in a current, is comparable to a cylinder which causes erosion as we know from chapter 4. Hence, increase of strength has to be the result of something else. The roots of trees and plants appear to function as a bottom protection. It is, however, not very clear yet how this works, because the question is whether the roots reduce the load around the stem or serve as armouring of the soil. Anyhow, roots appear to reduce the scour around trees. Dorst,1995, did some research on scour of a sand bottom with willow trees, with and without roots. He found a reduction of 40% of the scour depth with roots only directly around the stem, compared with the same stems without roots. With a root system extending much further from the stem (some trees develop a sort of root mats), the reduction was even larger ($\approx$ 75 %).

Vegetation also influences the resistance against sliding. The roots clearly armour the soil, see Figure 12-10 from CIRIA,1992.

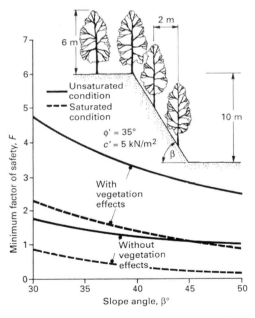

Figure 12-10 Roots as armour against sliding

Grass

Grass and other herb-like vegetation require special attention. Grass cannot grow under water, but can stand current and wave loads for many hours depending on the magnitude of the load. This section on bank protection treats current velocities. Wave loads are treated with shore protections in section 12.4.3.

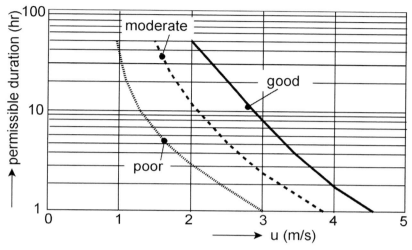

Figure 12-11 Permissible flow duration on grass cover

Figure 12-11 shows a graph with permissible flow velocities and their duration. Grass can stand very high velocities for a short period of time or lower velocities for several days. If the current velocity is low, the water flows through the grass. If the velocity increases, the leaves flatten on the soil and cover it, thus increasing the protection.

Section 12.4.3 elaborates on the relation between management and grass quality.

12.3.3 Vegetation with reinforcing mats

If the wave or current loads exceed those of the previous section, a fascine mattress or some other artificial mat can reinforce the slope. Figure 12-12 shows the concept: a (thin) fascine mattress, or similar structure such as a gabion or composite mattress, increases revetment's strength. Only a single layer of stone is applied in order not to hinder plant growth. The formulas of chapter 8 are used to calculate the dimensions, but a margin should be taken into account since only one layer is applied. When the plants are fully grown, the roots can serve to retain the soil, but at the beginning, some filter is needed. A degradable geotextile, as mentioned in section 12.2.2, can solve this problem.

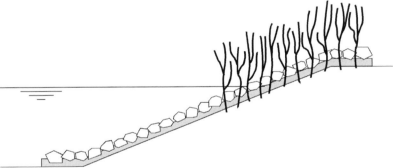

Figure 12-12 Reinforced vegetation

12.3.4 Load reductors

When the wave loads are definitely too great for vegetation to survive, a load reductor in front of the bank is necessary to create an environment in which aquatic plants can flourish. The strip behind the wave reductor then becomes a quiet area. To refresh the water in the strip, the exchange of water between canal or river and the strip should be possible. Figure 12-13 shows some examples of load reductors. (a) is a low protection (sheet piles or a stone dam), see chapter 7 for some numbers for wave reduction. (b) represents sheet piles with holes under water at regular intervals along the canal, (c) is a protection with its crest above water, but with interruptions along the length of the canal. This means that, locally, the wave attack on the bank will be high again. To prevent damage to the vegetation, a solution as presented in (d) is a possibility, asking again for slightly more space. Note that the cross-sections (a) and (b) and the plan views (c) and (d) do not match. They represent different solutions.

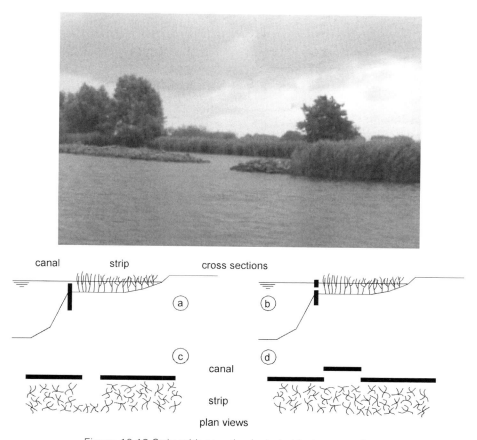

Figure 12-13 Strip with aquatic plants behind wave reductor

In a river or lake, wind waves and currents have to take care of the exchange of water in the strip. Since these phenomena also form the threat to the bank and vegetation stability, a subtle compromise between load and protection needs to be found and should always be tested in a real situation.

These wave reductors and strips can be conctructed in a navigation channel where the loads are caused by the ship waves. To refresh the water inside the strip, the energy of the passing ships can be applied. The main engine behind the exchange of water is the primary wave, see Figure 12-14. The waterlevel depression in the canal causes an outflow through the apertures in the protection, after which a gradual inflow occurs, refreshing some of the water in the strip. The apertures should be located at regular intervals (order of magnitude 50 – 100 m, also depending on the intensity of the navigation). See CUR-200, 1999 for more details.

Often, water exchange also causes sedimentation inside the strip which can hardly be prevented. A deep part near the apertures seems the most appropriate measure to deal with this aspect. Sedimentation mainly takes place in this part, making maintenence easier (Hooijmeijer, 1998).

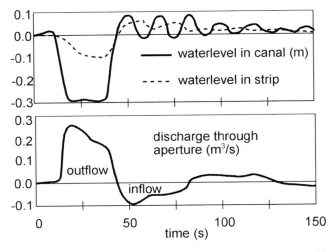

Figure 12-14 Exchange of water in strip due to passing ship

12.3.5 Groynes

Groynes can be found along many rivers. If the shipping intensity increases or flow velocities increase, erosion between the groynes might occur, see Figure 12-15. Instead of a stone revetment, some simple alternatives are possible. (a) shows the possibility of sediment supply. The supplement largely destroys the existing, small vegetation, but the natural vegetation can return by seeds that are washed ashore. (b) shows an extension of the groynes in landward direction. This leaves the natural processes intact to a large extent, but some land will be lost.

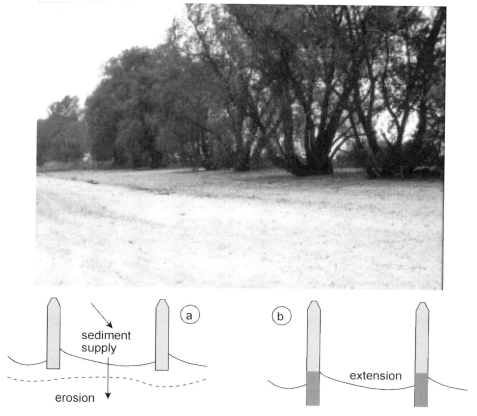

Figure 12-15 Erosion between groynes

12.4 Shore protections

12.4.1 Mangroves

Mangrove forests are the natural vegetation of many tropical coasts and tidal inlets, they form a higly productive ecosystem, a nursery for many marine species. Mangrove trees miraculously thrive in very dynamic circumstances. They can cope with salt water where as most other plants cannot. Seedlings have little opportunity to settle, so mangroves are viviparous, giving birth to an almost complete tree in a capsule (the propagule) that can travel with the tide and can turn into an upright standing young tree within a few days.

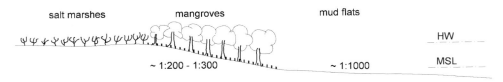

Figure 12-16 Cross section coast with mangroves

Figure 12-16 gives a typical cross section of a mud coast with mangrove vegetation. There is an increasing awareness of the importance and vulnerability of mangrove forests. These forests are also a natural coastal protection and where they are removed, for whatever reason, erosion and/or artificial protection is the price to be paid. Planting these vegetable breakwaters everywhere, however, is not a very promising concept. Mangroves can only exist on coasts with a moderate wave climate (Noakes, 1955). This is mainly because the seedlings cannot settle in highly dynamic conditions (Sato, 1985). Once the trees are grown up, however, they can even stand an occasional cyclone (Stoddart, 1965; Hopley, 1974).

There is a wealth of literature on mangroves, but little from a physical or coastal engineering point of view. From a hydraulic engineering point of view, an insight into the influence of mangrove forests on wave transmission and coastal stability is of great interest.

One of the most striking visible features of mangroves is the root system. Because mangroves usually live in anaerobic conditions in the mud soil, they improve their gas exchange with the atmosphere by means of aerial roots. Of the many mangrove species in the world, the two most important are Avicennia and Rhizophora. These species have completely different aerial roots. Rhizophora has "prop"-roots or "stilt"-roots, while Avicennnia grows "snorkel"-type pneumatophores, which emerge vertically from the bottom, see Figure 12-17. These roots play an important role in wave damping, probably even more than the trunks of the trees.

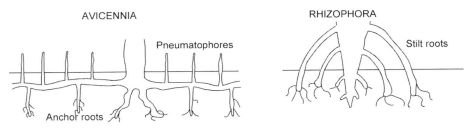

Figure 12-17 Mangrove roots

With this breathing system, mangrove trees need fresh air regularly and for this reason they can only live in the upper tidal zone, approximately above Mean Sea Level up to High Water (Spring) Level. They can stand storm surges, but after extremely long periods of flooding, drowning of mangroves has been reported

(Steinke and Ward, 1989) and even suffocation when large quantities of sediment have covered the root systems during a storm. Below MSL, the seedlings cannot settle and at higher levels, the mangroves cannot compete with other plant species. The coastline in front of mangroves often consists of mud flats with typical slopes of approximately 1:1000. Behind the mangroves, between HWS-level and the level of occasional flooding by the ocean, salt marshes can be found with halophytic (salt-loving) herbs and grasses, before the "normal" vegetation starts, see Figure 12-16. The width of these tidal forests (with typical slopes of about 1:200 - 1:300) is determined mainly by the tidal range (which can vary from a few decimeters to more than 5 meters along tropical coasts).

Figure 12-18 Young rhizophora mangrove trees

At Delft University of Technology a study concerning the wave transmission in mangrove forests was carried out, see Schiereck and Booij, 1995. Based on the hydrodynamical damping of waves around cylinders the effect of stems and roots of mangroves was estimated. Since there is a lot of variation in dimensions and density of the vegetation, three typical cases were discerned: a sparse, an average and a dense vegetation, see Figure 12-19 where the wave transmission is defined as H_t/H_i, the transmitted wave height divided by the incoming wave height.

With a wave energy model (Holthuijsen et al, 1989) the energy dissipation on several slopes was computed for the mangrove forests, as shown in Figure 12-19. If the energy dissipation per unit area is assumed uniform (as was done e.g. by Bruun, 1954, for beaches), slopes of 1:100 to 1:300 are found for mangroves and 1:1000 to 1:2000 for mud flats. These numbers are roughly in line with what is found in nature, see Figure 12-16.

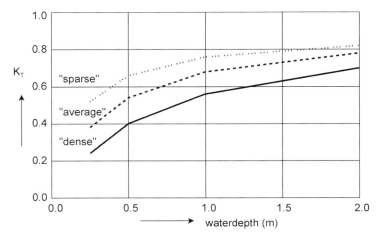

Figure 12-19 Wave transmission through 100 m of mangroves

The significance of these findingsfollows from Figure 12-16. When the mangroves are removed, the slope will become ~ 1:1000 leading to a regression of the coastline of many hundreds of meters. So, indeed, where a natural protection is present, one should preserve it. Reforestation is difficult and can require temporary protection, see Figure 12-20, see also next section on load reduction.

Figure 12-20 Mangrove reforestation with wave load reductor

12.4.2 Load reduction

When the loads on a shore cause too much erosion, a load reductor in front of the shore can sometimes be a solution. A rockfill dam is a hard protection, but the original shore behind it can stay intact and can serve as a recreational beach and/or the original flora and fauna can be maintained or introduced a second time. An example is a coast where mangroves were originally present, but have been removed. To restore the vegetation, a temporary protection is necessary. Chapter 7 mentions wave transmission across dams related to the relative cress height, regardless of the width and slope. Equation (12.4) describes the experimental results:

$$\frac{H_T}{H_I} = \left(\frac{B}{H_I}\right)^{-0.31} \left(1 - \exp(-0.5\xi)\right) F_{\mathrm{dam}} \tag{12.4}$$

F_{dam} in this equation represents the dam type for which Table 12-1 gives some values:

Table 12-1

Dam type	Rock	Gabions	Closed (asphalt or pitched blocks
F_{dam}	0.64	0.7	0.8

Figure 12-21 Wave reductor

Figure 12-22 shows some results of equation (12.4). (a) shows the results for various crest widths (the minimum width in equation (12.4) should be taken $B/H_I = 1$) and $\xi = 2$. A greater width, of course, gives a lower transmission. (b) shows the results for various ξ and $B/H_I = 1$; a larger ξ (steeper slope or less steep waves) gives more transmission.

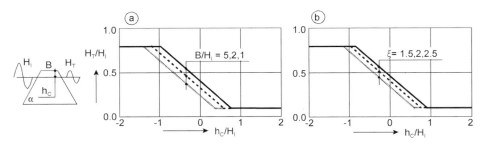

Figure 12-22 Wave transmission dams

In order to facilitate the exchange of water and animals, apertures at regular intervals are necessary, see Figure 12-23.

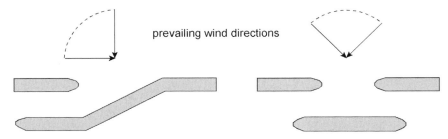

Figure 12-23 Water exchange possibilities

12.4.3 Grass dikes and revetments

Grass can stand amazingly high current velocities and waves, provided the duration of the loads is limited and the grass is well managed. In the Netherlands much research on the strength of grass has been carried out, which is summarized in TAW, 1998. Figure 12-24 shows grass on a clay layer. The upper layer with the roots is called turf. The strength of grass comes mainly from the roots. Evenly distributed and healthy grass roots provide a strong cover layer. The strength is greatly influenced by the degree of fertilization. Much manure gives "lazy" grass with weak roots, while grass that gets little manure has to work hard for its food and gets long and strong roots.

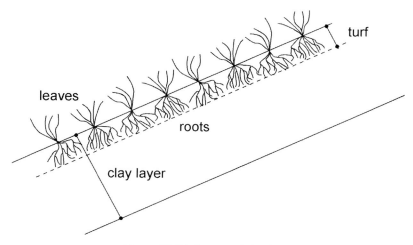

Figure 12-24 Grass and clay layer

The load and strength mechanisms are a combination of the features treated for asphalt layers and block revetments. The wave impact is absorbed by the deformation of the turf and clay layer, while the retracting waves cause waterpressures from inside as is the case for a block revetment, see Figure 12-25.

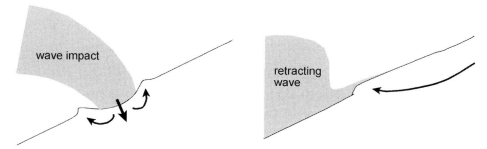

Figure 12-25 Load and strength of grass under wave attack

Erosion of grass in waves has been investigated, see TAW, 1998 and can be described with:

$$t_{\max} = \frac{d}{\gamma E} = \frac{d}{\gamma c_E H_s^2} \tag{12.5}$$

in which d is the turf thickness, E is the erosion speed and γ is a safety coefficient. c_E represents the quality of the grass:

Table 12-2

Grass quality	Good	Moderate	Bad
c_E (10^{-6} 1/ms)	0.5 – 1.5	1.5 – 2.5	2.5 – 3.5

Figure 12-26 shows the results of equation (12.5) for $d = 0.05$ m and $\gamma = 2$. Grass layers on dikes can stand waves with $H_s = 1.5$ m as long as the quality of the grass is good and the storm does not last longer than several hours. This makes grass an excellent cover for the higher parts of the dikes where waves occur only occasionally. When the waves are higher than 1.5 m, the zone of maximum wave attack should be covered with stone or asphalt, but grass still can be applied in the run-up zone. Good grass layers can stand waves lower than 0.4 m for several days, so in estuaries or rivers with a limited fetch, grass dikes can be a good alternative for the higher parts (grass can never live under water permanently).

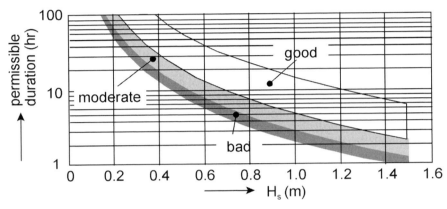

Figure 12-26 Permissible duration of wave load and grass quality

As previously mentioned, the quality of the grass, and more in particular the roots, depends heavily on the degree of fertilization. So, intensive grazing by cattle reduces the quality significantly. Moreover, cattle damage the grass layer with their hoofs. There is, however, no contradiction between nature and strength. Without manure, the grass is stronger and a larger biodiversity can be expected. Many herbs and flowers then flourish in grass. The biodiversity can be used as a measure of the quality of the grass, providing a visual tool for inspection (see TAW, 1998 and Sprangers, 1999).

13 CONSTRUCTION

Construction of beach groyne (courtesy Boskalis)

13.1 Introduction

This chapter treats some aspects of the construction of protections. Section 13.2 starts with an overview of some equipment, while the execution of bed, bank and shore protection works will be treated in sections 13.3, 13.4 and 13.5 respectively. Section 13.6, finally, deals with quality assurance.

All engineers are confronted with the practical sides of construction, but hydraulic engineers, like farmers and fishermen, also heavily depend on nature. Working in moving water and in changing weather conditions makes things more complicated. In tidal waters, the work is dominated by the rhythm of the tides, particularly in the zone between HW level and LW level. Waterborne operations are sensitive to wave conditions which influence the productivity, the construction method and even the design. Some parts are loaded during construction or repair in a completely different way than in the final stage, e.g. a filter layer may be unprotected for some time, vulnerable for wave attack or high flow velocities during springtide.

The circumstances during construction have such a large influence on the productivity and the final quality of the structure that it is necessary to have at least some statistical data on the conditions at the construction site. Frequencies of waterlevels, currents, waves and wind should be available or estimated before starting. During construction or maintenance, a good weather forecast is indispensable. Some activities are preferably not planned during the relatively rough season. In the Netherlands, for example, the execution of works involving sea defences is forbidden by law between october and april. In tropical countries, monsoons can create conditions lasting months, in which operations along a coast are risky or practically impossible.

Working in water and wind can not be learned from a book, experience is indispensable. Often adapting to the circumstances is necessary. Improvisation should be avoided, but that is not always possible, certainly not in remote areas. This chapter merely sums up the most important facets of the construction, including some of the difficulties hydraulic engineers may be confronted with. More information can be found in CUR,1995 which has been an important source for this chapter.

13.2 Equipment

13.2.1 General

Constructing bed, bank and shore protections means working on the transition area between land and water, hence the two main types of equipment are *land based* and *waterborne*. The choice between these two types in a particular case depends on the availibility of equipment and the possibility of working from the land side or from the waterside. For a large project, any existing equipment can be mobilized and the choice will depend chiefly on costs. For a mega-project it can even be profitable to develop and build special equipment which is written off for that project only. E.g for the Eastern Scheldt storm surge barrier in the Netherlands, special barges have been built to sink the block mats for the bed protections. For the breakwaters at Hook of Holland (Port of Rotterdam) side stone dumping vessels have been constructed and special barges have been built to place the concrete blocks of the armour layer.

For small projects, standard available equipment will be used. In a less developed area where very little is available and possible, creativity can be crucial to come to workable solutions. In the following sections, some widely used standard equipment will be mentioned.

For the application of rock, a distinction can be made between *direct dumping* and *controlled placement*. Some equipment can only be used for dumping of large quantities of stones while other machines are fit (or are only able) to place each stone individually, there are, of course, various possibilities in between. This distinction between the various types of equipment is not black and white, for each piece of equipment the possibilities will be described.

A third distinction could be made between *transportation* equipment, like trucks or cargo barges, and *handling* equipment like cranes or stone dumpers. Combinations are also possible and sometimes the difference is not so clear. So, here too, there is no black and white distinction and the characteristics will be mentioned accordingly when the equipment is described.

13.2.2 Land based equipment

Most land based equipment moves on wheels or tracks. Figure 13-1 shows some examples. Any truck can be used to carry stones, but special off highway *dump trucks* can carry from 30 tons up to 150 tons. They can be used for transport and for direct dumping. *Wheel* (or track) *loaders* have typical capacities of 5 to 15 tons and can be used for handling and loading on a construction site and for short distance transport on a construction site.

Shovels can do more or less the same, but with longer arms, they can reach farther. Typical capacities are 5 to 40 tons. Shovels move stones away and upwards to a position higher than the driver, while *backhoes* look very much the same but "scrape" the material upwards from a lower position towards the driver. *Bulldozers*, finally, only push material forward.

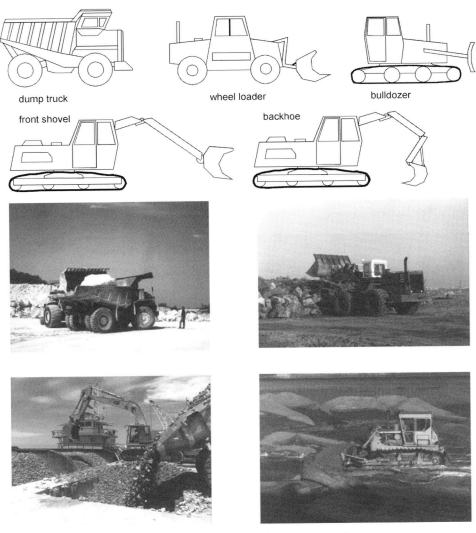

dump truck   wheel loader   bulldozer

front shovel   backhoe

Figure 13-1 Some examples of land based equipment

Loaders, shovels and backhoes can lift material several meters, but for larger distances *cranes* have to be used. Cranes are usually equipped with a clamshell or a grab, see Figure 13-2. *Clamshells* can be used for finer material like sand or gravel, while a *grab* can carry large stones, up to about 8 tons. An average operation speed for convential cranes is about 20 cycles per hour, while for hydraulic cranes, this can go up to 30-40 cycles per hour. The lifting capacity of a crane depends largely on the reach, see also Figure 13-2.

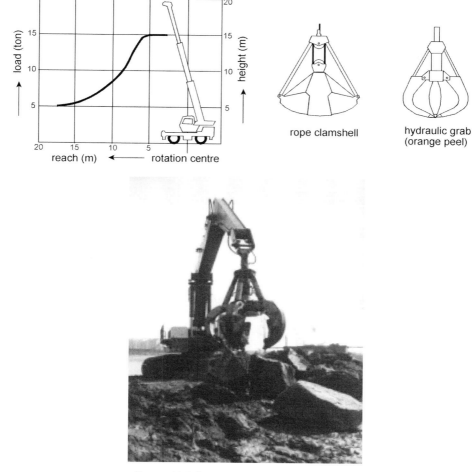

rope clamshell

hydraulic grab
(orange peel)

Figure 13-2 Crane capacity and accessories

Working conditions

Land based operations in hydraulic engineering know the same limitations as working outside on any construction site. Extra problems can be caused by wave run-up. Land based operations should therefore be planned well above possible wave run-up during construction or the downtime due to excessive wave run-up should be allowed for in the planning. When a breakwater is constructed with dump trucks, the crest on which the trucks drive is normally 1-1.5 m above normal (high) waterlevels. Generally, in a tidal area, on low lying parts of the construction site, oprations will be limited to low water periods.

A vehicle with caterpillar tracks can drive on larger stones than one with tyres. If the stones are too large for truck tyres, the large gaps on the surface can be filled with finer material. This is possible for rock up weighing up to 1 ton. But if the fine material has to be removed again because it has a negative influence on permeability and stability (see chapter 8), this becomes very costly.

13.2.3 Waterborne equipment

Figure 13-3 Split barge (Boskalis)

Figure 13-4 gives some examples of waterborne equipment. Starting with the idea that the only difference between land based and waterborne equipment is that it drives rather than floats, one can place any piece of land based equipment on a barge or *pontoon*, see (a). Capacities will be rather low, while the placing is reasonably well controlled. Individual stones can be placed quite accurately. A *side stone dumping vessel* (b) is the appropriate tool for many situations. Stones are transported on deck of the vessel and can be pushed into the water by moving beams. The loading capacity can be up to 1500 tons, while the dumping capacity is about 60 - 70 tons per minute. The dumping is reasonably well controlled, depending on the velocity and the waterdepth, see section 13.3. If no side stone dumping vessel is available, a pontoon and a bulldozer can form a substitute for a fairly uncomplicated job.

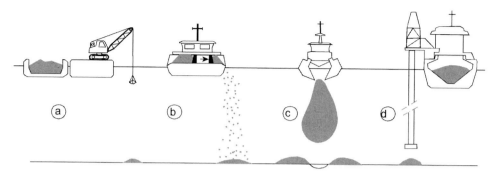

Figure 13-4 Some examples of waterborne equipment

A *split barge* (c) or a barge with bottom doors can be used when large quantities of stone have to be dumped, like in the core of a breakwater or any other dam structure. For accurate placing, the split barge is not a good alternative. When used for bottom protections on large depths, the result can even be a crater instead of a bulge, see Figure 13-4c. This is especially the case in deep water. If accurate placing on large depths is important, a *fall-pipe vessel* can be the answer (d). This is for example the case when a pipeline has to be protected.

Figure 13-5 Side stone dumping vessel (courtesy Boskalis)

Working conditions

Waterborne operations are usually more vulnerable to environmental conditions than land based operations. Tides are an obvious example. During low water, the depth can be insufficient for barges. On the other hand, high water can create ideal conditions for waterborne operations. So, in any case working time versus downtime heavily depends on the tides, which behave independent of normal working hours.

Current velocities of up to 1.5–2 m/s usually pose no special problems. Higher velocities make special measures necessary, either by means of strong anchoring possibilities or by means of engine thrust in all directions.

Waves can cause many kinds of problems. Short wind waves are usually less of a nuisance than swell. Most barges that have been designed to work in coastal waters can operate in wind waves of up to 1 – 1.5 m, while the same barges experience problems with swell waves of merely 0.5 m. Cranes mounted on a vessel are vulnerable to rolling. Tilting of just a few degrees can cause downtime for such a crane.

Navigation along the construction site, finally, can cause hindrance to the operations. With special warning or signs for the passing ships, they can be urged to slow down.

13.3 Bed protections

13.3.1 Loose rock

Constructing a loose rock bed protection can normally be done best with a side stone dumping vessel. Such a vessel sprinkles the stones rather than dumps them. Waterdepth and velocity influence the accuracy of placing the stones. Figure 13-6 shows possible deviations.

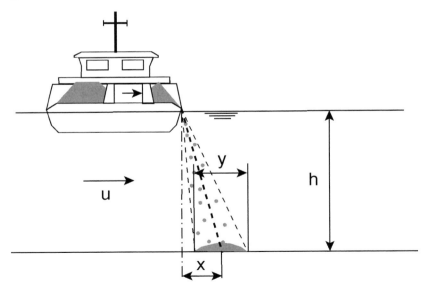

Figure 13-6 Deviations from dumping location with side stone damping vessel

The deviation of the centre of the mound of the stones compared with the location vertically under the ship can be estimated from the current velocity and the fall velocity of a stone in water. The latter can be found by equating gravity and flow resistance:

$$mg = F_D \quad \rightarrow \quad (\rho_s - \rho_w)d_{n50}^3 g = C' \rho_w u^2 d_{n50}^2 \quad \rightarrow \quad u_{Fall} = C' \sqrt{\Delta g d_{n50}} \quad (13.1)$$

where C' is a constant of proportionality depending, among other things, on the shape of the stones. The deviation, x, now can be computed simply from the tangent of the angle between fall velocity and current velocity:

$$x = C \frac{hu}{\sqrt{g \Delta d_{n50}}} \quad (13.2)$$

where C is a coefficient with typical values of around $0.7 - 0.8$.

The path to x in Figure 13-6 is the average path and the sinking stones show random variations around this average, leading to a width of the bulge, y. Empirically, this width was found to be (Van Gelderen, 1999):

$$y = K\sqrt{h} \quad (13.3)$$

in which $K \approx 1.9 \sqrt{m}$ for broken stone and $K \approx 2.1 \sqrt{m}$ for rounded stone.

Another problem with dumping can be segregation of fine and coarse material. To prevent this, it is usually sufficient to ensure a ratio $d_{n90}/d_{n10} < 5 - 10$.

13.3.2 Fascine mattresses

Fascine mattresses are preferably constructed on a slope near the water, see Figure 13-7. In a tidal area, the construction site is usually located between low and high water, making towing away the mattress quite easy but reducing the possible working time. On a site completely above water, work is possible all of the time, but a strong towboat is necessary to get the mattress into the water. The construction site in Figure 13-7 is too small for a complete mattress, so during construction, part of the mat has to be towed away. The finished wooden mat floats and is towed away to the place where it is wanted. A beam is connected to the mattress to distribute the pulling forces evenly. Typical dimensions of fascine mattresses are lengths and widths of several ten's of meters.

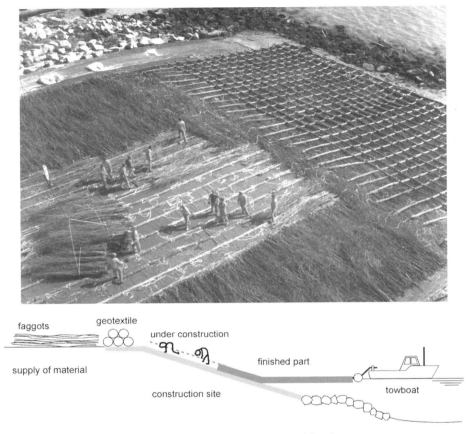

faggots   geotextile

under construction

supply of material

finished part

construction site   towboat

Figure 13-7 Construction and towing away of fascine mattress

The following is a description of a sinking method, often used in the Netherlands. Other methods are possible.

Upon arrival at the sinking location, the mattress is tied between two pontoons. In a tidal area, the sinking is done during the slack-water periods between ebb and flood or vice versa. A heavy metal beam causes one end of the mattress to sink. A side stone dumping vessel is manoeuvered between the pontoons and stones are dropped on the mattress, making it sink further. These stones give the mattress a provisional stability on the bottom (normally, light stones are used for this first cover: $150 - 200$ kg/m^2 of stones $10 - 60$ kg). The stone dumping vessel moves slowly with the tidal current direction, dumping at one or two sides. Once the mattress is completely on the bottom, the beams are retrieved and extra stones are dumped to finish the bed protection (depending on the size required for the loads after construction, normally $1.5 - 2 \cdot d_{n50}$ which, for stones $10 - 60$ kg comes down to around 500 kg/m^2).

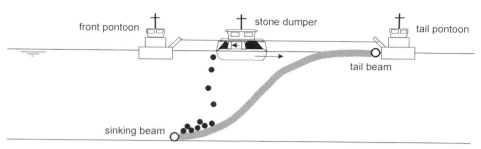

Figure 13-8 Sinking fascine mattress

Figure 13-9 Construction of bamboo mattress

13.3.3 Prefabricated mats

Prefabricated mats are transported and sunk with special equipment. For the Eastern Scheldt storm surge barrier, the mat dimensions were 200×30 m^2.

Figure 13-10 Sinking block mat in Eastern Scheldt (van Oord ACZ)

13.4 Bank protections

13.4.1 Revetments

Figure 13-11 Placing geotextile in simple revetment (Johan van der Ham)

The construction of revetments can be done from two sides, so both land based and waterborne equipment can be involved. Basically, a revetment consists of a top layer and a filter with some toe protection. The most simple version is a geotextile with stones on top. Figure 13-11, Figure 13-12 and Figure 13-13 show some possible construction methods. A geotextile is rolled down from the bank. To a very limited extent, this can be done manually, if necessary with the assistance of a diver. A fascine roll tied to the lower end can give some stiffness to ease the process. For larger structures, a pontoon can pull a steel pipe down the slope, around which the geotextile is reeled. A side stone dumping vessel then covers the geotextile with stones, upward from the toe to prevent the whole geotextile sliding down.

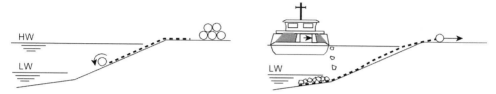

Figure 13-12 Construction of a simple revetment with side stone dumping vessel

Figure 13-13 Placing stones with floating crane and grab (Johan van der Ham)

When a fascine mattress is applied on a bank in a tidal area, or as a toe protection for a dike, it can be placed at high water after which it is anchored by placing stones on top, see Figure 13-14.

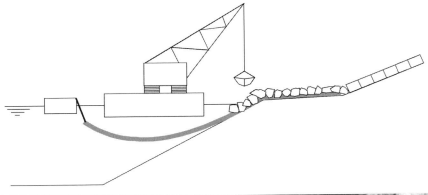

Figure 13-14 Placing fascine mattress on revetment

Above low water, a top layer of a revetment can be placed from the bank. For light stones (up to 10 – 60 kg) this can be done with bulldozers or wheel loaders and placing stones manually. For heavier stones, a crane with a grab is normally used, see Figure 13-15.

Figure 13-15 Placing of small and large stones from landside

It is also possible to place prefabricated mats on a bank which has the advantage of placing large areas in one action, Figure 13-16 shows an example. To hoist the mat and to prevent damage, an equator is needed.

Figure 13-16 Placing of prefabricated mat on revetment

13.5 Shore protections

13.5.1 Dikes

Dikes consist roughly of a body and a protection layer which is quite similar to a revetment. The body can be made of clay, which has a good resistance even after the protection layer has been removed, or of sand. In the last case, the protection layer fulfils a much more essential function than in the first case. New dike bodies are mostly constructed of sand nowadays, as clay is not always available in large quantities.

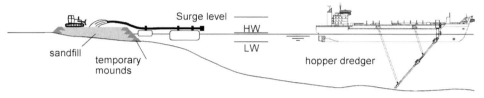

Figure 13-17 Construction new dike body of sand

Figure 13-17 shows a way to construct a new dike. Sand is dredged from a site which is near enough to keep the costs low and far enough to prevent morfological consequences for the coastline. The sand is pumped either directly from a stationary dredger or is unloaded onto a pontoon using a self-propelled hopper dredger. The sand-water mixture which is released from a pipe, creates very mild slopes. Under water 1:15 – 1:25 and above water 1:30 – 1:40. This will result in a dike body which is often too large or in much loss of material. To avoid this, small mounds of fine quarry stone, or other available flow resistant material are constructed to create steeper slopes. In between these mounds the sand is sprayed and spread with bulldozers.

When the body is finished, the protection layers are laid on the dike. To protect the dike against the every day loads during construction, the toe and under water parts are made first, e.g. by applying a fascine mattress with stone cover. After that, the dike revetment is constructed. Depending on the type of top layer, e.g. blocks, asphalt or grass, specialized equipment is used. Figure 13-18 shows the construction of a block revetment layer and figure 13-19 the construction of an asphalt layer.

Figure 13-18 Mechanical placement of Hydroblocks (courtesy Greenbanks)

Figure 13-19 Construction of an asphalt revetment (courtesy Netherlands pavement consultants)

13.5.2 Groynes and breakwaters

Coastal groynes and breakwaters are very much the same, the difference lies mainly in the dimensions. Both ususally have to be constructed under exposed conditions. The choice for land based or waterborne construction depends on many factors, but the availibility of equipment and the source of rock are probably decisive. When the quarry is inland of the considered coast, land based operation might be the first choice because the rock is transported with trucks, while waterborne operation is logical when the stones are delivered by barges. When transshipment of material is necessary anyhow, the choice depends on other factors. Figure 3-20 shows the construction of a large groyne.

Figure 13-20 Construction of a groyne using both land based and marine equipment

Land based construction

Trucks are important tools for land based construction of groynes and breakwaters. They can be used for direct dumping, but when more placing accuracy is needed, or when the required slope can not be acquired with direct dumping, additional cranes will be necessary. This will take much more time when every truck has to be unloaded on the spot. Sometimes it is possible to create a stock pile with direct dumping from where a crane can continue to work.

Passing possibilities for trucks are important for productivity, especially on long structures. When a crane is working, passing is even indispensible, see Figure 13-21. Backhoe cranes can be used for stones which are not too heavy (up to 1 – 2 ton) and reaches which are not too long. For the land based construction of a toe structure, long reach cranes are necessary. This is possibly easier with waterborne equipment.

Figure 13-21 Land based equipment on breakwater

Waterborne construction

When the breakwater becomes very long and traffic jams on the crest become a problem or when cranes have to reach too far, the use of barges becomes an option. For the core of a groyne or breakwater, a split barge can be very economic if large quantities have to be dumped. For toes and armour layers, a more accurate placing is needed and a side stone dumping vessel can be used. For large elements, a crane on a pontoon, or a specially designed barge, is the answer (see Figure 13-22).

Figure 13-22 Placing elements on a breakwater with crane ship (van Oord ACZ)

Combined land based and waterborne construction

Often a combination of the two main categories of construction equipment is used, certainly for larger structures. Figure 13-23 shows the construction of a large breakwater. First, the bottom is protected with mattresses where large filter gradients are expected and with gravel on other parts. On top of that, part of the core is made with quarry run dumped from split barges. This can be done up to a few meters under normal high water, depending on the draught of the barges. The rest of the core is provided and dumped with trucks, while the slopes are trimmed and part of the secondary armour layer is applied on the slopes with a crane on a pontoon.

The toe of the breakwater is placed with the crane on the pontoon and the secondary armour layer is completed, either with the same crane and pontoon or with a crane which operates fom the crest. The armour layer, consisting of very heavy concrete blocks, is placed with a heavy duty crane. The crown wall is cast in situ and the armour layer is finished.

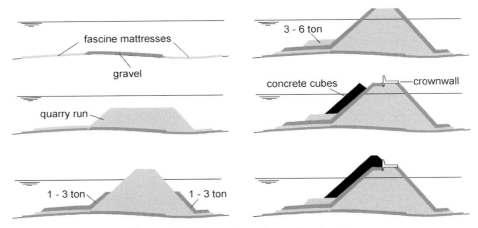

Figure 13-23 Execution scheme breakwater

13.6 Quality assurance

13.6.1 General

Quality assurance, quite normal in mechanical and electronic industries, is relatively new in a traditional business, such as hydraulic engineering. Of course, there has always been some kind of quality assurance implicitly incorporated in the craftmanship of the personnel, but it was never explicitly mentioned in contracts. Moreover, in contrast with cars or computers, the client (usually some public authority) was always present during the construction of his product.

Two factors are changing, increasing the necessity of a quality assurance system. Firstly, traditional methods and manual labour are being replaced by new mechanical methods for which less experience is available. Secondly, there is a shift in the relation between the client and the contractor. More and more "turn-key" or "design and build" contracts or whatever name they have, are signed worldwide. In those cases, the contractor has more responsibility for the quality of the product. Since it is impossible to rely only on an inspection of the final product (which is usually under water and covered with a lot of material), a good quality control system for the whole process is essential.

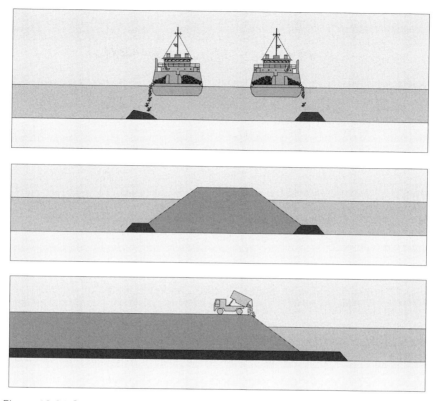

Figure 13-24 Combination of waterborne and land-based operations (van Oord ACZ)

Quality is defined as: *the degree to which the product, process or services complies with the functional requirements.* Meeting the required quality standards is mainly a matter of organisation rather than a technical issue. The realization of a product of sufficient quality is the result of good control of all activities, which concerns all parties involved (client, design engineer, contractor). The following activities play an important role in quality control assurance:

- Definition of existing needs
- Specification of functional requirements
- Design process
- Work preparation
- Construction
- Commissioning and testing
- Maintenance
- Evaluation

Quality assurance is not merely limited to the construction stage but involves all project stages. Reversely, the design should not only take into account functional requirements, but should also take into account practical construction requirements such as tolerances and construction feasibility using standard or specially built equipment. During all stages of a project, the quality should be safeguarded by a quality manager who reports independently to the project manager.

13.6.2 Tolerances

Land based operations

Using standard equipment, the tolerances can reach the following values:

Table 13-1

Material	Tolerance above water	Tolerance under water
Gravel	0.05 m	0.10 – 0.15 m
Rock	$0.25 – 0.5\ d_{n50}$ (min. 0.2 m)	$0.25 – 0.5\ d_{n50}$ (min. 0.2 m)

With good supervision and manual assistance even better values can be reached. Land based construction with materials like concrete blocks or asphalt can yield accuracies in the order of magnitude of cm's.

Waterborne operations

If the right equipment is used, with individual placing of stones, the same vertical tolerances as mentioned for land based operations should generally be attainable. For stone dumping activities (bulk placing), Table 13-2 gives some numbers.

Table 13-2

Waterdepth	$W_{stone} < 300$ kg	$W_{stone} > 300$ kg
Above water	± 0.2 m	+ 0.4 to – 0.2 m
< 5 m	+ 0.5 to – 0.3 m	+ 0.8 to – 0.3 m
5 – 15 m	approximately ± 0.5 m	+ 1.2 to – 0.4 m
> 15 m	approximately ± 0.5 m	+ 1.5 to – 0.5 m

Vertical tolerance of dredging activities, which are often a preparation for protection works, ranges from 0.1 to 0.5 m.

Horizontal tolerance depends on the propulsion of the equipment (bow and stern thrusters or other propellers that can be pointed in any direction) and the navigational equipment (e.g. based on Global Positioning System or Differential Global Positioning System, Dynamic positioning, etc.). If waves and currents are absent or negligible, a horizontal accuracy of approximately 1 m can be achieved.

APPENDIX A
Material properties

A.1 Block weight and size[1]

The relationship between size and weight of individual blocks may be defined in terms of the equivalent-volume cube (side D_n) or the equivalent volume sphere (diameter D_s) which, with weight density ρ_a and block weight W, give the following relationships:

$$D_n = 1.0 \cdot \left(\frac{W}{g \cdot \rho_a} \right)^{1/3}$$

$$D_s = 1.24 \cdot \left(\frac{W}{g \cdot \rho_a} \right)^{1/2}$$

$$D_n = 0.806 \cdot D_s$$

As indicated in the previous section, where graded rock materials are used, size and weight relations refer to medians or averages.

Where the particles are small enough (less than 200 mm) the statistical values are most conveniently derived from sieve analyses. The median sieve size (square openings) (D_{50}) on the percentage passing cumulative curve can be related to the median weight (W_{50}) on the percentage lighter by weight cumulative curve by a simple conversion factor ρ_a. (This dimension D_{50} is the same as the median value of the gross shape dimension z.)

---

[1] Based on CUR154 [1991]

The 50% passing nominal diameter D_{n50} is the size of the cube with equivalent volume to the block with median weight, and is given by

$$D_{n50} = \left(\frac{W_{50}}{g \cdot \rho_a} \right)^{1/3}$$

The conversion factors relating D_{50} to D_{n50} or W_{50} have been determined experimentally by various workers and can be summarised as:

$$\frac{D_{n50}}{D_{50}} = 0.84$$

or

$$F_s = \left(\frac{W_{50}}{g \cdot \rho_a \cdot D_{50}^3} \right) = 0.60$$

where F_s is the shape factor. F_s may vary from 0.34 to 0.72, while in studies using different shape classes of limestone fragments, F_s for all five classes fell between 0.66 and 0.70.

Samples containing blocks larger than 100 mm are difficult to analyse by sieving techniques and direct measurement becomes a more appropriate means of determining size distribution. The relationship between size and weight distributions have been noted above. With large block sizes (i.e. larger than sieve sizes of 250 mm) measurement of weight may prove more practicable than measurement of dimensions, and the size-weight conversion factor such as F_s can be used to determine the median sieve size and other geometric design parameters that may be expressed in terms of sieve sizes. W_{50} (or M_{50}) is the most important structural parameter in the rock armour design being related to the design parameter D_{n50}, as described above. Gradings of rock fulfilling the class limit specification described in the following section may be expected to have standard deviations in D_{n50}, varying from 1% for heavy gradings to 7% for wide light gradings.

A.2 Grading

In a sample of natural quarry blocks there will be a range of block weights, and G in this sense, all rock materials is, to some extent, graded. The particle weight distribution is most conveniently presented in a percentage lighter by weight cumulative curve, where W_{50} expresses the block weight for which 50% of the total sample weight is of lighter blocks (i.e. the median weight) and W_{85} and W_{15} are similarly defined. The overall steepness of the curve indicates the grading width, and a popular quantitative indication of grading width is the $\dfrac{W_{85}}{W_{15}}$ ratio or its cube root, which is equivalent to the $\dfrac{D_{85}}{D_{15}}$ ratio determined from the cumulative curve of the equivalent cube or sieve diameters of the sample. The following ranges are recommended for describing the grading widths:

$$\frac{D_{85}}{D_{15}}$$

or

	$(W_{85} / W_{15})^{1/3}$	W_{85} / W_{15}
Narrow or "single-sized" gradation	Less than 1.5	1.7-2.7
Wide gradation	1.5-2.5	2.7-16.0
Very wide or "quarry run" gradation	2.5-5.0+	16.0-125+

The term 'rip-rap' usually applies to armouring stones of wide gradation which are generally bulk placed and used in revetments. The phrase 'well graded' should generally be avoided when describing grading width. It merely implies that there are no significant 'gaps' in material sizes over the total width of the grading.

Standardisation of gradings

There are many advantages in introducing standard grading classes. These mostly concern the economics of production, selection, stockpiling and quality control from the producer's viewpoint. With only a few specified grading classes, the producer is encouraged to produce and stock the graded products, knowing that designers are more likely to provide them the market by referring to these standards wherever possible. The proposed standard gradings for armour are relatively narrow. This can result in increased selection costs, but these will often be completely offset by the possibility of using thinner layers to achieve the same design function. Standard gradings are not needed for temporary dedicated quarries supplying single projects where maximised utilisation of the blasted rock is required.

It is convenient to divide graded rock into:

- 'Heavy gradings' for larger sizes appropriate to armour layers and which are normally handled individually; 'Light gradings' appropriate to sheltered cover applications, underlayers and filterlayers that are produced in bulk, usually by crusher opening and grid bar separation adjustments;

- 'Fine gradings' that are of such a size that all pieces can be processed by production screens with square openings (i.e. less than 200 mm). For fine gradings, the sieve size ratio from the cumulative sieve curve ($\dfrac{D_{60}}{D_{10}} = U_c$, the uniformity coefficient) is often used to characterise the width of grading in a sample.

Standard gradings are more or less essential for fine and light gradings. However, for heavy gradings it is not difficult because of individual handling to define and produce gradings other than standard (see below). For example, if the 1-3 t grading is (just) too small for a particular application, choice of the first safe standard grading of 3-6 t will lead to an excessive layer thickness and weight of stone, and here use of a non-standard grading may well be appropriate. Again, ceiling sizes of stones in quarries arising from geological constraints may dictate an upper limit.

A consistent scheme for defining grading requirements for the four suggested heavy standard grading classes is given in Dutch standard NEN5180, see the table of Figure A-1. Box A1 explains the scheme in detail and in Box A2, the effective mean, W_{em} which is important for rapid assessment of grading is described.

Three light grading classes have been defined by weight in an identical way to the heavy gradings, as shown in the table. Corresponding requirements have been proposed when these gradings are specified using 'sieve' sizes and have been designated 200/350 mm, 350/550 mm and 200/500 mm classes. The test verification in the latter case requires both gauging of blocks and average weight determination. Both specifications for the light gradings are intended to produce approximately similar graded stone products.

Class designation (kg)	ELCL (y<2)	LCL (0<y<10)	UCL (70<y<100)	EUCL (97>y)	Requirements for range of effective mean weight, i.e. excluding pieces less than ELCL	Additional information Expected range for compliance with standard gradings	
Where y is the percent by weight lighter on the cumulative plot						W_{50}	W_{85}/W_{15}
Class limit definition by weight ,W_y (kg)							
Heavy gradings 300-1000 kg	200	300	1000	1500	540-690 kg	595-760 kg	2.3-3.8
1000-3000 kg	650	1000	3000	4500	1700-2100 kg	1800-2200 kg	2.2-3.6
3000-6000 kg	2000	3000	6000	6000	4200-4800 kg	4200-4800 kg	1.6-2.2
6000-10000 kg	4000	6000	10000	15000	7500-8500 kg	7500-8500 kg	1.4-1.8
Light gradings 10-60 kg	2	10	60	120	20-35 kg	26-46 kg	32.-7.7
60-300 kg	30	60	300	450	130-190 kg	150-220 kg	2.8-6.0
10-200 kg (wide)	2	10	200	300	30-90 kg	70-130 kg	5.0-11.4

	Class limit definition by square hole (mm)				Average weight retained on L hole, W_L (kg) for rock density, ρ_L (t/m³)		
					<2.5	2.5 – 2.9	>2.9
					Min. / Max.	Min. / Max.	Min. / Max.
200/350 mm	100	200	(350)	400	20 / 40	25 / 45	25 / 50
350/550 mm	250	350	(550)	650	115 / 180	130 / 200	145 / 240
200/500 mm (wide)	100	200	(500)	550	45 / 80	50 / 90	55 / 100

Note:

The 10-60 kg, 60-300 kg and 10-200 kg classes are approximately equivalent to the 200.350 mm, 350/550 mm and 200/500 mm classes, respectively. The sizes given in brackets are those of the equivalent UCL or U hole and are not requirements but for consistency are used for designating the class of grading.

Figure A-1 Standard gradings

Box A1 Explanation of class limit system of standard gradings
Rather than using envelopes drawn on a cumulative plot to define the limits of a standard grading, the proposed system refers to either:
1. A series of weights of stones and their corresponding ranges of the cumulative percentage by weight lighter than values which are acceptable; or
2. A series of 'sieve' size of stones and their corresponding ranges of cumulative percentage by weight passing values, together with the average weight of stones in the grading.
Each weight-standardised grading class is designated by referring to the weights of its lower class limit (LCL) and its upper class limit (UCL). In order to further define the grading requirements realistically and to ensure that there are not too many undersized or oversized blocks in a given grading class, it is necessary to set two further limits, the extreme lower class limit (ELCL) and the extreme upper class limit (EUCL). The standard grading scheme then uses percentage by weight lighter than values, y (equivalent to percentage by weight passing for aggregated), 0, 2, 10, 70, 97 and 100 to set the maximum and minimum percentiles corresponding with each of the four weight values given by ELCL, LCL, UCL, and EUCL (see below). Note that although straight-line segments have been drawn to indicate an envelope, grading curves can go outside these straight-section envelopes and still fulfil the requirements given by the four class limits. However, the further requirement

specifying that the effective mean weight, W_{em}, should fall within a set range will help to reassure the designer that there is an appropriate range of W_{50} values.

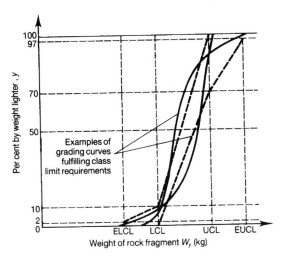

Size and average weight standardised gradings

Each size-standardised grading class is defined with reference to the 'sieve' sizes of its ELCL, LCL and EUCL as explained above (no UCL specified) together with the average weight, WL, of all rocks not passing the LCL hole. However, for consistency they are designated i.e. named using the LCL and UCL hole sizes. Apart from bulk weighing to determine WL, the test verification of size is limited to gauging of blocks using three square gauge holes of sizes as defined in the table and imposing the requirement consistent with the weight-grading scheme that:

1. Less than 2% by weight shall pass the ELCL hole, known as the EL hole;
2. Less than 10% by weight shall pass the LCL hole, known as the L hole;
3. At least 97% by weight shall pass the EUCL hole, known as the EU hole.

The size-limit specification of light gradings is only appropriate for underlayers applications whereas the weight-limit specification is always necessary for cover layers. Any of the three standard light gradings specified by size limits but tested and certified by their equivalent weight designated grading class would be deemed acceptable for all uses as the weight definition is taken as primary.

Box A2 Median and effective mean weight

A number of parameters in addition to the designated class limits may be required for design and specification purpose. An acceptance range for the important design parameter, W_{50} (median weight), is also useful for each grading class. However, in some circumstances an effective mean weight W_{em} for a particular consignment may be more easily obtained simply by bulk weighing and counting. In order to avoid including fragments and splinters in the, distribution, W_{em} is defined as the arithmetic average weight of all blocks in the consignment or sample, excluding those which fall below the ELCL weight for the grading class. An empirical conversion factor relating W_{em} to W_{50} allows an estimate of W_{50} without the necessity of weighing each piece to obtain a weight-distribution curve. As the grading becomes wider, so will W_{50} depart more from W_{em}. Approximate relations are as follows:

10-60 kg
$$\frac{W_{50}}{W_{em}} = 1.3$$

60-300 kg
$$\frac{W_{50}}{W_{em}} = 1.15$$

300-1000 kg
$$\frac{W_{50}}{W_{em}} = 1.10$$

1000-3000 kg
$$\frac{W_{50}}{W_{em}} = 1.05$$

3000-6000 kg and

6000-10000kg
$$\frac{W_{50}}{W_{em}} = 1.00$$

The weight envelopes for the standard light and heavy gradings are shown in Figure A-2 with appropriate size-weight conversion. In this figure, rocks of any density and weight can be related to the corresponding nominal diameter, D_n, sieve size, z, and minimum rock thickness, d, using the alternative horizontal scales on this chart. The 'sieve' size, z, enveloped for the standard tine and light gradings are shown in the following two figures, again with conversions to D_n and d. The standard fine and light gradings are produced by screens and grids and sometimes with eye selection to remove oversize material in the 60-300kg and 10-200 kg classes. The poor screening efficiency that occurs in practice means that a correction factor would be needed in addition to the given theoretical relationships for sieve size, z, and minimum stone thicknesses, d, should a producer wish to use Figures A-2 and A-3 to indicate the combination of screens and grids that would yield the standard gradings. For the 10-200 kg more widely graded class, experience in the UK quarries using simple grizzlies to remove undersize material, and eye selection to remove oversize material suggests that for rock of average (2.7 t/m^3) density, setting the grizzly clear spacing at 225 mm and using eye selection of material above 450 mm will be a good starting point. This advice must be tempered, however, by individual quarry managers' own experience at their particular quarry.

The grading envelopes become progressively narrower in the 'heavy grading' classes, consistent with design requirements and the geological constraints on producing large sizes of blocks. However, projects requiring blocks larger than 10 t should not make the (non-standard) grading class excessively narrow because of the producer's extreme difficulty in selecting accurately and the wastage from oversize block production. For example at $\rho_r = 2.7$ t/m^3, D_n for a 15 t block = 1.7 m, and D_n for a 20 t block = 1.95 m, a difference of about 10%, which is very difficult to select precisely except by individual weighing.

Wide or quarry run gradings may be economic to produce and will certainly be used for core. The amount of fine material may be difficult to control. One rapid method is to use a loader with a meshed bucket base such that material of less than 100 mm passes through. The operator's success in removing fines will depend on not overfilling the bucket, the type and dryness of the rock material. Other possibilities using grizzlies and divergators can give greater control of fines. Screening is undoubtedly necessary in the production of permeable core materials when a high percentage of tines are present in the blast pile.

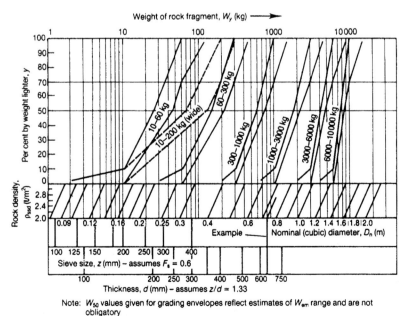

Note: W_{50} values given for grading envelopes reflect estimates of W_{em} range and are not obligatory

Figure A-2 Weight gradings and size relations for the standard light and heavy grading classes

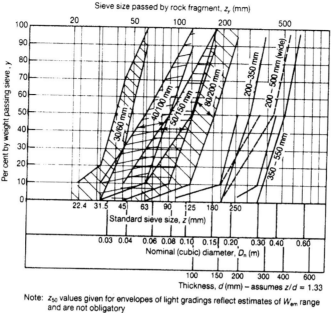

Note: z_{50} values given for envelopes of light gradings reflect estimates of W_{em} range and are not obligatory

Figure A-3 Size gradings and relationships for the standard fine and light granding classes

A.3 Geotextiles

Geotextiles are permeable textiles made from artificial fibres used in conjunction with soil or rocks as an integral part of a man-made project and were first used in the Netherlands. They are frequently employed as filter membranes and as the interface between differently graded layers. Geotextiles are also used as bed protection and can be loaded with concrete blocks (block mattress), bituminous bound crushed stone and sand (fixtone mattress) and geotextile tubes filled with gravel (gravel-sausage mattress). Gravel bags have also been used for special filter requirements.

The basic functions of geotextiles may be listed as follows:
- Separation: the geotextile separates layers of different grain size;
- Filtration: the geotextile retains the soil particles while allowing water to pass
- through;
- Reinforcement: the geotextile increases the stability of the soil body;
- Fluid transmission: the geotextile functions as a drain because it has a higher
- water-transporting capacity than the surrounding materials.

Geotextile manufacture
Geotextiles are manufactured from a variety of artificial polymers:

1. Polyamide (PA);
2. Polyester (PETP);
3. Polyethylene (low-density LDPE and high-density HDPE);
4. Polypropylene (PP);
5. Polyvinylchloride (PVC);
6. Chlorinated polyethylene (CPE).

The first four are the most widely used although many variations are possible. Additives are also employed in geotextile manufacture to minimise ageing, to introduce colour and as anti-oxidants and uv stabilisers.

Comparisons of properties of the four main polymer families are shown in Figure A-4. These are very broad, because there are many variants within each group. Some properties (such as strength) are also greatly influenced by the different processes of manufacture. A classification of geotextiles based on the type of production and the form of the basic elements is given in Figure A-5.

		polyester	polyamide	polypropylene	polyethylene
Strength		+++	++	+	+
Elastic modulus		+++	++	+	+
Strain at failure		++	++	+++	+++
Creep		+	++	+++	+++
Unit weight		+++	++	+	+
Cost		+++	++	+	+
Resistance to:					
UV-	Stabilised	+++	++	+++	+++
light	Unstabilised	+++	++	++	+
Alkalis		+	+++	+++	+++
Fungus, vermin, insects		++	++	++	+++
Fuel		++	++	+	+
Detergents		+++	+++	+++	+++

+++ high   + low

Figure A-4 Comparative properties of general polymer families

The basic elements used in geotextiles are monofilaments, multifilaments, tapes, weaving film and stable fibres. Monofilaments are single, thick, generally circular cross-sectioned threads with a diameter ranging from 0.1 mm up to a few millimetres. Multifilaments (yarns) are composed of a bundle of very thin threads. Yarns are also obtained from strips and from wide films. Tapes are flat, very long plastic strips between 1 and 15 mm wide with a thickness of 20-80 μm. A weaving film is sometimes used for the warp 'threads' in a fabric.

The basic fibre is a fibre of length and fineness suitable for conversion into yarns or non-woven geotextiles. For non-woven fabrics the length is usually about 60 mm.

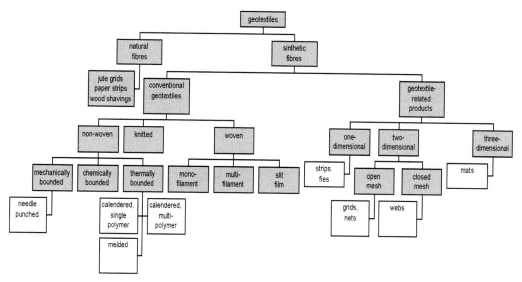

Figure A-5 Geotextile classification groups

Woven geotextiles

A woven fabric is a flat structure of at least two sets of threads. The sets are woven together, one referred to as the warp, running in a lengthwise direction, and the other, the weft, running across. Woven geotextiles can be categorised by the type of thread from which the fabric is manufactured.

Monofilament fabrics are used for gauzes of meshes, which offer relatively small resistance to through-flow. The mesh size must obviously be adapted to the grain size of the material to be retained. Monofilament fabrics are made principally from HDPE and PP.

Tape fabrics are made from very long strips of usually stretched HDPE or PP film, which are laid untwisted and flat in the fabric. They are laid closely together, and as a result there are only limited openings in the fabric.

Split-film fabrics are made from mostly fibrillated yarns of PP or HDPE. The size of the openings in the fabric depends on the thickness and form of the cross-section of the yams and on the fabric construction. These split-film fabrics are generally heavy. Tape and split-film fabrics are often called slit-films.

Multifilament fabrics are often described as cloth, because they tend to have a textile appearance and are twisted or untwisted multifilament yams. The fabrics are usually made from PA 6, PA 6.6 or from PETP.

Besides the above-mentioned monofilament fabrics, special mesh-type constructions are produced such as those with a monofilament warp and a multifilament weft which have outstanding water-permeability and sand-retention properties. Other examples include open meshes in which the woven or unwoven warp and weft threads are attached at crossing points by chemical or thermal bonding and other meshes constructed by using knitting techniques.

Non-woven geotextiles
A non-woven geotextile is a textile structure produced by bonding or interlocking of staple fibres, either monofilaments or multifilaments arranged at random, accomplished by mechanical, chemical, thermal or solvent means. Non-woven gauzes are structures with large meshes which are formed by placing threads or tapes at predetermined distances on top of one another and bonding them at the intersections by a chemical, thermal or mechanical process.

Geotextile-related products
These products are distinguished in one-dimensional (strips, ties), two-dimensional (gids, nets, webs) and three-dimensional (mats) products. Grids are lattices made from perforated and then stretched polymer sheets. Three-dimensional mats are produced by extruding monofilaments into a rotating profile roller, followed by coating so that the threads adhere to each other at crossings which are spatially arranged. The matting material itself occupies less then 10% of the mat volume. The mats are 5-25 mm thick and about 1-6 m wide.

Characteristics and properties
The geotextile properties stem primarily from their functional requirements. Since the geotextile can have a variety of functions, requirements are diverse. For reinforcement the emphasis is on mechanical properties such as E-modulus and strength, for filters it is on properties such as water permeability and soil tightness. The durability required will depend on the specific application and lifetime required. Geotextiles must also fulfil secondary functional requirements related to the execution of the work (e.g. a certain amount of uv-resistance is needed) or it must have resistance to mechanical wear and tear if construction equipment is to be driven over the fabric. The suitability of a geotextile should be checked against these functional requirements during the design phase of the project.

Although specification requirement tests need only be carried out once, the following quality control tests may be required during production: tearing strength, grab test strength, tensile strength, strain at breaking load, moduli and mass distribution. These tests should be made in both the length and width directions. The thickness, the mass per unit area and the bursting strength may also need to be checked, and in some applications water permeability and sand-retaining properties will be important. A large number of national and a few international standard test methods are available covering these requirements.

Dimensions

The maximum standard width available for both woven and non-woven fabrics is 5-5.5 m. The length is limited by the available transport facilities and ease of handling on-site. Depending on the mass per unit area, the length generally lies in the range 50-200 m.

Jointing is necessary to obtain greater dimensions. In practice, large areas can be covered by overlapping sheets. Where physical continuity is required without overlap then heat welding (some non-wovens) or stitching may be used. The seam forms the weakest link in the geotextile construction and should therefore be checked thoroughly against the specifications. The thickness of most geotextiles lies between 0.2 and 10mm when unloaded, although this may sometimes reduce under pressure.

In general, the mass of non-woven geotextiles lies in the range 100-1000 g/m^2, 100-300 g/m^2 being the most commonly used. Woven fabrics can be heavier and masses between 100 and 2000 g/m^2 are possible. The greater demand is for the lighter grades in the range 100-200 g/m^2. Generally, the lighter types of geotextiles are used as separators, the heavier woven fabrics for reinforcement and the heavier non-wovens for fluid transmission.

Mechanical properties

The mechanical properties of geotextiles depend on a number of factors: temperature, atmospheric conditions, the stress-strain history, the mechanical properties of the material and fibre structure, the structures of the yarn and of the geotextile, the direction of anisotropy, and the rate of loading and ageing. Most fabrics exhibit cross-contraction under loading. However, light tape fabrics and fabrics with so-called 'straight' warp construction do not exhibit cross-contraction and construction strain (strain due to fibre straightening). Test methods can be categorised into those that do not prevent cross-contraction (uniaxial) and those that do (biaxial). Methods commonly employed for tensile testing are strip tensile, grab tensile, manchet tensile, plain-strain tensile, wide width tensile and biaxial.

A great variation in both strength and stiffness exists. The strength varies generally between 10 and 250 kN/m. Non-woven and woven PP fabrics are not ideal in situations where high strength is combined with low strain because of the large elasticity of these geotextiles.

All meltspun synthetic polymers, as used in geotextiles, have visco-elastic behaviour, which means that the mechanical behaviour is time-dependent. This becomes manifest in creep and relaxation phenomena. Creep data for polymer materials can be presented in several ways. Often log ε (strain) is plotted against log t (time) for various levels of the ultimate short-term load U, i.e. 50% U, 25% U, etc. The sensitivity to creep of polymers increases considerably in the sequence PETP, PA, PP and PE. For geotextiles that are loaded for prolonged periods of time (10-100 years) the permissible load for polyester is the order of 50% of the tensile strength, for polyamide 40%, and for polypropylene and polyethylene below 25%.

Burst and puncture strength of geotextiles is important in coastal and shoreline rock structures. In these tests a circular piece of geotextile is clamped between two rings and loaded directly by gas or water pressure or by a physical object, Puncture tests can be used for investigating the resistance of a geotextile to puncturing by, for instance, falling stones. The other tests available include the California Bearing Ratio (CBR) plunger tests, the cone drop test and the rammer test (BAW). Test methods are also available for strength parameters such as tear strength, abrasion resistance and friction coefficient.

Hydraulic properties
The water permeability of a geotextile incorporated into a structure depends on the geotextile itself, the subsoil, the load imposed on the face of the geotextile, the hydraulic load, the blocking of the geotextile, the clogging of the geotextile, the water temperature and the composition of the water. The pore size and the number of pores per unit area of a geotextile primarily determine the permeability. The subsoil and imposed load determine how much water has to be discharged and the compression of the geotextile.

Flow through the fabric is normally laminar when the geotextile is embedded in the soil, but becomes turbulent when subjected to wave action (for example, under a coverlayer of rip-rap or blocks). Blocking of flow occurs when soil particles partly wedge into the pores. This normally only arises in situations of unidirectional flow rather than the oscillating flows, which frequently occur under wave action. Clogging of flow occurs when fine particles (for example, iron particles) settle on or in the geotextile or at the interface between the geotextile and the subsoil. Reductional factors in permeability of geotextiles due to clogging of the order of 10 have been found.

Permeability is usually measured in the laboratory using values of hydraulic gradient low enough for laminar flow. For thin, more permeable fabrics, the permeability test may be performed on several separated layers to increase the measurable water head and still maintain laminar flow, the equation for flow through the fabric is

$$k_g = \frac{q \cdot t_g}{a_g \cdot \Delta H}$$

where: k_g = permeability of geotextile (ms^{-1})
        q = rate of flow (m^{3}s^{-1})
        a_g = surface area of geotextile (m^2)
        ΔH = head loss (m)
        t_g = geotextile thickness (m)

The permeability may be expressed as permittivity (k_g / k_r), which is useful in comparing geotextiles because permittivity varies less than permeability when there is a change of normal stress applied to the geotextile. The permittivity can be found by dividing the permeability by the nominal fabric thickness, the permittivity of geotextiles varies between 10^{-2} s^{-1} and 10 s^{-1}.

Geotextiles will only be completely soiltight when the largest opening is smaller than the smallest particle of the subsoil. This is usually not required because of the filter function of the geotextile. When the openings are larger the soil tightness depends basically on the hydraulic loads, the soil grading and the geotextile characteristics. The soil tightness of a geotextile on a particular type of soil can be determined by reconstructing the situation in a model laboratory apparatus and then carrying out measurements using the hydraulic boundary conditions. In addition, the soil tightness can be characterized by a ratio of O / D, where O represents the diameter of holes in a fabric and D the diameter of particles (e.g. O_{90} / D_{50}). A number of methods have been developed in which an opening characteristic (e.g. O_{90} or O_{95}) is determined in terms of a sieve size. These tests are based on either wet or dry sieving.

Chemical properties

One of the characteristic features of synthetic polymers is their relative insensitivity to the action of a great number of chemicals and environmental effects. Nonetheless, each plastic has a number of weaknesses which must be taken into account in the design and application. Specifically, the life of geotextiles can be affected by oxidation and by some types of soil/water/air pollution. Many synthetic polymers are sensitive to oxidation. The end result of oxidation is that mechanical properties such as strength, elasticity and strain absorption capacity deteriorate and the geotextile eventually becomes brittle and cracks.

Investigations have shown that, provided the geotextile is not loaded above a certain percentage of the instantaneous breaking strength, the thermo-oxidative resistance will determine the theoretical life of the material. The allowable load for PETP is, at most, 50%, for polyamides it is somewhat lower, and for polypropylene and polyethylene it is about 10-30%. (This guidance only applies where the geotextile functions as a filter and is not withstanding mechanical loads).

Specific additives have been developed to counteract these processes. These can be grouped according to their protection function as either anti-oxidants or UV-stabilisers. In fact, the thermo-oxidative resistance of a geotextile is determined by a number of factors: the thermo-oxidative resistance of the polymer itself, the composition of the anti-oxidant packet, the effect of the thermo-oxidative catalytic compounds in the environment, the effects of processing on the long-term thermo-oxidative resistance, the resistance of the anti-oxidant additives to leaching by water and the practical site conditions.

A.4 Composite materials

Composite systems are normally adopted where improved stability of the rock, materials being used is sought or where ease of placing the protection system is important.

Gabions

A gabion is a box or mattress-shaped container made out of hexagonal (or sometimes square) steel wire mesh strengthened by selvedges of heavier wire, and in some cases by mesh diaphragms which divide it into compartments (see Figure A-6). Assembled gabions are wired together in position and filled with quarried stone or coarse shingle to form a retaining or anti-erosion structure. The wire diameter varies but is typically 2-3 mm. The wire is usually galvanised or PVC-coated. PVC-coated wire should be used for marine applications and for polluted conditions.

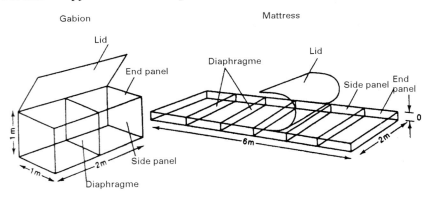

Figure A-6 Gabions

The durability of gabions depends on the chemical quality of the water and the presence of waterborne attrition agents. The influence of the pH on the loss of the galvanic zinc protection is small for pH values in the range 6-12 and there are examples of gabions with negligible loss over 15 years. Grouting of the stone-filled gabions or mattresses can give some protection to the wire mesh against abrasion and corrosion, but this depends on the type of grout and the amount used.

The dimensions of gabions vary, but typically range in length from 2 to 4 m (mattress, 6m), with widths about 1 m (mattress, 2m), and height 0.3-1.0 m (mattress, max. 0.3 m). The mesh size is typically 50-100 mm.

The units are flexible and conform to changes in the ground surface due to settlement. Prefabricated gabions can be placed under water. Gabions can thus be used in a wide variety of marine works: groynes, dune and cliff protection, protection of pipelines and cables, and as toe protection. Mattress-shaped gabions are flexible and are therefore able to follow bed profiles both initially and after any scouring which may take place. Gabions can also be piled up to the form retaining walls or revetments. In order to prevent migration of solid through the structure they may be used in conjunction with geotextile filter layers.

In certain applications the gabion structure needs impermeability or weight to counter uplift. To give these characteristics, the stone is grouted with mastic or a cement-bound grouting. The weight of the structure can also be influenced by the choice of the density of the stone blocks with which the gabion or gabion-mattress is filled.

Mattresses

Prefabricated mats for hydraulic constructions have been used extensively as a medium for combating bed and embankment erosion. The range of mat types have increased considerably during the last decade. Formerly, use was made of handmade mats of natural materials such as reed, twigs/branches (willow), rope, quarry stone and rubble. Today, geotextiles, bituminous and cement-bonded materials, steel cables and wire gauze are used as well as the more traditional materials.

Functions and principles

A distinction must be made between mattresses used for the protection of continually submerged beds and foreshores and mats employed for slope or front face protection of embankments. Bottom protection mats are concerned with:

- The mitigation of groundwater flow in the underlying seabed to prevent horizontal transport and erosion;
- Prevention of piping of soil particles;
- Assisting in the overall geotechnical stability of a larger structure;
- Offering resistance to loads imposed by anchors, trawler boards, etc. and thereby protecting pipes and cable lines; :
- Providing a protection system with the flexibility to allow for anticipated settlements.

Mattresses for embankment retention have similar functions, i.e:
- Prevention of soil transport;

- Resistance to wave action;
- Provision of flexibility to allow for settlement.

Additionally, they may be required to:
- Not trap refuse; and
- Support vegetative growth.

Mats are manufactured from the following elements:
1. The carrier: Fabric, gauze, cables and bundlings by which the necessary strength is obtained for its manufacture, transport and installation.
2. The filter: A geotextile which satisfies the conditions with respect to permeability, soil density, strength, etc. In some embankment-mat constructions the geotextile is applied separately.
3. The ballast: Concrete blocks, packaged sand, gravel, stone or asphalt necessary for the stability of the protection. Quarry stone can be used to serve as a material to sink the mat.
4. The connections: Pegs, wires, etc. by which the ballast material is fixed to the filter and the carrier materials.

In addition to the mat proper, a further deposit of gravel, aggregates, quarry stone, slag, etc. may be necessary when the ballast material stability in itself is insufficient or when extra protection is demanded.
The carrier and the filter can often be combined in the form of a single geotextile with the necessary mechanical and hydraulic properties. Sometimes the filter is combined with the ballast, which is then dimensioned as a granular filter.

Manufacture and placing
Fascine mats are typically up to 100 m long and 16-20 m wide. Historically, these mats were manufactured from osiers which were clamped between the so-called 'raster' of bundles of twigs/branches bound together to a circumference of 0.3 m.
'Plugs' were driven into the crossing-points of the bundles to serve as anchor points for towing and sinking the mats (for bed protection) or collar-pieces (for foreshore protection). Because of the destruction of the osiers by pile-worm attack, a reed layer was incorporated into the mats. These fascine mats were made on embankments in tidal areas or on special slipways.
The finished mat or collar-piece was then towed to its destination and sunk using quarry stones which were placed in position by hand, with the uppermost rasterwork of bundles serving to maintain the stone in place during sinking. After sinking, more and heavier stones were discharged onto the mat if this was found to be necessary.

Modern mats are made in accordance with the same principles. The twigs/ branches and reed are replaced by a geotextile into which loops are woven and to which the rasterwork of bundles are affixed. A reed mat may be affixed onto the geotextile in order to prevent damage to the cloth during the discharge of stones onto the mat. The geotextile extends out beyond the rasterwork with the aid of lath outriggers so that, on sinking, the mats overlap. Sometimes a double rasterwork is applied to give the mats a greater rigidity and more edge support to the quarry stone load. The mat construction can serve as a protection for the bottom, a foreshore and even as an embankment protector. In the last case the mat is hauled up against the embankment itself.

Modern alternatives to fascine mats include mats where the quarried stone is replaced by open-stone asphalt reinforced with steel mesh, or by 'sausages' of cloth or gauze filled with a ballast material such as sand, gravel, bituminous or cement-bound mixtures.

A.5 Physical properties of soil[2]

Figure A-7 summarises the important physical parameters of the soil. The soil characteristics which affect vegetation establishment and growth, their principal determinants and the ways in which they can be modified are given in Figure A-7. Soil contains water, air, fine earth, stones and organic matter.

Parameter	Definition	Assessment
Soil texture	Description based on proportions by weight of sand, silt and clay as percentages of fine earth fraction <2 mm in size	Field estimate or laboratory measurement
Stoniness, % vol	Proportion of large particles, >2 mm	Direct measurement, or field estimation
Dry bulk density (ρ_b) ton/m^3	Apparent density of soil in situ (on a dry basis)	Field measurement, either removal of undisturbed core or replacement method (sand or water)
Particle density (ρ_s) ton/m^3	Density of the soil particles	Laboratory measurement by displacement. Most soils are consistent with the value of about 2650 kg/m^3
Void ratio (e)	Ratio: volume of soil voids to volume of solids	$e = \rho_s/(\rho_b-\rho_s)$
Porosity (n) % (total pore space)	Volume of soil voids expressed as a percentage of total in situ soil volume (note: voids occupied by air and water)	$n = (1-\rho_b/\rho_s)\times100$
Soil erodibility factor	The risk of erosion by air of water due to the nature of the soil itself	Direct measurement or estimation based on soil texture
Packing density (L_d) (rooting potential)	A more reliable indicator of the effects of compaction that bulk density alone; allows for clay content	$L_d = \rho_b + (0.009 \times \%\ \text{clay})$

Figure A-7 Soil physical parameters (definitions related to plant growth)

[2] based on Coppin and Richards, 1990

Rooting potential indicates the resistance of the soil to root penetration, which depends mainly on the soil's bulk density and on mechanical strength. Roots have great difficulty penetrating soil with strengths greater than 2.0 to 2.5 MNm^{-2}, though higher limiting values have been suggested for coarse-textured soils. Generally, root growth is enhanced by greater moisture content and voids, and is retarded by higher bulk density and clay content. Critical dry bulk densities for soils, above which root growth is severely restricted, are about 1.4 mgm^{-3} for clay soils and 1.7 mgm^{-3} for sandy soils. As clay content is so important in determining the rooting potential, a term packing density (L_d) is often used to determine the maximum density to which a soil can be compacted and still permit root growth (see Figure A-8).

Rooting potential indicates the resistance of the soil to root penetration, which *Soil texture* describes the particle size distribution and gives an indication of the likely behaviour of a soil in respect of handling, root growth or drainage. Descriptions such as sandy loam or clay are based on measured proportions and mixtures of clay, silt and sand in the fine earth <2 mm) fraction, as shown in Figure A-9.

Important soil characteristics	Principal determinants				Modifiers					
	particle size	packing density	porosity	organic mater	vegetation cover	topography	cultivation	compaction	additions	time
Texture	●									
Soil structure	●		●	●			○	○		○
Rooting potential	●	●	●	●			○	○		
Soil water capacity	●		●	●				○	○	
Permeability and water acceptance			●		○	○	○	○		○
Ion exchange capacity	●			●	○				○	
Erodibility	●		●	●	○	○	○	○		
Ease of cultivation	●	●	●		○			○	○	

Figure A-8 Soil physical characteristics

Soil structure is a characteristic which describes the arrangement and size of particle aggregates or 'peds'. Structure develops over time, as fine soil particles aggregate into crumbs and blocks. This increases the number of large pore spaces and thus the permeability and rooting potential of the soil. The presence of organic matter and plant roots plays a major role in developing and maintaining soil structure. Structure is easily damaged by handling or cultivation during wet conditions, when the soil is weaker.

The usual particle grading curves prepared to BS 5930:1981 are familiar to most engineers. Soils are described according to the British Soil Classification System for Engineering Purposes.

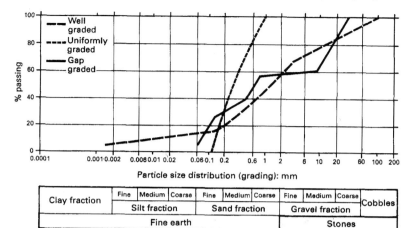

Clay fraction	Fine	Medium	Coarse	Fine	Medium	Coarse	Fine	Medium	Coarse	Cobbles
	Silt fraction			Sand fraction			Gravel fraction			
Fine earth							Stones			

For soil survey work, texture descriptions are based on the fine earth fractions, that is <2 mm size. The overall proportion of gravel and cobbles defines the *stoniness*. The proportions of sand, silt and clay in the fine earth soil matrix define the texture classes as given in the triangular diagram.

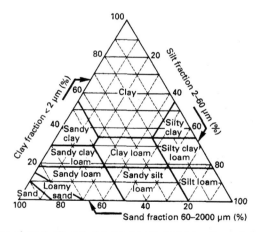

The proportions of sand, silt and clay can be obtained from the particle grading curve, calculating the quantity of each size fraction as a percentage of the <2 mm fraction.

Figure A-9 Soil texture

Soil potential

The physical, water and chemical characteristics of the soil can be combined into an overall assessment of soil potential for plant growth. A scheme for this is given in Figure A-10. In the table class A is the highest quality and suitable for situations where good quality fertile topsoil is necessary. However, class C, whilst of poorer quality, would still be suitable for many situations. In many cases it would be possible to modify or manage a class B or C soil to improve its quality.

Parameter	Unit	Suitability class			Unsuitable
		A	B	C	
Soil type					
Texture	Description[1] and % clay	fLS,SL SZL,ZL	SZL,CL, ZCL,LS	C<45% SC,ZC,S	C>45%
Stoniness	% vol	<5	5-10	10-15	>15
Available water capacity	% vol	>20	15-20	10-15	<10
pH		5.7-7.0	5.2-5.5 7.0-7.3	4.7-5.5 7.3-7.8	<4.7 >7.8
Conductivity	mmho/cm	<4	4-8	8-10	>10
Pyrite	%weight	-	<0.2	0.2-3.0	>3.0
Soil fertility					
Total nitrogen	% weight	>0.2	0.05-0.2	<0.05	
Total phosphorus	mg/kg	>37	27-37	<27	
Total potash	mg/kg	>360	180-380	<180	
Available phosphorus	mg/kg	>20	14-20	<14	
Available potassium	mg/kg	>185	90-185	<90	

Note 1: f=fine, S=sand, C=clay, L=loam, Z=silt

Figure A-10 Assessment of soil potential

Selection of soil materials

The simple system for assessing soil potential described in Figure A-10 can be used as a general guide to classify all material which the engineer intends to use as soil, regardless of origin, according to its potential for plant growth. Classes are allocated as follows.

Soil type

Class A   Highest growth potential, important when it is necessary for soil to have minimal restrictions on plant growth. Suitable for final soil covering or topsoil.

Class B   Where growth potential is not critical, but reasonable growth is still required. Also suitable for subsoil layers beneath Class A.

Class C   Will still support good growth if managed properly, but susceptible to handling problems which may restrict growth. Can also be used as subsoil layers.

Suitability cannot be based on soil, texture alone and the classes proposed in Box A-6 have only general application. For some uses special soil characteristics may be required; examples are given in Figure A-11.

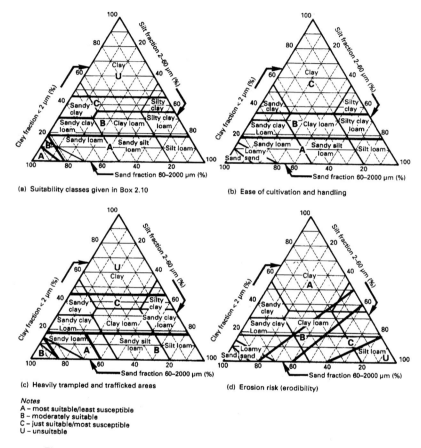

(a) Suitability classes given in Box 2.10

(b) Ease of cultivation and handling

(c) Heavily trampled and trafficked areas

(d) Erosion risk (erodibility)

Notes
A – most suitable/least susceptible
B – moderately suitable
C – just suitable/most susceptible
U – unsuitable

Figure A-11 Spoil suitability for specific purpose based on texture

Soil fertility

Class A   Highly fertile, will produce dense vigorous growth, requiring higher maintenance and leading more quickly to successional changes. This is not always necessary nor desirable, however, and the group is best used for intensively managed areas and for grazing.

Class B   Moderate fertility; fertilisers may be required to support very productive growth.

Class C   Minimal fertility, suitable for low-maintenance vegetation but fertilisers will be necessary. Swards should have good legume component.

Soil physical properties and fertility are defined separately so that different classes can be selected for each, depending on the situation.

APPENDIX B
Examples

In this appendix three cases are described:

- Bank protection along a river mouth
- Bed protection under a closure dam
- Breakwater as part of a shore protection

These cases are intended to give an idea of a preliminary design. They are necessarily simplifications of real life situations. The used boundary conditions are as realistic as possible. The reader should be aware that design is not a mathematical science. There is always more than one solution and even the magnitude of the various parameters in the formulas is often ambiguous.

The reader should be aware that the following examples are fictional and are only intended to familiarize him/her with a number of practical design problems that may be encountered during a feasibility study. In the case of a project of some importance and/or size, this preliminary phase is followed by detailed computations.

B.1 River bank

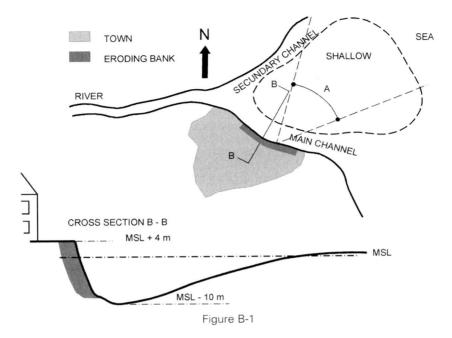

Figure B-1

B.1.1 Introduction

The waterfront of a town, located on the bank of a river mouth, is subjected to erosion. The ground level of the town is high enough to prevent flooding, but the erosion is threatening the road along the waterfront, see Figure B-1.

The average depth of the river near the eroding bank location is about 5 m below MSL, but near the town, in the outer bend, it is 10 m below MSL. The erosion process does not appear to stop and is shifting the main channel continuously to the south. The main channel is 500 m wide at MSL.

Waves enter the area from sector A. Their height is limited by a shallow area in the river mouth, which lies approximately 0.5 m above MSL. The river bed and banks consist of sand grains with a diameter of 300 μm.

The following table gives the boundary conditions for this case.

Exceedance frequency (1/year)	Waterlevel (MSL + m)	Wave height Hs (m)	Wave period Tp (s)	Maximum ebb flow velocity during tidal cycle (Q/A in m/s)
Average HW	1.0			
Average LW	−1.0			1.5
1	2.0	0.75	4	2.0
0.1	2.5	1.0	5	2.5
0.01	3.0	1.25	6	3.0
0.001	3.5	1.5	7	3.5

Note:

1. The waves are correlated with the extreme waterlevels (a high waterlevel is also caused by the wind).

2. The given velocities are the discharges divided by the area (Q/A) and are not correlated with the high waterlevels (the location is very close to the sea and a high river discharge or high tidal range is independent of the average sea water level). In a tidal cycle these velocities occur around MSL.

B.1.2 Design frequency

There is no danger of flooding, the bank protection can easily be inspected and repaired and damage to the road will not lead to great danger of the buildings at the other side of the road immediately. With inspection every year and a accepted chance of exceedance of 10%, this leads to a design frequency of 0.1/year. Hence, design conditions are: water level MSL + 2.5 m in combination with waves: $H_s = 1$ m, $T_p = 5$ s and, separately, $u = 2.5$ m/s.

B.1.3 Choice of protection type

There are several ways to protect the river bank against erosion, e.g. by means of groynes or with a revetment made of rock. First, we will roughly compare these two alternatives.

For the lay out of groynes, see section 2.6. To keep the main flow from the bank, the groyne should have a length of at least 1/5 of the distance between the groynes. A distance of 250 m and a groyne length of 50 m are used. For the slopes of the groyne 1:2 is chosen (rather steep to save material). The crests of the groynes are equal to the waterlevel of the design flood to prevent high velocities directly along the bank. The slope of the (unprotected) bank between the groynes is assumed 1:3.

For the dimensions of a revetment, we need the width and the thickness. The width is determined by the slope. An important parameter for the slope of the revetment is the possible seepage. Due to rainfall and/or to fluctuating waterlevels in the river, seepage almost always takes place. In that case, section 5.4.2 simply gives: $\alpha \leq \phi/2$. From figure 3-7 we find for sand with $d_{50} = 0.3$ mm: $\phi \approx 30°$ (river sand is rather rounded). From this we find: $\alpha \le 15°$, which means a slope of 1 : 3.5. The thickness depends on the diameter which still has to be determined. For now, a safe estimate is 0.5 m thick.

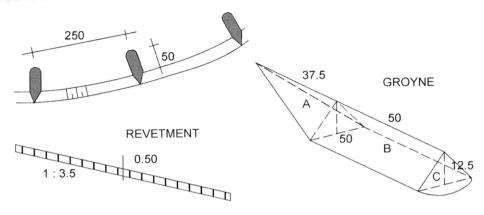

Figure B-2

It is difficult to compare all aspects of the two solutions, groynes and a revetment. Costs and the construction method, for example, depend heavily on local circumstances. Therefore, the comparison is made, simply based on quantities of material. The design waterdepth is 10 + 2.5 m. For the groyne, the volume of stone for 250 m bank length (one groyne for 250 m bank) is: A + B + C = 1/2 × 1/2 × 12.5 × 50 × 37.5 + 1/2 × 12.5 × 50 x 50 + π × 25^2 × 12.5 / 6 = 5860 + 15625 + 4090 = 25575 m^3. For the revetment, the volume of stone per 250 m bank length then becomes: 12.5 × $\sqrt{(1^2 + 3.5^2)}$ × 250 × 0.5 = 5690 m^3.

Groynes need over 4 times more stones (the volume of stone is quadratic with the depth for groynes and linear for revetments, so only in shallow water groynes will need less stone), but for now we will forget about the material and concentrate on the slope of the revetment, 1:3.5, and the extra protection at the toe of the revetment to withstand continuing erosion. If bottom protection will be necessary at the head of the groyne, it can be expected that the construction costs will also be higher for groynes. Moreover, we will see that wave attack is an important load on this bank, so between the groynes an additional protection will be needed. So, a revetment is considered preferable (groynes are applied in a shallow river when a navigation channel with a certain depth is wanted).

B.1.4 Top layer

The river bank is attacked both by flow and by waves. To start with the flow, we need to know the design velocity near the bank. The values given as boundary conditions can not be used directly, since they represent an average in the cross section (Q/A). Since the bank is located in an outer bend of the river, the velocity will be higher than this average. As a first estimate we increase the velocities by 40%, leading to a design velocity of 1.4 * 2.5 = 3.5 m/s (measurements are, of course, preferable and not so difficult to obtain!). We further assume that this velocity occurs at the toe of the slope (see section 3.3), so, the water depth is about 10 m (maximum velocity occurs in the tidal cycle around MSL, see note with boundary condition table).

Using the Shields equation, with $\psi = 0.03$ and $\Delta = 1.65$, we find iteratively with $C = 18*\log(12*h/k_r)$ and $k = 2*d_{n50}$: $d_{n50} = u^2/(C^2*\psi*\Delta) \approx 0.1$ m. This is, however, for uniform flow and with a horizontal bed. The flow in a river is more turbulent than in a uniform flow, due to variations in the cross section and the fact that the stones lie on a slope of 1:3.5. To take this into account we apply two correction factors.

K_v takes the extra load due to turbulence into account and is assumed 1.2, which is a rather conservative estimate. K_α takes the decreased strength into account and is calculated with: $K_\alpha = \sqrt{(1 - \sin^2\alpha/\sin^2\alpha)}$. For $\alpha = 16°$ and $\varphi = 40°$ this leads to $K_\alpha = 0.9$. d_{n50} then becomes: $K_v^2*d_{n50\text{-Shields}}/K_{0\_} = 0.16$ m (K_v is an amplification factor for the velocity, hence K_v^2 is used for the diameter; $K_{\alpha\_}$ is defined as a correction factor for the diameter). From the curves in Appendix A we find that stone class 80/200 mm is too light and we use stone class 10 - 60 kg.

For the wave attack, we will use the Van der Meer formula (for a rough first estimate Hudson can also be used, but for revetments, Van der Meer is preferred, since the permeability of a revetment is quite different from that in a breakwater, for which Hudson was derived).

With $H_s = 1$ m, $\cot\alpha = 3.5$, $P = 0.1$ (relatively short waves hardly penetrate into the subsoil, hence it is considered "impermeable"), $T_m = 0.85 * T_p = 4.25$, S = 2 (threshold of damage, since the chosen boundary conditions already have a rather high probability of exceedance: 10%) and N = 7000 (maximum value, for the same reason), we find: $\xi = 1.5$, while for the transition between the two van der Meer formulae $\xi = 2.2$, so we are dealing with the formula for plunging breakers. With $\Delta = 1.65$ this leads to a $d_{n50} = 0.38$ m. From the curves in Appendix A we find that this requires rock 60 - 300 kg.

It is obvious that the wave attack is dominant. The wave attack, however, does not act on the whole slope. The region of wave attack lies approximately 2*Hs around the waterlevel. For the final design we also have to take into account lower waterlevels (with lower wave heights). A practical measure (also for construction) is applying 60 -300 kg from Mean Low Water (MSL – 1 m) up to MSL + design waterlevel + wave run-up (see equation 7.15 in section 7.4.2) = MSL + 2.5 + 1.5 $\gamma_{r\_}$ H_s = MSL + 2.5 + 1.35 ≈ MSL + 4 m.

Depending on the height of the bank, a practical transition to what lies above has to be made. In this case we are dealing with a city waterfront with a road along the river, so, one can think of a wall of masonry. Below the low waterlevel, the attack on the slope is from the flow only. So, the stone 10-60 kg can be applied down to the bottom.

B.1.5 Filter

Between the stone revetment and the original bank material, a filter will be necessary to resist the turbulent flow, but most of all to withstand the wave-action. In that case the geometrical rules should be followed:

$$d_{15F} < 5 \times d_{85B}$$

From appendix A, figure A-2 we find for the small particles of the top layer ($d_{n50} \approx$ 0.38 m, stone of 60 - 300 kg): $d_{15F} \approx 0.3$ m. The large particles of the second layer (d_{85B}) should be larger than 0.06 m. In that case, the smaller particles of the second layer will have a size of around 3 cm. This is still too large to lay on top of the original material (0.3 mm). So, more layers in between are necessary, leading to several filter layers. In that case, a geotextile can be a good solution. From the table in section 6.3.2 on geotextiles we find that the maximum opening in the textile (O_{90}) should be smaller than $2*d_{90}$ of the base material, in this case about 0.8 mm.
The following is a practical solution:
Apply a geotextile onto the whole slope, together with a layer of stones 10-60 kg. If these are not too sharp, they can act as a weight for the geotextile, without damaging the textile. In the zone of wave attack, rocks 60 - 300 kg are applied.

B.1.6 Scour and toe

As was already mentioned in the introduction, the erosion process in the outer bend of the main channel continues, so at the toe of the slope, scouring will also go on. Since the shift southward is now blocked, the erosion will cause deepening of the bottom at the toe. With morphological computations, the course of this process in the years to come can be estimated. Several measures can deal with this problem (see chapter 11):
- excavate down to the expected scour depth and continue the revetment
- apply a falling apron
- continue the revetment on the bottom of the river, assuming that the geotextile and the stones on top of it are able to sink along with the erosion process (hanging apron).

Both the falling apron and the hanging apron require regular inspection, e.g. every year or after a high discharge or extreme tide. The third solution has been selected in this case and Figure B-3 shows the "final preliminary" design. This is only a rough sketch without any detail. These can vary, depending on the construction method.

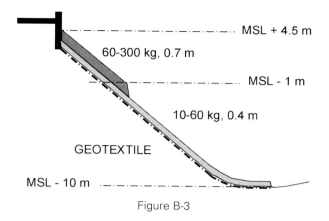

60-300 kg, 0.7 m

MSL + 4.5 m

MSL - 1 m

10-60 kg, 0.4 m

GEOTEXTILE

MSL - 10 m

Figure B-3

B.2 Caisson closure

B.2.1 Introduction

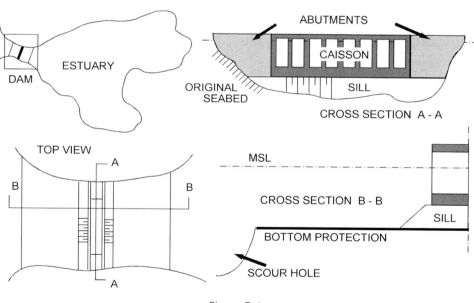

Figure B-4

A tidal basin is to be closed off from the sea, see Figure B-4. The closure will be realised with caissons on a sill. The mouth of the basin, where the closure is going to take place, is 100 m wide and 10 m deep. The bed of the estuary consists of loosely packed sand with a diameter of 0.25 mm.

The site is located in a country with summer and winter periods. During winter, activities have to be interrupted, due to the weather conditions. The closure cannot be completed in one summer season with the available equipment. At first, the bottom is protected against scour by means of mattresses. Then, a sill (blocking 25% of the cross section) is constructed as a foundation for the caissons (Phase 1). These activities are carried out during the summer season of the first year. The next summer, two (closed) abutments are constructed, together blocking another 25% (50% blocking in total, Phase 2). After that, a caisson with sluice gates, is placed to finish the closure. These sluice gates are opened immediately after the caisson has been positioned. The purpose of these gates is to reduce the head differences over the caisson until it is fully ballasted and stabilized and, in this case most of all, to be able to adjust the waterlevel inside the estuary before it is closed. After placing this final caisson and opening the sluice gates, the total blockage is 60% of the original cross-section (Phase 3). The situation with only the sill in the channel lasts for 250 days. The situation with two abutments lasts about 30 days and the phase with the final caisson placed, about 10 days. This represents a very rough schematization of the closing procedure. For a real closure, more phases are usually taken into account, but this schematization is good enough for demonstration purposes.

B.2.2 Boundary conditions

The most important boundary conditions are the tidal levels and current velocities. Of course, wave data are important for the stability of the caissons during transport from the building location to the dam area and during the closure operation, but these aspects are not considered here.

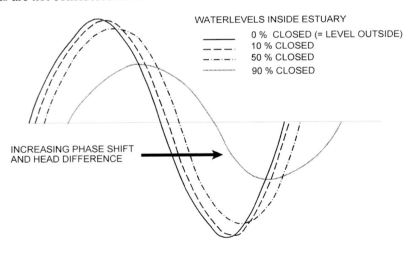

Figure B-5

When a tidal basin is closed, the tide in the basin gradually decreases. Figure B-5 shows what happens to the waterlevel inside the estuary. When the estuary is still completely open, the waterlevels at both sides of the proposed closure dam are practically the same. When the closure dam is under construction, the resistance of the dam leads to a head difference across the dam with a phase shift and a lowering of the tide inside. This can be calculated with a relatively simple tidal calculation. In the first phases of the closure, the combination of head difference and phase shift is such, that there is hardly any lowering of the tide inside. Figure B-6 roughly shows the relation between the decrease of the maximum discharge during a tidal cycle and the diminishing cross section of the tidal channel. When the channel is already half closed, one sees that the discharge is still 90 % intact. This means that the velocity is almost twice as high as in the original situation and this goes together with the head difference across the dam.

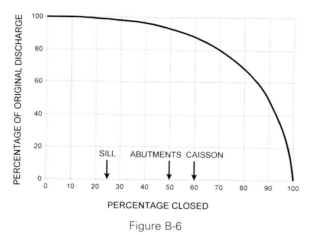

Figure B-6

The relation sketched above is in principle found with all tidal closures. The determination of discharges, waterlevels and velocities is nowadays done with numerical tidal models. In this example we will use schematized boundary conditions. The relation in Figure B-6 is used and the discharge as a function of time is supposed to be a sinus. For all situations, the velocities can be calculated using the continuity equation.

For scour computations, average springtide conditions are used, as they are assumed to be representative for the process during closure. The maximum discharge during a normal spring tide cycle in the original situation (no closure dam) is 1600 m^3/s (figure obtained from measurements), while the cross-section area $\approx 100 \times 10 = 1000$ m^2 (waterlevel fluctuations neglected). This leads to a maximum velocity during a normal springtide cycle, averaged over the cross-section, of 1.6 m/s. With this, the velocity in all phases can be calculated from the continuity equation.

For the stability of the sill, a discharge with a frequency of occurence of 0.1/year is used (accepted probability of exceedance in the critical period of about one month of 1%). This discharge is 15% higher than during normal springtide (from measurements). For stability, more extreme conditions are chosen than for scour. Scour is a morpholigical process where the time dimension plays a role, while the stability of the sill depends on one extreme condition.

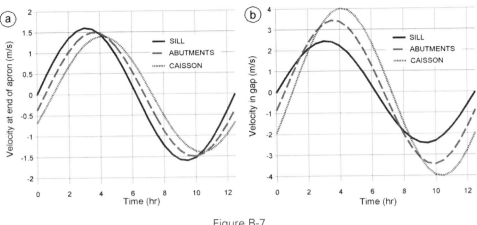

Figure B-7

The velocities to be used in the scour computations, at the end of the bottom protection, are presented in Figure B-7a. They have been calculated, by simply applying the continuity equation:

When the sill is constructed, the discharge still has about the same value as in the situation without sill, see Figure B-6. The cross-sectional area at the end of the apron is the same as in the centre-line of the closure dam, hence the maximum velocity during a normal springtide is also 1.6 m/s. After construction of the abutments (50 % closure), the discharges are about 90 - 95 % of the original values, leading to a velocity at the end of the apron of about 1.5 m/s. The same procedure leads to 1.4 m/s after placing the caisson (60 % closure, giving a little less than 90 % of the original discharges).

The velocities used for the stability conditions of the top layer of the sill, under the caisson, are given in Figure B-7b. The procedure is essentially the same as for the velocities at the end of the bottom protection, but now the cross-sectional area is no longer constant. Furthermore, the velocities are determined for a tide 15 % stronger than normal springtide.

After construction of the sill, 25 % of the original cross-section is closed, while the tidal discharges are still practically the same. In the narrowest cross-section this leads to a velocity of: 1.6 * 1.15 * 1/0.75 = 2.45 m/s on top of the sill. After completion of the abutments (50 % closure), this becomes: 1.5 * 1.15 * 1/0.5 = 3.45 m/s and after placing the caisson (60 % closure): 1.4 * 1.15 * 1/0.4 = 4 m/s in the narrowest gap of the closure dam (the apertures of the caisson).

B.2.3 Top layer sill

Figure B-7b gives the design velocities for the three building stages. With the sill, the velocities are a little less than 2.5 m/s, while the whole width of the channel is still available for the flow. After construction of the abutments, the velocity increases up to almost 3.5 m/s, see Figure B-7, while about 30% of the width is blocked, leading to flow separation in the gap resulting in an extra load on the sill. The velocities after placing the caisson mount to 4 m/s, but this is inside the caisson. The sill itself is now protected by the caisson bottom. So, for the stability of the top layer of the sill, the situation with abutments is normative.

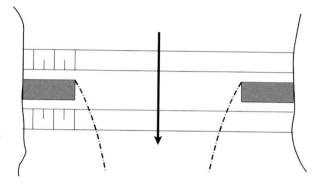

Figure B-8

From the Shields formula with $u = 3.45$ m/s, $\Delta = 1.65$, $\psi = 0.03$ and $k = 2*d_{n50}$ and a waterdepth of 7.5 m above the sill, we iteratively find: $d_{n50} = 0.11$ m. This is true for uniform flow, so we have to make a correction for the flow separation in the gap. From section 3.4.3 (Table 3.1) we find for a constriction with a vertical board a K_v-value of 1.7, using the average velocity in the gap. This leads to $d_{n50} = 1.7^2 * d_{n50\text{-shields}} = 0.32$ m. This means rock of 60 - 300 kg is required, as we can see from the curves in Appendix A.

B.2.4 Top layer bottom protection

Past the sill the flow spreads again over the whole width and depth of the gap, leading to lower velocities. The stones directly beyond the sill are in an unfavourable position, because of the deceleration behind the vertical step, see figure 3.13. With a sill height of 25% of the depth and equation 3.16, a K_v-value of 1.33 is found. But there is still also the horizontal constriction, which causes an amplification of the load. It must be said that this combination of a vertical and a horizontal constriction is complicated and for an important closure work, investigation with a scale model will be necessary. Here it is assumed that, applied to the average velocity beyond the sill, the total K_v, including the effects of both constrictions, simply is the product of the two coefficients: $1.7*1.33 = 2.3$. The average velocity beyond the sill, is 50% of the value on top of the sill, hence $3.45/2 \approx 1.75$ m/s. With a waterdepth of 10 m, we find $d_{n50\text{-Shields}} = 0.015$ m and with $K_v = 2.3$: $d_{n50} = 2.3^2 * 0.015 = 0.08$ m. This means much smaller stones than on the sill and the question rises again whether the followed procedure is correct. The answer can be either to do model tests or use larger stones, e.g. 10-60 kg near the sill. Further downstream, smaller stones can be applied, since the flow spreads and the intensity of the turbulence becomes much lower. At the end of the bottom protection, gravel with a diameter of a few centimeters will probably suffice.

B.2.5 Filter in sill

The top layer of the sill will be made of rock 60 - 300 kg with a d_{n50} of 0.38 m, while the original bed consists of sand with $d_{50} = 0.25$ mm. When these large stones are dumped on the sand, the flow through the sill will erode the bed, as there will be a large velocity over the sill and/or a large head difference across the sill. This will result in settlement of the sill and the caissons.

The maximum load will occur when the caisson is placed and the gates of the open caisson are closed. The level inside the estuary will be constant, while the tide at sea goes up and down. The head difference across the dam causes porous flow in the sill and a filter will be needed to prevent the sand being washed away. This filter can be designed with the classical, geometric filter rules, but this would lead to a conservative design, since such a filter could theoretically stand any head difference. Therefore we apply hydraulic criteria, see section 6.2.3. The porous flow is parallel to the interface, so we use equation 6.2.

At first we have to decide on the design head difference. The situation with the closed caisson lasts only a few days (directly after placement of the final caisson, the dam is covered with sand), so the period to consider in the determination of the probability is very short, say 4 days. This phase in the closure however is very critical, so a very low probability has to be taken, say 0.1%. This leads to a design frequency of 1/4000 days or 0.1/year. A design head difference of 2 m is found for that situation (from measurements). With a caisson width of 10 m, a hydraulic gradient of 20% serves as a first estimate (the seepage length through the sill is somewhat longer, but the gradient at the inflow and outflow points under the caissons will be higher than the average value).

Next, we have to determine the critical shear velocity from the Shields-curve (figure 3.2). With the viscosity $v = 10^{-6}$, $\Delta = 1.65$ and $d = 0.25$ mm we iteratively find: $u_{*c} = 0.013$ m/s and $\varphi = 0.04$. Now, with equation 6.2, $I_c = 0.2$ and the porosity of the filter material $n_F = 0.4$, we find $(n_F * d_{15F}/d_{50B}) = 8$ and from there $d_{15F} = 0.005$ mm. Assuming $d_{50}/d_{15} \approx 1.5$ this leads to $d_{50F} = 0.008$ m or fine gravel as a first layer on top of the sand.

From this layer we go on with the same equation, now using 0.008 as d_{50B}. This leads to a second layer with a d_{50} of about 0.12 m, hence coarse gravel, small rock or slag material. Applying this rule for the third time we find that this material will be stable under the top layer of 60 - 300 kg.

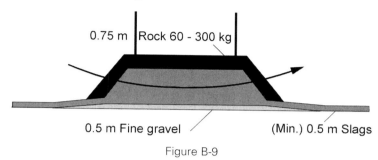

0.75 m Rock 60 - 300 kg

0.5 m Fine gravel   (Min.) 0.5 m Slags

Figure B-9

Figure B-9 shows the final build up of the sill, taking into account the height of the sill (2.5 m), the minimum thickness of the various layers, the price of the materials and the possible height until which the materials can be used in the flow (check with Shields). It seems practical to use slag both for the sill and the bottom protection. The construction scheme is then as follows:

1 dump fine gravel under the sill area
2 dump slags over the whole bottom area
3 dump rock 60 - 300 kg to complete the sill (pay attention: the top of the sill needs to be made as flat as possible for the placing and foundation of the caissons). Of course, the bottom protection can include a geotextile, making the fine gravel superfluous. The question is, whether this is economical in this case.

B.2.6 Scour

To get an idea of the scour during closure, we make some calculations using equation 4.10. For the parameters in this equation, we take: $\Delta = 1.65$, $h_0 = 10$ m and $u_c = C*u_{*c}/\sqrt{g} = 97*0.013/\sqrt{9.81} = 0.4$ m/s (with $C = 18*\log(12*10/2*0.00025) = 97\sqrt{m}/s$).

α is determined from model tests for a closure of some dimensions. We will make a first estimate for α in the three building phases. The length of the bottom protection is assumed $10 * h_0 = 100$ m at each side of the axis of the dam. (In fact the scour calculations serve to determine the length of the bottom protection, but in order to calculate the scour, a length has to be assumed. $10*h_0$ is a reasonable value to start with).

Figure B-10 shows flow lines for the various building stages. During phase 1 there is only a vertical constriction of 25% due to the sill. The flow separation beyond the sill spreads down to the bottom long before the end of the bottom protection. Horizontally, the flow covers the whole width of the channel. Figure 4.14 indicates a value of $\alpha = 2$ for this case. In phase 2 there is a horizontal constriction of 25% in combination with the sill. Now the flow separates in the horizontal constriction and is still not fully spread at the end of the bottom protection. A rough guess from the same figures leads to $\alpha = 4$ for that phase (for the deepest scour holes in the vortex streets between the main flow and the stagnant zone). In the third phase, the total constriction is more, but the caisson gives a better spreading of the flow than in the case with only the abutment caissons. Here we take $\alpha = 3$.

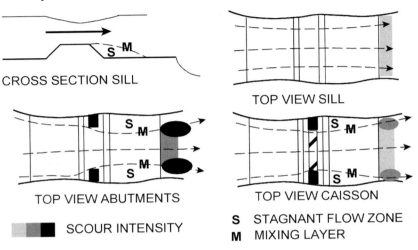

CROSS SECTION SILL

TOP VIEW SILL

TOP VIEW ABUTMENTS

TOP VIEW CAISSON

SCOUR INTENSITY

S   STAGNANT FLOW ZONE
M   MIXING LAYER

Figure B-10

In combination with the velocities from the boundary conditions, we now have all the necessary parameters to compute the scour depth as a function of time. But first we have to use equation 4.14 to compute a tidal average of the scouring process. For phase 1 this is shown in Figure B-11:

First, $\alpha u - u_c$ is determined for every half hour, it is then raised to the power 1.7 and the average is calculated over the *whole* tidal range. This results in a value of 1.42 for phase 1. This is the value to be used in equation 6.4 to replace $(\alpha u - u_c)^{1.7}$. This calculation has to be carried out for ebb and flood flow separately. (In Figure B-10 only one side of the scour is drawn; here both sides are identical, because of the sinusoidal shape of the velocity curves, but for real tides they will be different. Moreover, the variation in depth during the tide is neglected here).

We want to use days for the time, while in equation 6.4, t is in hours, so we have to multiply t with 24 in the formula. For each phase, the scour depth as a function of time is calculated, see Figure B-12.

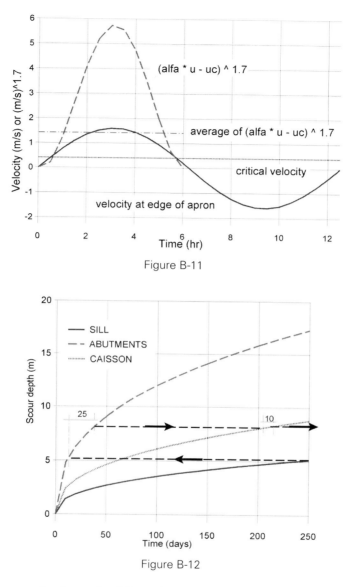

Figure B-11

Figure B-12

The total scour can be determined from these curves as follows:

The sill construction phase lasts 250 days; after that the scour depth is a little more than 5 m. Then the phase with the abutments starts, beginning with the scour from the previous phase and lasting 25 days. In that period the hole deepens 3 m. The caisson phase finally, scours less than phase 2 and lasts only 10 days. There is hardly any extra scour. So, the final depth is about 8 m.

The question is now whether the bottom protection of 100 m at each side of the dam axis is long enough. The introduction said that the sand is loosely packed. This means that the danger of flow slides exists, which would result in very gentle slopes (order of magnitude 1:15). With a slope of the scour hole itself of 1:2, this would mean that a flow slide could damage the sill with the caissons (8×15 > 100 + 8×2). The damage could be prevented by lengthening the protection or the slide could be prevented by dumping slag or gravel on the slope of the scour hole adjacent to the apron. After phase 1, a hole of 5 m is expected and there seems to be more than enough time to cover the slope before the caissons are placed. After 200 days the maximum scour depth due to the sill is nearly reached, see Figure B-12, so the most critical parts of the scour hole should be covered in 50 days, which seems a realistic possibility.

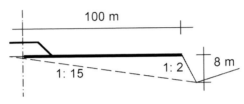

Figure B-13

This could, however, become a disappointment. In these calculations we have neglected the sediment transport in the tidal channel. Sediment transport reduces the scour depth, so, one could expect the above mentioned calculations to be pessimistic, but the opposite is true. Reduction due to sediment transport is only effective when the eroding power is relatively weak; hence, in the situation with only a sill. After that the scour will reach more or less the same values, but now in a much shorter time. This is demonstrated below.

The sediment transport is assumed to be: S = 3 m³/day/m (from some sediment transport formula or measurements). The effect of this transport is a reduction of the scour depth. We can calculate this reduction with equation 4.16 through 4.18. To do so, we have to make assumptions for the slopes of the scour hole. Here we take 4 and 40 for β_1 and β_2 respectively. Equation 6.13 then becomes: $h_{sm\ red} = \sqrt{(22*h_{sm\ unred}^2 - S*t)}/\sqrt{22}$, with t in days. With this we can make calculations for the reduced scour depths for the three phases, see Figure B-14.

The result is hardly any scour due to the sill (less than 1 m), while the final scour depth is almost the same as in the case where the reduction was neglected. The scouring in phase 2 is so violent that reduction hardly plays a role. This means that covering of the slopes of the scour hole cannot start until the abutments are completed (before the scour reaches values of about 5m, covering is of little use). But then, there are only a few days available to execute the covering job, so now there is more danger of the scour holes to collapse than without scour reduction! If one is afraid that the bottom protection may be too short, the only solution is to make it longer.

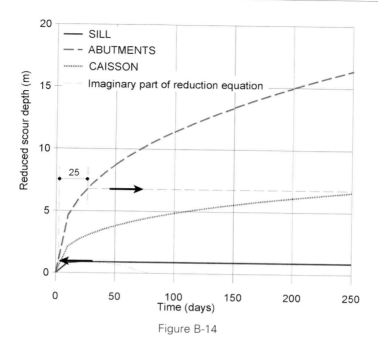

Figure B-14

B.3 Breakwater

B.3.1 Introduction

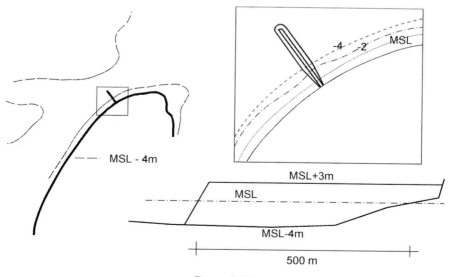

Figure B-15

On a sandy coast, a 500 m long breakwater will be constructed to influence the longshore sediment transport, see Figure B-14. The crest will be made at MSL + 3 m with a width of 5 m, for reasons of accessibility for maintenance and fishing. The subsoil consists entirely of sand of 200 µm, which is densly packed.

The manager of this coastline carries out yearly inspections and does not want there to be too much maintenance, so the design will be made for circumstances with an exceedance frequency of 0.01/year.

B.3.2 Boundary conditions

In the following table and figure, the most important boundary conditions are presented:

Exceedance frequency (1/year)	Waterlevel (MSL + m)	Wave height Hs (m)	Wave period Tp (s)
Average HW	0.8		
Average LW	-0.9		
1	2.05	2.52	11.3
0.5	2.25	2.62	11.5
0.2	2.45	2.72	11.7
0.1	2.65	2.82	12.0
0.05	2.85	2.92	12.2
0.02	3.05	3.02	12.4
0.01	3.25	3.12	12.6
0.005	3.45	3.22	12.8
0.002	3.60	3.30	12.9
0.001	3.75	3.38	13.0

The waterlevels come from observations and extrapolation.

The wave height is calculated as follows: the location where the breakwater will be constructed, is sheltered from the prevailing winds by shallow areas with a depth of about MSL - 2.5m. In the future, due to morphological changes, a lowering to MSL − 3m is possible. Wave heights under extreme conditions will therefore be depth-limited and are estimated from: $H_s \approx 0.5 * (3 + waterlevel)$.

The wave periods come from observation in deep water. The assumption is that the peak period of the wave spectrum is not changed by the shallows. This is on the safe side, since the shallow areas will generate higher harmonics in the waves, decreasing T_p somewhat, while here a longer period is unfavourable in the stability calculations.

Under average conditions, the waves are not depth limited. So, for the construction period, wave data are necessary. It appears that, during the summer season (in which the breakwater will be built), 20% of the time waves are higher than $H_s = 1$ m with T_p

≈ 5 s occur; 10% of the time $H_s > 1.5$ m with $T_p ≈ 6$ s and 1% of the time $H_s > 2$ m with $T_p ≈ 8$ s.

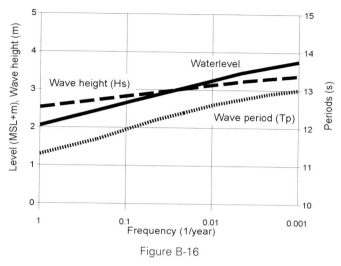

Figure B-16

B.3.3 Top layer - cross section

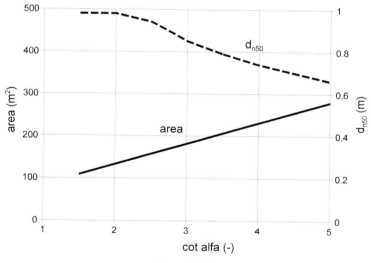

Figure B-17

At first, the slope of the breakwater trunk will be determined. The gentler the slope, the smaller the stone that is needed for stability in waves, but the larger the total quantity of stones. Figure B-15 shows the result for a dam with a 5 m wide crest at MSL + 3m and a bottomdepth of MSL - 4m. The stability was calculated with the Van der Meer formula for 0.01/year conditions. From a slope of 1:1.5 and 1:2, there is a transition between surging and plunging breakers, hence the same calculated diameter. In general, the decrease of the diameter will not compensate the larger area: the price per ton for these stones will not differ very much. In this case, we also assume there is no significant difference in the price of placing the stones. A slope of 1:2 is chosen although 1:1.5 is cheaper. This is a preliminary design which should not be approached too optimisitic for a first estimate of costs.

Warning:
The conclusion: "the steeper the cheaper" can change completely in other countries, when rock of a certain diameter, or the equipment to place it, is not available in the area. In that case the available diameter or equipment determines the slope. Another reason for avoiding larger stones can be when the equipment used to place the stones has its limitations.

B.3.4 Top layer - stone diameter

To calculate the necessary diameter of the stones in the top layer of the breakwater, the formula of Van der Meer is used for conditions with a frequency of 0.01/year. From the boundary conditions we find: waterlevel = MSL + 3.25 m, H_s = 3.12 m and T_p = 12.6 s. In the Van der Meer formula, the period is not T_p but T_m, so a correction is made: $T_m \approx 0.85 T_p$ = 10.7 s.

Other parameters in the formula are:
1. P (permeability): for a breakwater with a permeable core a value of 0.5 is recommended,
2. Δ (relative density): use is made of basalt with ρ = 3000 kg/m3. With seawater density = 1020 kg/m3, this leads to Δ = 1.95.
3. S (damage): some damage from the design waves is allowed, but removal of the top layer should be avoided, so a value of 4 is used.
4. N (number of waves): the maximum number is used (N = 7000) since storms in the area can last for several days.

From all this we find: ξ_m = 3.79, while the transition between the formulae for plunging and surging waves is given by equation 8.9: ξ_m = 3.54, so we have to use the equation for surging breakers (8.8b), leading to d_{n50} = 0.98 m. A reduction factor can be applied, since the waterlevel is higher than the crest level of the breakwater, see section 8.3.3. Since the crest is hardly under water, we will use equation 8.10 and we assume the maximum effect of this equation: reduction of 0.8 on the diameter, leading to d_{n50} = 0.78 m.

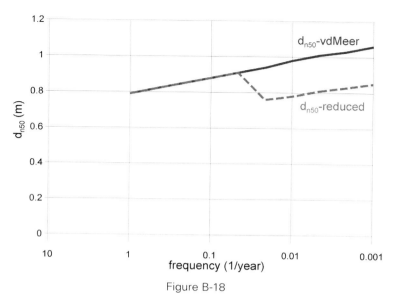

Figure B-18

Now, however, the question arises as to whether this wave condition is dominant, since the breakwater crest is under water and with lower waterlevels and waves, the reduction can not be applied, possibly leading to the use of larger stones. Figure B-18 shows the result for all waterlevel and wave conditions. It appears that up to a frequency of 0.05/year, there is no reduction, while for more extreme values, the crest disappears under water leading to the maximum reduction. So, the 0.05/year conditions are dominating leading to a $d_{n50} = 0.91$ m. From the curves in appendix A, we find that this equivalent with rock 1000-3000 kg.

B.3.5 Toe

For the support of the rocks at the toe, a berm is recommended (otherwise, stones at the toe will easily move under wave attack). To determine the necessary diameter for the berm, we look at figure 8.15a. For waterdepths of 6-8 m and a berm height of 1.5 m, we find values of $H/\Delta D_n$ between 5 and 7, leading to a necessary d_{n50} of around 0.25 m. A reason for using larger stones is that the diameter of stones of a berm, in order to provide sufficient side support, should not be less than one class lower than the top layer, hence with 1000-3000 kg as a top layer, the berm should be made of at least 300-1000 kg. For practical reasons, in order to avoid all different kinds of stones, we will use use the same stones as those in the top layer (1000-3000 kg).

B.3.6 Head

The head of the breakwater is attacked more heavily by the waves than the trunk. To account for this effect, there are two possibilities: a gentler slope or a larger stone. Again, a steep slope is cheaper, provided the stone weight does not become prohibitive, but means an extra class of stone. Here for the head, rock of 3000-6000 kg will be used instead of 1000-3000 kg.

B.3.7 Core

Filter rules

The main function of the breakwater is to interrupt the sediment transport locally. Therefore, finer material inside the dam should be used, since rocks with a diameter of around 1 m will cause much turbulence inside the dam, making the sandtightness questionable. Another reason for the use of finer material can be that it is cheaper (when there is a large difference in diameter); on the other hand it is more complicated to build the dam from different materials, leading to more different actions.

Applying geometrical filter rules (dynamic loading!), we find from the curves in appendix A for the 1000-3000 kg rock a d_{15} of 0.7 m. For stability, this leads to a minimum d_{85} for the next layer of 0.14 m corresponding with stones of 80/200 mm as an absolute minimum. For the construction and friction between the layers use of a somewhat larger stone would be preferable, leading to 10-60 kg.

However, the final choice for the cross-sectional build-up of the breakwater will be made after the construction of the breakwater is considered.

Construction

The stability of the core material in the construction phase should be considered, since uncontrolled losses of dam material will cost a lot of time and money. In the section on boundary conditions, it was said that waves of 1 to 1.5 m can occur during construction.

Figure B-19 shows the damage number in the Van der Meer formula as a function of the wave height. In order to calculate S, the Van der Meer formula was rewritten in the shape: $S = f(H,P,N, ...$ etc.). For the number of waves $N = 2000$ was taken and T_m = 6 s (a summer storm of 3 - 4 hours). A damage number of 25 already indicates considerable flattening of the slopes.

From this result we can see that the core material is only stable at very small wave heights and that even rock of 10-60 kg is easily damaged with wave heights between 1 and 1.5 m. Therefore, 60-300 kg will be used on top of a core of stones of 80-200 mm. The dumping of the core material will be followed immediately by the second layer. After that it seems possible to leave the dam for some weeks before constructing the top layer.

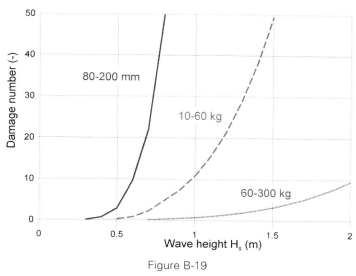

Figure B-19

3.3.8 Bottom protection

Width

The functions of the bottom protection are mainly:
- to keep scour holes away from the dam
- to make a transition between the large rocks of the top layer and the sandy bottom (filter)

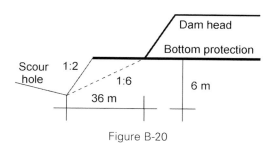

Figure B-20

For the first function, the bottom protection should be wider than the dam itself. Scour can come from tidal currents or rip currents and is not easy to determine. A first rough estimate can be made, by considering the dam as a blunt abutment, leading to a scour depth of about 1.5 times the waterdepth. Maximum flow due to the tides usually occurs around MSL, giving a scour hole of 6 m deep. With an assumed slope of 1:2 and an end slope after possible sliding (densely packed sand), the protection should be at least 24 m from the toe of the dam.

B.3.9 Top layer

The second function can be fulfilled by making a geotextile, reinforced with a fascine mattress and covered with stones. These stones are necessary to keep the mattress on the bottom before it is covered with the final structure. In the final stage, the top layer of the bottom protection should withstand the loads, in particular the wave shear and the tidal flow around the head.

The necessary stone size can be calculated from the modified Shields curve by Sleath, see Figure 8.6. To calculate the shear stress, we use the linear wave theory with the approach of Jonsson, see table 7.1 and figure 7.7. For 0.01/year conditions we find:

H_{max} = 1.75 * 3.12 m ≈ 5.6 m (maximum wave in a wave train of 1000 waves with Rayleigh distribution), T_p = 12.6 s and h = 4 + 3.25 = 7.25 m. From linear wave theory a wave length of 103 m in this waterdepth is found, with a bottom velocity of 3 m/s and a particle displacement of 6.15 m. Iteratively it is found that a_b/k_r = 9.55 (with k_r = 2 * d_{n50}), c_f = 0.073 (in equation 7.3) and d_{n50} = 0.32 m (with ψ = 0.056 in Figure 8.6). This means that rock of 60-300 kg is needed as a top layer on the bottom protection.

B.3.10 Final dimensions

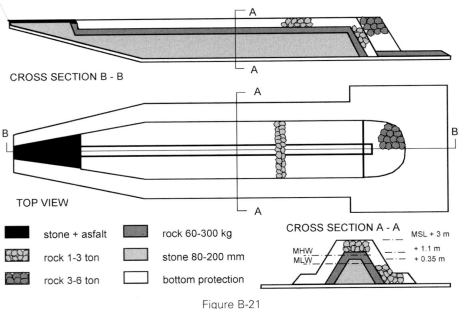

Figure B-21

Figure B-21 shows the "final preliminary" design. The various layers get a thickness of about 2*d. This results in 1.9 m for the 1000-3000 kg rock and 0.75 m for the 60-

300 kg rock. The top layer on the beach is made of stones 80-200 mm penetrated with asphalt in order to facilitate crossing with cars.

At the head, the rock 3000-6000 kg is constructed as a homogeneous layer and acts as an additional protection of this head.

Note:

It will be clear that there is still a lot to be optimized and investigated before the final design can be constructed. Details like the transition between the beach and the dam need attention. The construction method itself has to be worked out. An average HW-level of MSL + 0.8 m, a LW-level of MSL - 0.9 m and a bottomlevel of MSL - 4 m make both land-based and water-borne operations necessary. The core will be dumped by barges and completed with land-based equipment during the low water periods of the tidal cycle. The second layer of stones 60-300 kg will be constructed with land-based equipment in the same manner. The top layer can be completed above high water level.

List of symbols

Basic units: kg, m, s, $\in$
Other units: $N = kg{\cdot}m/s^2$ $Pa = N/m^2$
                  $J = kg{\cdot}m$ $W = J/s$

Symbol	Definition	Unit
a	wave amplitude ($= H/2$)	m
a_b	wave amplitude at bottom	m
A	area	m^2
A_c	area of canal's cross section	m^2
A_s	area of ship's cross section	m^2
b	width of canal, jet or mixing layer	m
B	width of ship, vegetation or (half) width of outflow in plane jet	m
c	1. concentration	kg/m^3
	2. coefficient in Sellmeyer's piping formula	-
	3. cohesion	N/m^2
	4. wave celerity	m/s
	5. stiffness coefficient soil (with wave impacts)	Pa/m
c_f	friction coefficient (between flow and wall)	-
C	Chezy coefficient	$m^{1/2}/s$
C_D	drag coefficient	-
C_L	lift coefficient	-
C_{creep}	coefficient in piping formulas	-
d	1. grain diameter	m

	2. layer thickness	m
	3. propeller diameter	m
d_n	nominal grain diameter (= $V^{1/3}$)	m
d_{n50}	median nominal diameter (= $(m_{50}/\rho_s)^{1/3}$)	m
d_x	grain diameter where x % of the grain mass has a smaller diameter	m
d_*	dimensionless grain diameter	-
D	1. diameter of cylinder	m
	2. protrusion abutment	m
	3. height of sill or dam	m
	4. thickness of soil layer	m
	5. draught ship	m
	6. diameter ship's propeller	
	7. energy loss in bore	J/sm^2
	8. damage in risk calculation	€
e	thickness geotextile	m
E	elasticity protection layer	Pa
f	1. friction coefficient (between materials)	-
	2. wave frequency (= $1/T$)	1/s
F	1. stability factor in slip circle computations	-
	2. fetch	m
F_f	flow force	N/m^3
g	acceleration of gravity	m/s^2
h	Waterdepth	m
h_0	original waterdepth	m
h_c	dam height relative to bottom	m
h_s	scour depth	m
h_{se}	equilibrium scour depth	m
h_t	waterdepth above toe of breakwater	m
H	1. wave height	m
	2. total head	m
	3. excess "pressure" against impervious layer	m
H_I	incoming wave height	m

H_R	reflected wave height	m
H_s	significant wave height (average height of 33% highest waves)	m
H_T	transmitted wave height	m
i	Gradient	-
i_c	critical gradient in filter	-
I	1. slope	-
	2. volume of scour hole per m width	m^2
k	1. permeability in porous flow	m/s
	2. wave number $(= 2\pi/L)$	1/m
	3. spring constant	N/m
k_r	(equivalent) roughness of bottom	m
K_D	damage coefficient in Hudson's formula	-
K_R	reflection coefficient $(= H_R/H_I)$	-
K_s	shape coefficient in pile scour formula	-
K_T	transmission coefficient $(= H_T/H_I)$	-
K_u	velocity coefficient in live-bed scour	-
K_v	velocity coefficient in stone stability	-
K_α	correction coefficient for stone stability on slope	-
L	1. wave length	m
	2. bottom protection length	m
	3. seepage length	m
L_0	deep water wave length $(= gT^2/2\pi)$	m
L_e	erosion length ("depth") in beach	m
m	mass	kg
m_0	area of wave spectrum	m^2
M	momentum	kg/s^2
n	porosity of soil (=volume of voids/volume of soil)	-
N	1. number of waves in Van der Meer formula	-
	2. number of stalks or stems of vegetation	$1/m^2$
O	size of openings in geotextile	m
p	1. pressure	N/m^2
	2. coefficient in parabolic beach profile	$m^{0.22}$

	3. probability density	-
P	1. permittivity geotextile	1/s
	2. permeability in Van der Meer formula	-
	3. power	W
P_F	failure probability	-
q_s	sediment transport per m width	m^2/s
Q	discharge	m^3/s
Q_b	percentage broken waves	-
r	1. relative turbulence	-
	2. radial distance from centre of jet	m
R	1. hydraulic radius	m
	2. strength ("résistance")	depends
R_c	crest height dam, relative to water level	m
R_u	wave run-up	m
R_d	wave run-down	m
s	1. wave steepness (= H/L)	-
	2. distance from ship's sailing line	m
s_0	wave steepness with deep water wave length (= $H/(gT^2/2\pi)$)	-
S	1. sediment transport	m^3/s
	2. damage level in Van der Meer formula	-
	3. load ("sollicitation")	depends
t	time	s
t_e	time to reach equilibrium scour	s
T	1. wave period	s
	2. turbulence period (averaging period)	s
T_m	mean wave period	s
T_p	peak wave period (where spectral energy is maximal)	s
T_s	significant wave period (average period of 33% highest waves)	s
u	velocity in x-direction	m/s
$\bar{u}$	velocity averaged in time or space	m/s
u_c	critical velocity	m/s
u_f	filter velocity	m/s

u_m	maximum velocity	m/s
u_b	bottom velocity	m/s
u_r	1. return flow velocity	m/s
	2. resulting velocity with waves and current	m/s
u_w	wind velocity	m/s
u_0	outflow velocity, start velocity	m/s
u'	turbulent velocity fluctuation in x-direction	m/s
u_*	shear velocity $(= \sqrt{\tau/\rho})$	m/s
v	velocity in y-direction	m/s
v_s	ship's speed	m/s
v'	turbulent velocity fluctuation in y-direction	m/s
V_l	limit speed of ship	m/s
w	1. velocity in z-direction	m/s
	2. width of soil layer in slip circle computation	m
w_s	fall velocity sediment	m/s
w'	turbulent velocity fluctuation in z-direction	m/s
W	1. weight	N
	2. relative width between piles in wave reduction formula	-
x	distance along horizontal axis (in flow direction)	m
y	1. distance along horizontal axis (perpendicular to flow direction)	m
	2. eccentricity ship in canal	m
z	1. distance along vertical axis	m
	2. water-level depression in primary ship's waves	m
z_b	distance propeller axis – bottom	m
$z_{\max}$	height of stern wave in primary ship's waves	m
Z	value reliability function	depends
α	1. slope angle	-
	2. coefficient in Breuser's scour formula	-
	3. coefficient in Sellmeyer's piping formula	-
	4. influence coefficient in reliability function	-
β	1. slope of scour hole	-
	2. angle of wave attack	-

	3. relative distance from average value in normal distribution	-
	4. coefficient in wave impact – strength equation	-
δ	thickness of boundary layer	m
Δ	relative density $[= (\rho_s - \rho_w)/\rho_w]$	-
ϕ	1. angle of repose	-
	2. velocity potential	m^2/s
	3. piezometric head	m
	4. angle between waves and current	-
	5. angle between ship's sailing line and cusp location	-
γ	safety coefficient	-
γ_b	breaker index $(= H/h)$	-
γ_B	berm coefficient in wave run-up	-
γ_β	wave angle coefficient in wave run-up	-
γ_f	foreshore coefficient in wave run-up	-
γ_r	roughness coefficient in wave run-up	-
η	water level in wave	m
κ	von Kármán constant (≈ 0.4)	-
λ	1. relaxation length in turbulence	m
	2. length of floating breakwater	m
Λ	leakage length	m
μ	1. dynamic viscosity	kg/ms
	2. discharge coefficient	-
	3. statistic average	depends
ν	kinematic viscosity $(= \mu/\rho)$	m^2/s
ρ_m	density material	kg/m^3
ρ_s	density of sediment	kg/m^3
ρ_w	density of water	kg/m^3
σ	1. total stress in soil	N/m^2
	2. standard deviation	depends
	3. stress in impervious layer due to wave impact	N/m^2
σ'	grain stress in soil	N/m^2
τ	shear stress	N/m^2

τ_c	critical shear stress	N/m^2
ω	angular frequency in waves ($= 2\pi/T$)	1/s
ξ	breaker parameter ($= \tan\alpha/\sqrt{s}$)	-
ξ_m	breaker parameter related to mean wave period	-
ξ_p	breaker parameter related to peak wave period	-
ψ	mobility parameter ($= \tau/\Delta gd$)	-
ψ_c	Shields (stability) parameter ($= \tau_c/\Delta gd$)	-
ζ	geometry coefficient ship	-
$\propto$	proportional to	-

Greek alphabet

lower case	CAPITAL	Name	lower case	CAPITAL	Name
α	A	alpha	ν	N	nu
β	B	beta	ξ	Ξ	ksi
γ	Γ	gamma	o	O	omikron
δ	Δ	delta	π	Π	pi
ε	E	epsilon	ρ	P	rho
ζ	Z	zeta	σ	Σ	sigma
η	H	eta	τ	T	tau
θ	Θ	theta	υ	Y	ypsilon
ι	I	iota	$\phi\,(\varphi)$	Φ	phi
κ	K	kappa	χ	X	chi
λ	Λ	lambda	ψ	Ψ	psi
μ	M	mu	ω	Ω	omega

References

Ahmed, F. & N. Rajaratham (1998) Flow around Bridge Piers. Journal of Hydraulic Engineering, vol. 124, no. 3, pp. 288-300

d'Angremond, K. (2001) Bed, Bank and Shore Protection II. Delft University of Technology

d'Angremond K., T. van der Meulen & G.J. Schiereck (1996) Toe Stability of rubble Mound Breakwater. 25th ICCE, Orlando, ASCE, New York

Ariens (1993) Relatie tussen Ontgrondingen en Steenstabiliteit van de Toplaag (in Dutch). M.Sc. Thesis, Delft University of Technology

Ashida (1973) Initiation of Motion and Roughness of Flows in steep Channels, Papers IAHR-Congress, Istanbul

Battjes, J.A. (1974) Computation of Set-up, longshore Currents, Run-up and Overtopping due to wind-generated Waves. Dissertation Delft University of Technology

Battjes, S.A. & K.W. Groenendijk (2000), Wave height distributions on shallow foreshores, Coastal Eng. (40) pp. 161-182

Battjes, J.A. en J.P.F.M. Janssen (1978) Energy Loss and Set-Up due to breaking of random Waves, Proceedings 16th ICCE (Hamburg). ASCE, New York

Bezuijen, A., M. Klein Breteler en A.M. Burger (1990) Placed Block Revetment. Coastal Protection, Pilarczyk, ed. Balkema, Rotterdam

Bijker, E.W. (1967) Some Considerations about Scales for coastal Models with moveable Bed. Delft Hydraulics, publ.nr. 50

Blom, P. (1993) On the shallow Water Equations for turbulent Flow over Sills. Delft University of Technology

Boer, de G. (1998) Transport van Stenen van een granulaire Bodemverdediging M.Sc. Thesis, Delft University of Technology

Booij, R. (1992) Turbulentie in de Waterloopkunde, college ct5312 (in Dutch). Delft University of Technology

Booij, R. (1998) Erosie onder geometrisch open Filter. Delft University of Technology, rep 2/98

Booij, N., R.C. Ris & L.H. Holthuijzen (1999) A third-generation Wave Model for coastal Regions. J.Geophys.Res. C4 (104) 7649-7666

Boutovski, A. (1998)Stabiliteit van gestortte steen, M.Sc.-thesis Delft University of technology

Breugel, van R.H. & T.D. ten Hove (1995) Steenstabiliteit bij horizontale Vernauwingen. M.Sc. Thesis, Delft University of Technology

Breusers, H.N.C., G. Nicolet, H.W. Shen (1977) Local Scour around cilindrical Piers. J. Hydr.Res. IAHR., 15(3):211-252

Breusers, H.N.C. en A.J. Raudkivi (1991) Scouring. Balkema, Rotterdam

Buffington, J.M. (1999) The Legend of A.F. Shields. J.Hydr.Eng. 124(4): 376-387

Buffington, J.M. & D.B. Montgomery (1997) A systematic Analysis of eight Decades of incipient Motion studies, with special Reference to gravel-bedded Rivers, Water Resources Research, 1999-33(8) 1993-2029

Burger, G. (1995) Stability of low crested Breakwaters. Msc. Thesis, Delft University of Technology

Calle, E.O.F. & J.B. Weijers (1994) Technisch Rapport voor Controle op het Mechanisme Piping bij Rivierdijken. TAW-DWW-Delft

Chiew,-Yee-Meng; Parker,-G. (1994) Incipient sediment motion on non-horizontal slopes J. Hydr. Res, IAHR vol. 32, no. 5, pp. 649-660

CUR 161 (1993) Filters in de Waterbouw. CUR, Gouda

CUR 146 (1991) Methode voor periodieke Sterktebeoordeling van Dijken. CUR, Gouda

CUR 162 (1992) Construeren met Grond. CUR, Gouda

CUR 154 (1991) Manual on the Use of Rock in coastal and shoreline Engineering. CUR, Gouda

CUR 155(1992) Handboek voor Dimensionering van gezette Taludbekledingen (in Dutch). CUR, Gouda

CUR 169 (1995) Manual on the Use of Rock in Hydraulic Engineering, CUR, Gouda

CUR 200 (1999) Natuurvriendelijke oevers, aanpak en toepassingen, CUR, Gouda

DHL (1988) Evaluatie Ontgrondingsonderzoek, report Q635

DHL (1969) Begin van beweging van bodemmateriaal, report S159-1

DHL (1970) Begin van Beweging, report M 1048

DHL (1985) Hydraulic desgn criteria for rockfill closure of tidal gaps, vertical closure, report M1741

DHL (1986) Hydraulic design criteria for rockfill closure of tidal gaps, evaluation report M1741

Dorst, C.J. (1995) Wilgen als oeverbescherming in kribvakken, M.Sc-thesis, Delft University of Technology, Delft

Fredsoe (1992) Mechanics of coastal Sediment Transport. Advanced Series in Ocean Engineering, Volume 3, World Scientific

Gelderen, F.J.G. (1999) Het enkele steenmodel, een verificatie op basis van modelproeven, M.Sc-thesis, Delft University of Technology, Delft

Gent, van, M. (1993) Stationary and oscillatory Flow through coarse porous Media, Comm. on Hydraulic and Geotechnical Engineering. Delft University of Technology, 93-9

Geodelft (1993) User's Manual MSeep

Gerressen, B., (1997) Stabiliteit van steenzettingen, beschouwing van een dijkbekleding als verend ondersteunde buigligger, M.Sc.-thesis, Delft University of Technology

Graf, W.H. (1971) Hydraulics of Sediment Transport. McGraw-Hill, New York

Grauw, de A.A.F.(1983) Design Criteria for granular Filters. Delft Hydraulics Publication 287

Grauw, de A.A.F. en K.W. Pilarczyk (1981) Model Prototype Conformity of local Scour in non-cohesive Sediments beneath an Overflow Dam. XIX IAHR Congress, New Delhi

Groot, de M.B., A. Bezuijen & A.M. Burger (1988) The Interaction between Soil, Water and Bed or Slope Protection. SOWAS-1988 Delft, Balkema, Rotterdam

Grune en Kohlhase (1974) Wave Transmission through a vertical slotted Wall, proc. 14th ICCE, III-pp. 1906-1923, ASCE, New York

Gunst, de M. (1999) Steenstabiliteit in een turbulente Stroming achter een Afstap. Msc. Thesis, Delft University of Technology

Hinze, J.O. (1975) Turbulence. McGraw-Hill, New York

Hoffmans, G.J.C.M. (1993) A hydraulic and morphological Criterion for upstream Slopes in local-scour Holes. Ministry of Transport, Public Works and Water Management, Road and Hydraulic Engineering Division, Delft

Holthuijsen, L.W. et al (1989) A prediction model for stationary chart crestal waves, Coastal Engineering (13) pp 23-54

Hooijmeijer, R.H. (1997) Sedimentatie in de natte strook, M.Sc-thesis, Delft University of Technology, Delft

Hopley, D. (1974) Coastal changes produced by tropical cyclone Althea in Queensland, Australian Geographer XII (5) pp 445-456

Hudson (1953) Wave Forces on Breakwaters, Proceedings-Separate ASCE, no 113

Iribarren (1938) Una Formula para el Calculo de los Diques de Escollera. M. Bermejillo-Pasajes, Madrid, Spain

Izbash, S.V. (1935) Construction of Dams by dumping of Stone in running Water. Moscow-Leningrad

Izbash, S.V. & K.Y. Zkhaldre (1970) Hydraulics of River Channel Closure. Butterworth, London

Jansen (1979) Principles of River Engineering, The non-tidal alluvial river. Pitman

Jonsson (1966) Wave boundary Layers and friction Factors, proc. 10th ICCE Tokyo, ASCE, New York

Jorissen, R. & N. Vrijling (1989) Local Scour downstream hydraulic Constructions, IAHR-Congress, Ottawa

Klok, P.K. (1996) De verborgen kracht van riet, M.Sc-thesis, Delft University of Technology, Delft

Lammers, J.C. (1997) Shields in de Praktijk. Msc. Thesis, Delft University of Technology, Delft

Lemmens, R.J.M. (1996) Natuurvriendelijke verbetering van de zanddichtheid van klassieke zinkstukken, M.Sc-thesis Delft University of Technology

Linden, van der (1985) Golfdempende Constructies (in Dutch). Msc. Thesis, Delft University of Technology

Meer, J. van der (1988) Rock Slopes and Gravel Beaches under Wave Attack. PhD. Thesis, Delft University of Technology

Meer, J. van der & K. d'Angremond (1991) Wave Transmission at low-crested Structures, Proceedings ICE Conference Coastal Structures and Breakwaters, Thomas Telford, London, UK

Mehaute, Le, B. (1969) An Introduction to Hydrodynamics and Water Waves. Springer, New York

Melville, B.W. & A.J. Raudkivi (1977) Flow Characteristics in local Scour at Bridge Piers. Journal of Hydraulic Research, vol. 15, no. 4, pp. 373-380

Nakagawa, H. & I. Nezu (1987) Experimental Investigation on turbulent Structure of backward-facing Step Flow in an open Channel. J. Hydr.Res., IAHR, 25(1):67-88

Noakes, D.S.P. (1955) Methods of increasing growth and obtaining natural regeneration of mangrove types in Malaya, Malayan Forester, (13) pp 23-30

Os, van P. (1998) Hydraulische Belasting op een geometrisch open Filterconstructie. Msc. Thesis, Delft University of Technology

Paintal (1971) Concept of critical shear Stress in loose boudary open Channels. Journal of Hydraulic Research, 9-1

Petschacher, M. (1994) VdP 1.5 for Windows, User Manual, ETH Zurich, Switzerland (see also www.Petschacher.at)

Pilarczyk, K.W.(1990) Coastal Protection, Proceedings of short Course on coastal Protection, Delft University of Technology, Balkema, Rotterdam

Rajaratham, N. (1976) Turbulent Jets. Elsevier, Amsterdam

Rance P.J. & N.F. Warren (1968) The Treshold of Movement of Coarse Material in oscilatory Flow. Proc. 11th ICCE London, ASCE, New York

Raudkivi, A.J. & R. Ettema (1985) Scour at cylindrical Bridge Piers in armoured Beds. J.Hydr.Eng. ASCE 111(4):713-731

Reynolds, A.J.(1977) Turbulent flow in engineering, Wiley, London

Ridder, de, H.A.J. (1999) Collegedictaat Methodisch Ontwerpen

Rijn, van, L.G. (1984) Sediment Transport. J. Hydr. Eng. (110) 10 1431-1456 (110) 11 1631-1641 (110) 12 1733-1754.

Rijn, van L.C. (1986) Mathematical Modelling of suspended Sediment in non-uniform Flows. Comm. nr. 365, Delft Hydraulics, Delft

Rouse H.(1958) Advanced mechanics of fluids, Wiley, London

RWS (1985) The Use of Asphalt in hydraulic Engineering. Rijkswaterstaat Communication, no. 37

RWS (1987) The Closure of Tidal Basins. Delft University Press

RWS (1990) Waterbouw, Rekenregels voor waterbouwkundig Ontwerpen (in Dutch). Bouwdienst Rijkswaterstaat

RWS/DHL (1988) Aantasting van Dwarsprofielen in Vaarwegen (in Dutch). Report M1115 XIX

Sato, K. (1985) Studies on the protective functions of mangrove forests against erosion, Science Bull. Univ. Ryukyos, pp 161-172

Schiereck, G.J., H.L. Fontijn, W.V. Grote & P.G.J. Sistermans (1994) Stability of Rock on Beaches, 24th ICCE, Kobe, ASCE, New York

Schiereck, G.J. & N. Booij (1995) Wave transmission in mangrove forests, Copedec IV, Rio de Janeiro

Schiereck, G.J. & H.L. Fontijn (1996) Pipeline Protection in the Surf Zone, 25th ICCE, Orlando, ASCE, New York

Schijf, J.B. (1949) Influence on the Form and Dimensions of the Cross-Section of the Canal, of the Form, of the Speed and the Propulsion System of Vessels. XVIIth PIANC, section 1, Subject 2, Lisbon

Schlichting, H. (1968) Boudary-layer Theory. McGraw-Hill, New York

Sellmeijer, J.B. (1988) On the Mechanism of Piping under impervious Structures. Ph.D. Thesis, Technical University of Delft, 111 pp.

Simons (1957) Theory and design of stable channels in alluvial channels, Colorado State Univ. rep. CER571DB

Sleath (1978) Measurements of bed Load in oscillatory Flow. ASCE Journal of the Waterway Port coastal and Ocean Division, Volume 104, NoWW4

Sorensen (1973) Water Waves produced by Ships. ASCE Journal of Waterways, Harbors and Coastal Engineering Division

Sprangen, J.T.C.M. (1999) Vegetative dynamics and erosion resistance of sea dyke grassland, PhD-thesis, Wageningen University

SPM (1984) Shore Protection Manual. US Army Coastal Engineering Center

Steinke. T.D. & C.J. Ward (1989) Some effects of the cyclones Domoina and Imboa on mangrove communities in St Lucia, S. Afr. Tydskr. Plantk. 55 (3) pp 390-398

Stoddart, D.R., (1965) Re-survey of hurricane effects on the British Honduras reefs and cays, Nature (207) pp 589-592

Stive, M.J.F. (1984) Energy Dissipation in Waves breaking on gentle Slopes. Coastal Engineering 8, pp. 99-127

Stoker, J.J. (1957) Water Waves. Interscience, New York, 595 pp.

Suiker, A.S.J. (1995) Inklemeffecten bij steenzettingen op dijken. M.Sc.-thesis, Delft University of Technology

TAW (2000), Technisch rapport asfalt, RWS-DWW, Delft

TAW (2002), Technisch rapport Golfoploop en Golfoverslag bij Dijken, RWS-DWW, Delft

TAW (1998) Technisch rapport Erosiebestendigheid van grasland als dijkbekleding, RWS-DWW, Delft

Tukker, J. (1997) Turbulence Structures in shallow free-surface mixing Layers. Comm. on hydr. and geotechn. eng., Delft University of Technology

Van Gent, M. (1992) Formulae to Descirbe porous Flow. Comm. on Hydr. & Geotech. Eng, Delft University of Technology, 92-2

Veldhuijzen & Zanten, van (1986) Geotextiles and Geomembranes in Civil Engineering. Balkema, Rotterdam

Vellinga (1986) Beach and Dune Erosion during Storm Surges. Dissertation, Delft University of Technology

Ven te Chow (1959) Open Channel Hydraulics. McGraw-Hill, New York

Verruijt, A. (1999) Grondmechanica, college ct2090, Delft University of Technology

Vrijling, J.K., H.K.T. Kuiper, R.E. Jorissen & H.E. Klatter (1992) The Maintenance of hydraulic Structures. Coastal Engineering 1992, pp.1693-1705

Wal, M. van der (2000), Kribben in Moderne Rivierkunde, Stichting PAO, Delft

Wang, X.K. & H.L. Fontijn (1993) Experimental study of the hydrodynamic Forces on a bed Element in an open Channel Flow with a backward-facing Step. J. of Fluids & Structures, 7, pp. 229-318

Weiden, van de (1989) General Introduction and hydraulic Aspects. Short Course on design of coastal Structures, AIT Bangkok

Wemelsfelder, P.J. (1939) Wetmatigheden in het Optreden van Stormvloeden. De Ingenieur nr. 9

Wiberg, P.L. & J.D. Smith (1987) Calculation of the critical shear Stress for Motion of Uniform and heterogeneous Sediments. Water Resources Research, 23(8), pp. 1471-

1480
Wijgerse, F. (2000) Shields op de Helling? (Het Ontwerpen van een Steenbestorting op een Oever onder Stromingsaanval). Msc. Thesis, Delft University of Technology

Wörman (1989) Riprap protection without filter layers. J. of Hydraulic Eng, ASCE, vol 115 (12)

Xie Shileng (1981) Scouring Patterns in front of vertical Breakwaters and their Influences on the Stability of the Foundations of the Breakwaters. Delft University of Technology

Zdravkovich, M.M. (1997) Flow around circular Cilinders; comprehensive Guide trough Flow Phenomena, Experiments, Applications, mathematical Models, and Computer Simulations. Vol. 1. Fundamentals. Oxford University Press, Oxford

Zuurveld, J. (1998) Hoofdstroming contra Menglaag (De Invloed van een Menglaag op het Begin van Bewegen van Bodemmateriaal). Msc. Thesis, Delft University of Technology

Index

Reminder II

Things you should remember after studying this book

1. $r = \dfrac{\sqrt{\overline{u'^2}}}{\overline{u}}$ (relative turbulence = root mean square fluctuations / average velocity)

     uniform flow: $r\overline{u} \approx 0.1$, mixing layer: $ru_0 \approx 0.2$, jets: $ru_m \approx 0.3$

2. $u_{max} \approx 2\overline{u}$ (maximum velocity around cylinder, relative to approach velocity)

3. $\left.\begin{array}{l} \dfrac{u_{*c}^2}{\Delta g d} = \psi \approx 0.03 \\[2ex] u_* = \overline{u}\dfrac{\sqrt{g}}{C} \end{array}\right\} \rightarrow \dfrac{u_c^2}{\Delta d} = \psi C^2$ stability in uniform flow, for other situations

           correction necessary: $K_v \approx \dfrac{1 + 3r}{1 + 3r_{un.\,flow}}$

4. $\dfrac{h_s}{D} \approx 2\tanh\dfrac{h_0}{D}$ (scour depth around cylinder)

5. $F_f = \rho_w g i \rightarrow i_c \approx 1$ (critical local vertical gradient in sand for fluidisation)

6. $\alpha < \dfrac{\phi}{2}$ (safe slope angle when horizontal seepage is present)

7. $\dfrac{d_{15F}}{d_{85B}} < 5$ (stab.) $+ \dfrac{d_{60}}{d_{10}} < 10$ (int.stab.) $+ \dfrac{d_{15F}}{d_{15B}} > 5$ (perm.)   geometrically closed filter

     rules, to be obeyed if large vertical gradients or dynamic loads can be expected

8. $\dfrac{h}{L} > \dfrac{1}{2} \rightarrow$ deep water waves $\dfrac{h}{L} < \dfrac{1}{20} \rightarrow$ shallow water waves

9. $\left.\begin{array}{l} \dfrac{H_b}{h} \approx 0.8 - 0.9 \\[2ex] H_{max} \approx (1.5 - 2)H_s \end{array}\right\} \rightarrow \dfrac{H_s}{h} \approx 0.4 - 0.6$ (H_s on shallow foreshore $\approx h/2$)

10. $\xi = \dfrac{\tan\alpha}{\sqrt{s}}$ (Iribarren number, slope steepness versus wave steepness)

11. $\left.\begin{array}{l} \dfrac{R_u}{H} \approx \xi \\[2ex] H_{2\%} \approx 1.5 H_s \end{array}\right\} \rightarrow R_{u2\%} \approx 1.5\xi H_s$

12. $H_T \approx \dfrac{H_I}{2}$ (wave transmission for dams with crest around waterlevel, $R_c = 0$)

13. $\dfrac{H_{sc}}{\Delta d} \approx 1.5 \sqrt[3]{\cot\alpha}$ (stability in waves, Hudson)

14. $P = 1 - \exp(-fT)$ (Poisson distribution)

     for small fT: $P \approx f$,   for $fT = 1$: $P \approx 0.63$